TROPICAL ENVI

Tropical environments comprise a major sub-system of the world ecosystem and events occurring in these systems necessarily have a global impact. Therefore an understanding of how tropical ecosystems function is becoming increasingly urgent for environmental issues and economic development.

Tropical Environments presents a comprehensive introduction to the complex systems of the tropics. Covering a broad, cross-regional range of humid through to semi-arid tropical climate zones, the book features a wealth of case studies drawn throughout the tropical world. The book tackles the major problems facing the tropics, from complex biological interactions and soil nutrient deficiencies to problems facing human populations, offering a balanced integration of biophysical and human management issues.

Contents: The History of Tropical Environments, Tropical Climates and Moisture Regimes, Soils and Nutrient Limitation in Tropical Ecosystems, Tropical Biotas, Tropical Forests and Forestry, Tropical Savannas, Traditional Tropical Agricultural Systems, Intensified Tropical Agriculture, Animal Production and Utilization in the Tropics, Land Use Planning, Rehabilitation and Research.

Martin Kellman is Professor of Geography and Biology at York University, Ontario, Canada. **Rosanne Tackaberry** is a Senior Science Advisor with the Government of Canada's Scientific Research and Experimental Development Program.

A VOLUME IN THE
ROUTLEDGE PHYSICAL ENVIRONMENT SERIES
Edited by Keith Richards
University of Cambridge

The Routledge Physical Environment series presents authoritative reviews of significant issues in physical geography and the environmental sciences. The series aims to become a complete text library, covering physical themes, specific environments, environmental change, policy and management as well as developments in methodology, techniques and philosophy.

Other titles in the series:

ENVIRONMENTAL HAZARDS:
ASSESSING RISK AND REDUCING DISASTER
K. Smith

WATER RESOURCES IN THE ARID REALM
E. Anderson and C. Agnew

ICE AGE EARTH:
LATE QUATERNARY GEOLOGY AND CLIMATE
A. Dawson

MOUNTAIN WEATHER AND CLIMATE, 2ND EDITION
R.G. Barry

SOILS AND ENVIRONMENT
S. Ellis and A. Mellor

GEOMORPHOLOGY OF DESERT DUNES
Nicholas Lancaster

TROPICAL ENVIRONMENTS

The functioning and management
of tropical ecosystems

Martin Kellman
and Rosanne Tackaberry

London and New York

To Augustine, Ramon, Vivencio, and their families, and the many millions of others who live and work in the rural landscapes of the tropics.

First published 1997
by Routledge
11 New Fetter Lane, London EC4P 4EE

Transferred to Digital Printing 2003

Simultaneously published in the USA and Canada
by Routledge
29 West 35th Street, New York, NY 10001

© 1997 Martin Kellman and Rosanne Tackaberry

Typeset in Garamond by
Florencetype Ltd, Stoodleigh, Devon

British Library Cataloguing in Publication Data
A catalogue record for this book is available from the British Library

Library of Congress Cataloging in Publication Data
Kellman, Martin C.
Tropical environments : the functioning and management of tropical ecosystems / Martin Kellman and Rosanne Tackaberry.
p. cm. – (Routledge physical environment series)
Includes bibliographical references and index.
1. Ecology—Tropics. 2. Tropics. 3. Ecosystem management—Tropics. I. Tackaberry, Rosanne II. Title. III. Series.
QH84.5.K44 1997
577'.0913—dc21 97–387

ISBN 0–415–11608–2 (hbk)
0–415–11609–0 (pbk)

CONTENTS

CONTENTS

CONTENTS

FIGURES

TABLES

PREFACE

This book had its origins in a senior undergraduate/graduate course that one of us has taught at York University for several years. The book is based on the assumption that there is value in a more synthetic view of the world, in this case the tropical world. This assumption flies in the face of much of what is done in modern science, where fragmentation into detailed part-knowledges is the prevailing pattern. While this fragmentation is essential to advances in scientific knowledge, we believe that a more synthetic view is also important, not only to an educated citizenry, but especially to those who must develop policy about land use and environmental management. The book seeks to span the gulf between the earth and life sciences, and assumes that readers will be familiar with basic ideas in environmental science, especially those of ecology and physical geography.

Preparation of a book of this sort has required a highly selective use of the available literature; the alternative would have been a volume of monumental proportions. Selectivity has been required both in what topics to encompass and in what to include when discussing these topics. In selecting at each of these levels, we have been guided by the need to present a coherent picture of the tropics, even if this proves to be a temporary one. In selecting particular material to cite as examples we have been guided by a desire to use information with which we are most familiar and hence least likely to be used in error. In doing this we are acutely aware that alternative material often exists which could have provided examples of equal or greater value, and we hope that users of the book will make reference to examples with which they are familiar. Our emphasis throughout the book has been on developing generalizations about how tropical ecosystems function and can best be managed, and within which particular observations can be seen to be situated. We hope that readers will view the synthesis about the tropics that we have presented as essentially temporary, to be evaluated, criticized and ultimately transformed as more information emerges.

We are donating all royalties from this book to the Planned Parenthood Federation of Canada for use in its international programmes. Planned Parenthood works both in Canada and in the developing world to improve people's access to information and services related to their reproductive and sexual health. One result of this work is a reduction in population growth, which we believe to be the single largest threat to the survival of tropical ecosystems and the well-being of human populations in the tropics.

ACKNOWLEDGEMENTS

Our research in various tropical regions has been facilitated by the assistance of an enormous number of people. Among these, we would especially like to thank: Alexander Barreto, Alberto Briceño, Julia Carabias, Gus Carty, Milton Flores, Janet Gibson, Sergio Guevara, Susan and Alan House, Augustine Howe, John Hudson, Ernesto Medina, Patricia Moreno, Irene Pisanty, Lesley Rigg, Oscar Rosado, Judith Rosales, Paul Seigel, Steve Sherwood, Ramon Silva, Bernard Triomphe, Vivencio Tubig and Donald Walker. The funding for much of this fieldwork has been provided by grants from the Natural Sciences and Engineering Council of Canada and the National Geographic Society.

Maps and diagrams have been prepared by Carol Randall and Carolyn King of the Cartography Laboratory, York University. The staff of the Resource Sharing Department, York University Libraries, have pursued many obscure references for us with diligence and good humour. Peter Harrison, Sarah Lloyd, Valerie Rose and all at Routledge have provided support and guidance during the editorial production process. Tila Kellman has read many parts of the manuscript and made improvements in content and style. To all of these persons we offer our sincere thanks.

The authors and publisher wish to thank the following for permission to reproduce copyright material. Every effort has been made to contact all copyright holders. However, if any one has not been contacted they should contact the publisher in the first instance.

Academic Press – Fig. 3.4; Blackwell Scientific Publications – Figs 6.2, 7.7; American Association for the Advancement of Science – Figs 7.11, 9.2, 9.3; American Geophysical Union – Fig. 3.15; Cambridge University Press – Figs 2.2, 6.14, 8.4, Tables 5.2, 5.3; Chapman & Hall – Fig. 2.4; Elsevier Science – Fig. 2.3, Tables 4.4, 5.4, 9.8; Food and Agriculture Organization of the United Nations – Table 6.5; Macmillan Magazines Ltd – Fig. 2.6; Oxford University Press – Fig 2.5, 6.13, Tables 6.2, 6.3, 6.6; Plenum Publishing – Fig. 4.3; Systematic Zoology – Fig. 5.7.

1

INTRODUCTION

Two conditions make tropical environments of special interest to both natural and social scientists: they cover a large proportion of the earth's total surface area; and they are home to a large proportion of the world's people, most of whom live in poverty.

The geometry of a rotating globe, such as the earth, requires that zones close to the equatorial plane of rotation make up a disproportionate share of the total surface area. Areas lying between the tropics of Cancer and Capricorn (23.5° N and S) make up 40 per cent of the total global surface; this rises to 58 per cent if the latitudinal zone is extended to 35° N and S. As a consequence, tropical zones play a large role in global energy budgets and the general circulation of the atmosphere. While the accidents of plate tectonics have resulted in low latitudes being under-represented in terms of land area, the extensive oceanic areas of the tropics nevertheless have a profound global impact. For example, the heat storage capacity of water coupled with the extent of the tropical Pacific Ocean results in events in this water body having a global impact on climate, notably by way of the El Niño Southern Oscillation (e.g. Graham *et al.* 1994).

Approximately half of all tropical land surfaces can be classified as humid, with rates of annual rainfall equalling or exceeding those of evapotranspiration. The remainder is subhumid or semi-arid (National Research Council 1993). About 60 countries occur entirely or partly within the humid tropics and in 1991 these contained, in aggregate, over 2 billion persons (National Research Council 1993). Per capita GNP for these countries averaged $1,262 and rates of natural population increase varied from 1.1 to 3.8 per cent per annum. These rates translate into population doubling times of 62 and 18 years, respectively. Thus these countries are not only generally poor, but likely to grow rapidly more so at the present rates of population increase.

MYTHS AND GENERALIZATIONS

Throughout history the term 'tropics' has been a fertile source of popular myths to the residents of temperate areas. Much of this mythology has had

1

its roots in ignorance about tropical regions and their inhabitants. Until recently, inadequate information has also led to a set of scientific myths about this region of the earth, many of which persist in some form to the present. Two of the most pervasive (and contradictory) myths about the tropics relate to the region's assumed bountifulness and to its fragility. The myth that the tropics offer unlimited agricultural potential served, in part, to fuel European colonial expansion in these areas, and in recent decades has helped justify further agricultural colonization in such places as Amazonia. The converse myth, that most tropical soils have no agricultural potential and should never be subject to cropping, became popular in the past half-century (e.g. Gourou 1958), and more recently has been generalized to hold that all tropical eco-systems are 'fragile' and should be protected from human activities of all kinds. As with most myths, both statements contain some kernels of truth, but were founded on a thin empirical base. The fact that both have persisted in some form in the face of accumulating contradictory evidence in recent decades can probably be attributed to the durability of myths once entrenched (Botkin 1990), and to the convenient justifications that each provides for either 'development' or 'preservation' in the tropics.

In recent decades there has been a virtual explosion of new empirical data on the tropics. While this should allow a thorough re-evaluation of the early generalizations and encourage reformulation of new, more accurate ones, this is made difficult both by the sheer volume of material now available and by the many contradictory conclusions that these data yield. The practice of science leads inexorably to reductionism and the compartmentalization of specialized part-knowledges, with synthesis and generalization rarely attempted or left in the hands of non-scientists. We consider this trend understandable but unfortunate, as most policy decisions on management affect whole systems, rather than just their parts, and policy formulation is best achieved by those informed by a broader view.

We present this book as an attempt to achieve some contemporary synthesis about how tropical ecosystems function. However, we also emphasize that there exists no broadly agreed-upon synthetic methodology in science, in contrast to analytical method (e.g. Popper 1965). This leaves synthesis as more of an art than an established scientific procedure. Moreover, synthesis assumes that some generalization is possible, and some would argue that this is no longer achievable about the tropics. Recent data have shown that tropical environments are highly variable from place to place, even at local scales, making it unlikely that sweeping, grand generalizations about these environments and their management will be either very accurate or useful. However, we believe that certain lesser generalizations about how tropical ecosystems function do exist, and can provide useful points of departure for establishing more site-specific solutions to management. These generalizations seek to specify dominant processes, and how these vary from place to place to provide different environmental 'recipes'. We suggest that these can then provide

2

guidance about what questions are most appropriately asked, what observations are most useful, and what experiments are most likely to be productive, when site-specific solutions are sought to the many pressing problems facing human populations in the tropics.

SOME DOMINANT PROCESSES

In this book we treat the functioning of natural tropical ecosystems and managed derivatives of these that result from forestry, agriculture and animal husbandry. These systems comprise the mosaic of rural tropical landscapes. In both natural and managed ecosystems in the tropics either or both of two processes dominate, and will receive special attention throughout the book: a complexity of biological interactions; and a deficiency of plant mineral nutrients in the soil. Unlike in temperate areas, where winter temperatures constrain biochemical processes and the uninhibited growth of biological populations, in the tropics populations can potentially grow until limited by some other factor. Moisture acts as a primary limiting resource in those parts of the tropics experiencing seasonal drought, but elsewhere limitation can only be achieved by interactions among biological populations, exhaustion of soil nutrient resources, or both. We would expect humid tropical environments to support productivities approximately three to four times those of temperate areas if climate were the dominant limiting factor upon this process. The fact that tropical productivity often shows a broad overlap with that of temperate areas indicates the importance of other limiting factors in the tropics.

Biological interactions include competition, herbivory, predation and parasitism. While these sorts of interactions are common to all ecosystems, whether tropical or temperate, contrasts between the two regions are especially apparent when they involve short-lived organisms such as insects, whose populations are eliminated seasonally in temperate winters. In temperate areas, insect populations often spend most of the short summer growing seasons recovering from winter elimination, while in the tropics they are free to expand and persist at levels set by food or host availability. As a consequence, tropical host populations, especially plants, are likely to be subject to much more intense and continuous herbivory, predation and parasitism than are their temperate counterparts. Tropical environments are also characterized by much more varied assemblages of biological populations, and the degree to which this may have been facilitated by climatic conditions will be discussed in Chapter 5. Irrespective of their origin, the existence of rich biotas, among which intense biological interactions occur, is a dominant characteristic of most natural tropical ecosystems. This biotic complexity not only makes it difficult to manage these systems in a semi-natural state, such as in forestry, but creates special problems when simpler artificial systems are intruded into the natural milieu. The pest and disease problems that are a constant, but

3

controllable, irritant in the agricultural ecosystems of temperate areas often become overwhelming in the tropics.

The absence of a dormant winter period also makes insufficient soil nutrients a further potential factor limiting plant growth in the tropics. This effect is exacerbated by the intensity of chemical weathering in the humid tropics and persistent deep percolation of water within the soils. The former process results in continual transformation of primary geological minerals, and the plant nutrients that they contain, to stable secondary minerals or water-soluble constituents, while the latter results in persistent gradual leakage of these constituents from the soil. The problem is further exacerbated by the fact that a large proportion of tropical land surfaces derive from the geologically stable plate fragments of Gondwanaland, which have not experienced the rejuvenation of soil parent materials precipitated by orogeny or volcanic and glacial activity. Natural selection among plants in this soil milieu has favoured conservatism in nutrient usage, often at the expense of rapid growth and productivity. Because rapid growth and high productivity are the very characteristics usually sought in managed ecosystems, a persistent incongruity exists between these systems and the soil substrates on which many of them depend.

ENVIRONMENTAL CARRYING CAPACITIES

Biological populations cannot continue to grow indefinitely when the resources available to support them are finite. As limits to a habitat's ability to sustain members of a population are approached, a variety of so-called 'density-dependent' mechanisms often come into play which reduce birth rates, increase death rates, or both, and result in an equilibration of the population at the environment's carrying capacity. In other instances populations may temporarily exceed the sustainable carrying capacity of the habitat. The result of such population overshoots is usually an over-consumption of one or more limiting resources, followed by a collapse of the population. Carrying capacities may be reduced at least temporarily after such an overshoot. Carrying capacities may also vary through time in response to such non-biological processes as rainfall variability. Places subject to this variability, such as large parts of the semi-arid tropics (see Chapter 3), will have sustainable carrying capacities set by conditions during extreme drought years, when only relatively small populations can be supported. Populations which develop during intervening favourable years may be unsustainable during extreme drought periods, and suffer high mortalities then.

Until recently, the inevitability of a finite environmental carrying capacity for human populations has escaped most members of these populations. This has probably been engendered both by the availability of new resources, in the form of unoccupied terrain and fossil energy sources, and by the ability of people to devise more efficient agricultural life-support systems. However,

4

even the most efficient agricultural system is limited by the biochemistry of the photosynthetic process, and the solar energy, moisture and nutrient resources upon which it relies. Consequently, it is inevitable that limits to human population sizes will have to be confronted by members of these populations. The population doubling times of most tropical countries indicate that, in the tropics, this confrontation must occur within the next several decades.

'Success' in biological terms is often equated with the sustenance of large numbers of organisms of a particular species, irrespective of their quality of life. In human populations, such a numerical definition of success would be judged ethically unacceptable by most people, and an acceptable quality of life for members of the population is usually assumed to be a necessary criterion of success. While 'quality of life' is an elastic term that cannot be equated simply with access to more resources, it is clear that a human population which is sustained at more than minimal levels of biological perpetuation is likely to place greater resource demands on its habitat than one that is not. This may be expected to lower the numerical carrying capacity of an environment for people.

Environmental carrying capacities across the earth's surface are not uniform, although the naive faith in technology that has pervaded much of western thinking throughout the twentieth century has often made it unpopular to emphasize this fact. Carrying capacities of all habitats have usually been seen as subject to changes that are exclusively upward, as a consequence of human ingenuity at devising ever more effective ways of transforming natural resources to human use (e.g. Brookfield and Padoch 1994). However, the deteriorating state of many tropical landscapes that are now occupied by human populations (Chapter 11) suggests that population overshoots may be widespread and that sustainable carrying capacities will be far lower in the future than those now prevailing. In this book we will argue that many tropical environments are subject to special sets of conditions that place relatively low ceilings upon the human population densities that they can sustain at acceptable standards of living. We believe that these special problems must be recognized and appropriate site-specific solutions devised if we are to avoid even more widespread environmental degradation. However, we also remain optimistic that this is a realistic goal to strive for, provided that human populations do not continue to grow indefinitely.

PRESERVATION, UTILIZATION AND SUSTAINABILITY

The debate about management of tropical resources is pervaded by a conflict between those most interested in preserving the many unique species and ecosystems of the tropics, and those promoting further utilization of the terrain occupied by these for the support of human populations. Preservationists can argue, with much empirical support, that occupation of more terrain and destruction of their natural ecosystems has brought only temporary relief to the problems of human over-population in the tropics, and served only to support more people in poverty rather than to raise material standards of living. In contrast, those favouring more resource utilization appeal to the moral repugnancy of allowing human populations to starve in the interests of preserving natural ecosystems and can argue that, where human populations can be stabilized or reduced, greater use of natural resources can raise incomes and standards of living.

This debate can be seen as part of a larger issue that ultimately must be confronted by the global human population: the limits to exploitable earth resources. While the world's human population is fed on plant and animal products that are, in theory, renewable resources, the production of these has come to rely increasingly on non-renewable 'industrial' inputs, most notably those provided by fossil fuels. Much of the recent expansion of the global human population, and the dramatic increase in the material well-being of residents of industrialized countries, can be attributed directly or indirectly to consumption of non-renewable fossil fuel resources. The eventual exhaustion of these will bring the need to devise systems that can utilize renewable resources, notably those of solar energy, and that can be sustained indefinitely.

This is likely to require not only a much-reduced global human population, but also a fundamental shift in human perceptions. Throughout much of human history, an expanding population and increased material standard of living have been sustained by the exploitation of ever more resources: terrain, fossil fuels and other mineral products. Indeed, in most industrial societies economic prosperity is largely equated with more rapid resource consumption. The transition to a society based on steady-state use of renewable resources is likely to require major transformations in the human psyche. The human psyche has evolved in a milieu of steady, if often slow, demographic growth and an ever-expanding consumption of resources. Most societies view change as necessary to survival and often equate change with expanded resource use. How to ensure change and prevent stagnation, within the constraints of steady-state ecosystem processes, is a major challenge for future generations.

2

THE HISTORY OF TROPICAL
ENVIRONMENTS

In this chapter we ask the question: 'For how long and over what extent have relatively warm climatic conditions, and the biological communities characteristic of these, existed on the earth's surface?' The widely used Koppen–Geiger climate classification system arbitrarily defines tropical climates as those having no month with mean temperatures below 18°C. While in Chapter 3 we choose to define tropical climates in a more dynamic way, the Koppen–Geiger definition provides a convenient first approximation of where tropical climates exist today; this is generally within (although not universal to) the latitudinal zone enclosed by the tropics of Cancer and Capricorn (Figure 3.1). As we shall see, this pattern has changed considerably through time. These changes raise questions about presumptions of stability and fragility for tropical ecosystems that have been based on an assumed lack of climate change in the tropics. This historical record also sheds new light on the enigma of the origins of tropical biotic diversity.

We concentrate upon the history of tropical climates since the Cretaceous (65–144 million years before present [My BP]; Table 2.1), which is the period during which the flowering plants (Angiospermae or angiosperms) began a rapid diversification and domination of the earth's vegetation. Today the angiosperms provide the basic, and diverse, biological ingredients of all tropical terrestrial communities, and their fossils provide a relatively interpretable signal of environmental conditions during this period.

PALAEOENVIRONMENTAL DATA

Reconstructions of ancient environments must draw upon a variety of data that are often incomplete in spatial coverage and lack prolonged temporal sequences. As a result, reconstructions are usually based on an accumulation of partial, but corroborative, data. Fortunately, the development of a variety of radiometric dating techniques over the past half-century has allowed increasingly reliable chronological control of historical data, and greatly facilitated palaeoenvironmental reconstructions.

Table 2.1 A simplified geological table

Era	Period	Epoch	Approximate time span (millions of years before present)
Cenozoic	Quaternary	Recent (Holocene)	0.01–present
		Pleistocene	2.8–0.01
	Tertiary	Pliocene	11–2.8
		Miocene	25–11
		Oligocene	36–25
		Eocene	54–36
		Palaeocene	65–54
Mesozoic	Cretaceous		144–65
	Jurassic		180–144
	Triassic		220–180
Palaeozoic	Permian		280–220
	Pennsylvanian		310–280
	Mississippian		335–310
	Devonian		405–335
	Silurian		425–405
	Ordovician		500–425
	Cambrian		600–500
Precambrian			Preceding 600

With the exception of temperature-sensitive stable isotopes of oxygen, which are of use primarily in indicating oceanic palaeotemperatures, most palaeoclimatic data are indirect and make use of a variety of indicators, such as fossil organisms, soil material or even landforms, to infer past climatic conditions. Use of these indirect data must invoke the geological principle of uniformitarianism: that present-day processes and interrelationships also prevailed in the past, and that their products, when present, can be used to infer the existence in the past of other correlated conditions. Use of these forms of data is clearly dependent on how reliable is the uniformitarian assumption, and upon how close and well understood are the connections between the indicators and their climate correlates. Their use to infer tropical palaeoclimates assumes that reasonably consistent indicators of these conditions exist.

Biological fossils are very widely used indicators of past climates, and the inability of perennial plants to avoid unfavourable climatic conditions by way of migration or hibernation makes fossils of these especially useful as climate indicators. Plant fossils comprise either 'macro-fossils', such as leaves, fruit and stems, or 'micro-fossils' made up primarily of pollen, spores and the remains of aquatic micro-organisms. While macro-fossils provide reliable indicators of the community composition at the location where preservation took place (often swampy sites), they usually provide only a limited spatial perspective. In contrast, micro-fossils provide less exact taxonomic information, but a better indicator of regional vegetation.

Fossils have been used as climate indicators in two ways. In the first, correlations between climatic conditions and the morphological features of contemporary organisms, irrespective of their taxonomic lineage, have been used to infer climatic conditions in the past from the same features in fossil assemblages. In this way, Wolfe (1985) has used such morphological features of fossil leaves as size, shape and thickness to reconstruct climates of the Tertiary period. The assumption made in these reconstructions is that natural selection of certain morphological traits is sufficiently pervasive to override peculiarities deriving from a particular ancestry. The second method is to use fossil taxa as indicators of the presence or absence of certain conditions in the past, assuming that ecological tolerances of the taxa do not differ greatly from those of their contemporary descendants. For example, contemporary massive carbonate reef deposits exist only where ocean water temperature exceeds 21°C, and the presence of fossilized reefs at high palaeolatitudes has been taken to indicate tropical conditions at these places in the past (Barron 1983). The interpretation of past climatic conditions from assemblages of fossil pollen and spores has been especially widely used for the Quaternary period, which comprises the past 2.8 million years of earth history (Table 2.1).

The use of soil, mineralogical and geomorphological data to interpret the history of tropical environments assumes that certain sorts of material are either specific to or excluded by particular conditions. Some data of this sort provide unequivocal climate signals; for example, the presence of sand dunes beneath existing closed vegetation clearly indicates previously more arid conditions (Figure 2.1). However, other signals are far more equivocal. For example, the existence of unweathered minerals in Amazon deltaic sediments of Pleistocene age was once taken to indicate a treeless watershed (Damuth and Fairbridge 1970), but has recently been reinterpreted as a product of increased erosion of unweathered rock exposures during river incision required by lower sea levels (Irion 1989). As a consequence, data of this sort normally require corroboration by other forms of paleoenvironmental material.

PLATE TECTONICS

Any interpretation of palaeoenvironments is complicated by the movements of continental plates over the earth's surface. During the period under consideration here, the major plate movements have involved separation into eastern and western fragments of the former northern (Laurasia) and southern (Gondwana) land masses, as a result of an expanding Atlantic Ocean. At the start of separation (c. 80 My BP), Laurasia and Gondwana were themselves separated by the narrow Tethys Sea, whose remnant persists as the

Figure 2.1 The Apure sandplain of southern Venezuela. This savanna-covered land-scape is underlain by a sand sheet (middle ground) traversed by elongated parallel fossil dunes (background) oriented NE–SW, the presumed wind orientation during sand movement. Palaeosols beneath the sand have been dated at 11,100–12,300 yr BP, indicating relatively recent sand deposition

Mediterranean (Figure 2.2). The major exceptions to this longitudinal pattern of movements have been Australia, Antarctica and India. These plates separated from Gondwana about 180 My ago, with India drifting northwards. Australia and Antarctica first drifted southward into high southerly latitudes, with Australia later separating and drifting northwards into more equatorial latitudes (Figure 2.2).

Rapid diversification of the angiosperms in the Cretaceous coincided with the early period of major plate fragmentation, and it is therefore not surprising that large numbers of angiosperm families are common to all plate fragments. Intercontinental differentiation of angiosperm floras is primarily at the generic level, although even here there exist frequent commonalities between ancient plate fragments. For example, Africa and South America share many genera, suggesting that isolation was not complete for a long period while taxa were differentiating at the generic level.

CRETACEOUS PALAEOENVIRONMENTS

A variety of data indicate that the earth's climates during the Cretaceous were both generally warm and lacking steep latitudinal temperature gradients.

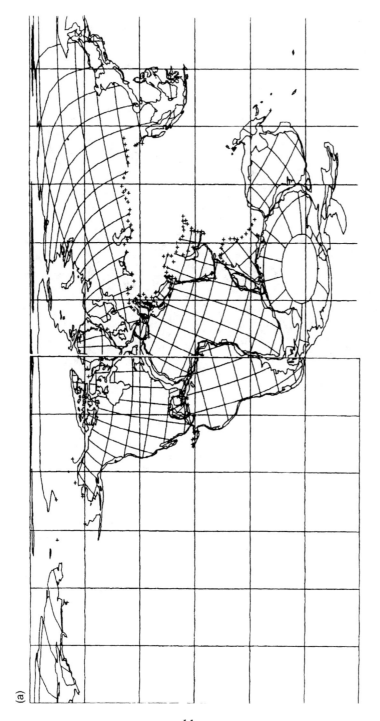

(a)

Figure 2.2 Continental plate positions at (a) 180 and (b) 80 million yr BP

Source: Smith *et al.* 1981: Maps 21 and 41

11

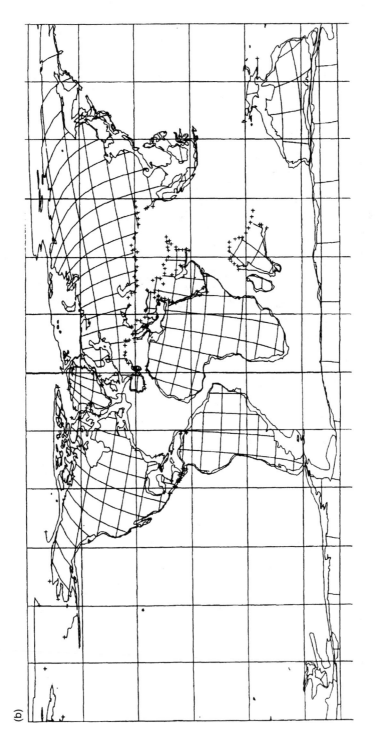

(b)

Figure 2.2 (b).

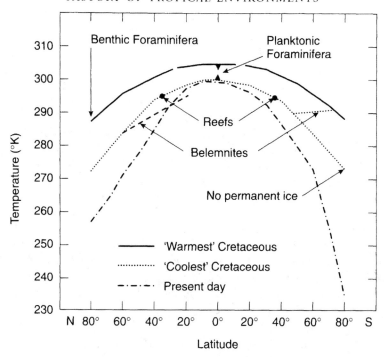

Figure 2.3 Summary of estimated mean annual surface temperatures at different latitudes during the mid-Cretaceous, using a variety of palaeo-temperature indicators. Present-day surface temperatures are shown for comparison

Source: Barron 1983: Figure 3

Tropical latitudes possessed temperatures that were as warm as those of the present, and may have been 3–5°C warmer, while polar latitudes experienced temperatures that appear to have been between 15 and 24°C warmer than at present (Figure 2.3). Rapid proliferation of the angiosperms took place at this time, and quantitative analysis of Cretaceous fossil floras (primarily from the Northern hemisphere) indicate that by *c*. 90 My BP these had achieved widespread dominance of terrestrial vegetation (Lidgard and Crane 1988). The taxa most affected by this expansion were ferns and cycads (primitive gymnosperms), rather than conifers (Lidgard and Crane 1988), and the interaction has usually been interpreted as one of competitive displacement of these reproductively and physiologically less efficient groups by the much more efficient angiosperms (Knoll 1984). During the diversification, angiosperm pollen first became important in low palaeolatitudes, with importance subsequently increasing in higher latitudes (Crane and Lidgard 1989). By the end of the Cretaceous, angiosperm pollen comprised >80 per cent of the taxa to 40° N palaeolatitudes, and generally 30–50 per cent of the taxa at higher latitudes (Crane and Lidgard 1989). Thus, while the site of origin of the angiosperms remains controversial, their earliest proliferation appears to have been in low latitudes.

The end of the Cretaceous is dated stratigraphically by the sudden disappearance of several animal taxonomic lineages from the fossil record, most notably the dinosaurs. This Cretaceous–Tertiary event at 65 My BP is now widely interpreted as the consequence of a major meteoric impact in the northern hemisphere, with the northern Yucatán peninsula recently identified as the most likely impact site (Hildebrand *et al.* 1995). While the event caused major extinctions among animals, its long-term effects on plants appears to have been less severe. A widespread 'spike' of charcoal fragments and fern spores in stratigraphic sequences following the event has been interpreted as a secondary succession following fire which, while pan-hemispheric, did not result in wholesale elimination of major plant taxonomic lineages (Tschudy *et al.* 1984; Nichols *et al.* 1986; Saito *et al.* 1986), although it may have selectively eliminated many broad-leaved evergreen species in the northern hemisphere (Wolfe and Upchurch 1986). However, the world entered the Tertiary with climatic conditions that were as warm and equable as was the preceding Cretaceous.

TERTIARY PALAEOENVIRONMENTS

Throughout the early Tertiary and into the early Eocene (i.e. 65–50 My BP), the world continued to be a warm and equable place, but with precipitation greater than that during the Cretaceous. Upchurch and Wolfe (1987) suggest that in North America this warm, wet period provided the most extensive humid tropical climate conditions available since the appearance of angiosperms, and promoted the development of closed multistratal tropical rain forest. In this forest, they believe, the angiosperms underwent rapid diversification into the many functional groups that characterize this vegetation type today; that is, tall trees, lianas, epiphytes and shade-tolerant seedlings. Shade tolerance normally requires large seededness (see Chapter 6), and the animals needed to disperse large seeds are assumed to have evolved concurrently. In North America, where plant fossil sequences are especially complete, there is evidence of diverse tropical rain forest to 50° N palaeolatitudes, and a form of tropical forest to 60–65° N (Wolfe 1985). In southern Alaska, the forest contained such tropical indicators as palms and mangroves. Palaeobotanical data from other regions are less complete, but confirm a pattern of world vegetation dominated by tropical forests. For example, in the early Eocene, the tropical swamp palm *Nypa* existed at 55–60° S palaeolatitude in southern Australia (Kemp 1981), suggesting sea surface temperatures of 20–25°C. *Nypa* is also recorded in Europe at a similar time (Moore 1972), confirming the enormous extent of tropical climate conditions in both hemispheres within this period (Figure 2.4). Neither palaeobotanical (Wolfe

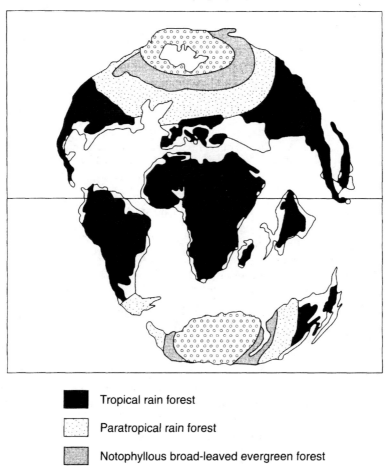

■ Tropical rain forest

▢ Paratropical rain forest

▨ Notophyllous broad-leaved evergreen forest

▢ Polar broad-leaved deciduous forest

Figure 2.4 Generalized global distribution of forest types during the early Eocene
Source: Tallis 1991: Figure 9.2

1985) nor geomorphological (Street 1981) data show any evidence for wide-spread arid zones during the early Tertiary. Subtropical desert climates, which are a dominant feature of the earth's present climatic zones (Chapter 3), appear to have emerged only in the Miocene (*c.* 25 My BP), coincidentally with a much steepened latitudinal temperature gradient (Wolfe 1978).

Subsequent developments in Tertiary climates involved a major cooling in the late Eocene (*c.* 35 My BP), which resulted in tropical forests being restricted to zones equatorward of 15°, in contrast to present limits at 20–25°

15

(Wolfe 1985). By the early Miocene (*c.* 20 My BP), intensified subtropical high-pressure systems and their associated desert climates had emerged; subsequently tropical environments would be confined to the equatorial zone between these high-pressure systems. The appearance of drier conditions within low latitudes in the early Miocene also coincides with the appearance of widespread savanna communities for the first time (Wolfe 1985). In Central America, elements of a dry forest flora appear to have existed in scattered pockets since late Eocene times, but only coalesced into a recognizable regional vegetation type in the Pliocene (2.8–11 My BP) (Graham and Dilcher 1995). Increased drought during the Miocene had a particularly profound effect on the vegetation of Australia. Tropical and subtropical forest was widespread on that continent in the early Miocene, but by the mid-Miocene this had shrunk to scattered pockets on the moister eastern seaboard, where it persists today (Figure 2.5). Elsewhere on the continent, rapid evolution of a unique drought-adapted flora took place, with trees of the genus *Eucalyptus* becoming especially prominent components of many communities.

In summary, the Tertiary began with warm, moist, forest-covered environments dominating much of the earth's surface. These were subject to progressive shrinkage and fragmentation throughout the period as a consequence of cooling in higher latitudes and the appearance of subtropical arid zones. Tropical moist climates, and the forest communities that these support, are therefore not only very ancient, but also of previously far greater extent. In contrast, dry tropical environments and communities are geologically recent, as are many of the temperate environments that appeared in higher latitudes in the late Tertiary. The increased cooling and drying that occurred throughout the late Tertiary culminated in a period of severe cooling that began about 2.8 My BP, to initiate the geological period known as the Quaternary (Demenocal 1995). Traditionally, this has been subdivided into a Pleistocene epoch prior to 10,000 yr BP and a Holocene epoch since 10,000 BP (Table 2.1).

THE TROPICS IN THE QUATERNARY

Many more data with which to interpret palaeoclimatic conditions exist for this period than for the earlier Tertiary and Cretaceous periods, although these data are often patchy in coverage. Continuous stratigraphic records for the Pleistocene are generally best-preserved from the latter part of this period, and radiocarbon dating of organic material is effective to approximately 50,000 years, allowing a particularly good chronological control over this

Figure 2.5 Distribution of rain forests in eastern Australia. The rain forests of northeastern Queensland are species-rich and similar to those of other tropical moist forests. Floristic richness declines southward in the subtropical rain forests of New South Wales, and is low in the temperate rain forests of Tasmania and Victoria.

Source: Adam 1992: Figure 3.8

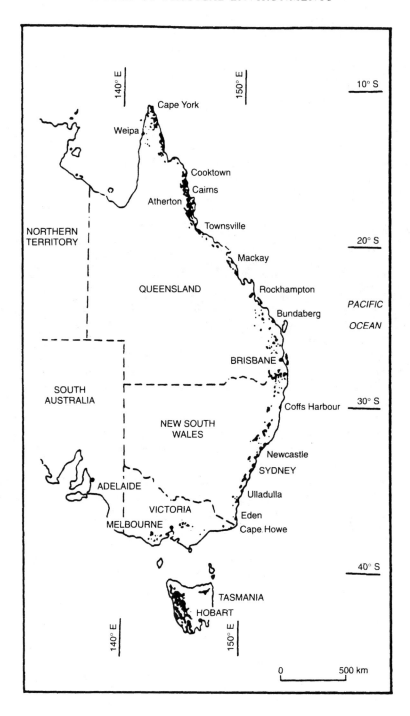

time span. This period encompasses the last major cold phase and glacial advance in temperate latitudes (termed the Wisconsin in North America and the Wurm in Europe), which began about 80,000 BP and reached a peak at approximately 18,000 BP. Areas that did not experience glaciation during the Pleistocene, such as the tropical lowlands, generally have many fewer small sedimentary basins in which fossil sequences can accumulate than do glaciated areas. As a consequence, extensive discontinuities exist in the spatial coverage of such data for the tropics, with the Amazon basin standing out as a particularly large gap. The use of fossil pollen data is further complicated in the case of tropical forests by the fact that many of the diverse pollen types of this vegetation remain unknown, and that pollen production by tropical forest trees is generally low because most are pollinated by insects, rather than by wind.

Early interpretations

Periodic cold temperatures during the Pleistocene were originally indicated by the widespread signs of glaciation in north-western Europe and North America. This reliance on glacial evidence led to an early de-emphasis on climate change in non-glaciated areas and the tacit assumption that these were relatively little affected. An accumulation of radiometrically dated pollen sequences from non-glaciated areas in Europe and North America over the past half-century has shown that these areas were also subject to extensive climate change during periods of glaciation, and supported vegetation covers very different from those of today (Huntley 1990; Webb 1988). However, until the past quarter-century, tropical areas were assumed to have remained relatively little affected by these changes at higher latitudes, an assumption based largely on an absence of palaeoenvironmental data.

The earliest evidence for colder conditions in the tropics came, as in temperate latitudes, from glacial features. All tropical regions possess some mountains of sufficient altitude to support glaciers today, and all show evidence of these glaciers having advanced to lower altitudes in the past (Clapperton 1993; Flenley 1979). Palynological evidence later confirmed that appreciable depression of vegetation zones also occurred in tropical mountain areas during the period 20,000–15,000 BP, with estimates varying from 650 to 1,700 m, depending on region and altitude (Flenley 1979). As world sea levels fell by only some 100 m during the last glacial maximum, there was clearly an insufficient altitudinal gradient within which a wholesale displacement of vertical climate and vegetation zones could take place; lowland environments must have been 'squeezed' altitudinally in some way. However, until very recently there existed no continuous pollen sequences from tropical lowlands for the critical period during and since the last glacial maximum.

In this void, some early scattered data led to the development of what, in retrospect, can be seen to be a premature theory about Pleistocene events in the lowland tropics, especially the New World tropics (Neotropics). In the late 1960s and early 1970s, scattered stratigraphic data for northern South America first suggested that substantial environmental change may have taken place in these tropical lowlands. For example, Bigarella and Mousinho (1966) interpreted geomorphological evidence from southern Brazil as indicating an alternation of dry and wet periods during the Pleistocene, while Damuth and Fairbridge (1970) found evidence in Amazon deltaic sediments of feldspar-rich materials that are not currently produced from the deeply weathered soils of the watershed. This too was interpreted as a sign of previous lack of forest cover and aridity. Concurrently, evidence for what seemed to be spatially congruent subspecific differentiation in some taxonomic groups (e.g. Haffer 1969; Vuilleumier 1971) led to the development of what came to be known as the 'refuge theory' (Prance 1982). Briefly stated, this proposed that the forest cover of the Amazon basin had been fragmented into a series of forest refugia in areas of higher rainfall, in which species were thought to have differentiated as a result of prolonged geographic isolation. Some even went so far as to suggest that this explained much of the species diversity of Neotropical tropical forest biotas (e.g. Simpson and Haffer 1978). Similar refugial patterns were postulated for the African tropics by Hamilton (1976), although this author found little evidence for speciation among forest organisms as a consequence of the assumed isolation.

The refuge theory has been presented in two multi-authored volumes (Prance 1982; Whitmore and Prance 1987), and is now thoroughly embedded in the textbook literature. Unfortunately, it is distressingly lacking in widespread support from palaeoenvironmental data. Endler's (1982) demonstration that boundaries between taxa did not conform to the predictions of the refuge theory and were more closely correlated with existing environmental discontinuities was largely ignored by his co-authors in the first volume, presenting the refuge theory (Prance 1982). Moreover, the palaeobotanical record clearly shows that diverse tropical forest communities pre-dated the Pleistocene. While some embellishment of this diversity could have occurred in the Pleistocene, this period certainly cannot be regarded as its primary source. Damuth and Fairbridge's (1970) feldspar-rich sediments have been reinterpreted as a result of river bed downcutting in response to lower sea levels (Irion 1989). However, the most convincing contradictory evidence has come in the form of palynological data from putative forest refugia which show an absence of lowland tropical forest in these places under full glacial conditions (Bush and Colinvaux 1990; Bush et al. 1990; Leyden 1984; Liu and Colinvaux 1985).

Current interpretations

While the central assumptions of the 'refuge theory' are no longer credible, the question remains as to what happened in the lowland tropics during periods of high-latitude climate cooling and glaciation. A gradual accumulation of palaeoenvironmental data over the past two decades permits some tentative answers to this question, although many gaps remain.

Isotopic analyses now generally confirm a drop in tropical sea surface temperatures of approximately 5°C at the time of maximum high-latitude glaciation (Guilderson et al. 1994), and a recent analysis of palaeotemperatures in north-eastern Brazil, using the noble gas content of fossil ground waters, indicates a temperature depression of a similar magnitude there (Stute et al. 1995). Temperature depressions of this magnitude are likely to have had wide environmental effects. Initiation of tropical hurricanes requires critically high sea surface temperatures (Chapter 3), and hurricanes are likely to have been completely absent during these periods. More importantly, tropical rainfall is primarily convective in origin (Chapter 3), and a lowered sea surface temperature would necessarily have resulted in less rainfall in many areas. The critical issue is whether this was sufficiently low to induce the widespread deforestation assumed by the refuge theory. The answer provided by an increasing number of palaeoecological studies in the lowlands seems to be 'yes' in some areas but probably 'no' in others.

Unequivocal evidence for deforestation during the last glacial period exists for a number of sites that today support tropical moist forest. For example, in the Peten area of Guatemala, an area now covered by evergreen tropical forest, and postulated to have been a Pleistocene forest refuge by Toledo (1982), Leyden (1984) found pollen indicating savanna and dry forest at 11,000 BP, which later changed to a forest of temperate (high-altitude) taxonomic affinities before the appearance of tropical forest taxa. She concludes that the evergreen tropical forest now covering this area is essentially a Holocene phenomenon. In the Lago Valencia basin of northern Venezuela, Bradbury et al. (1981) have found evidence for an arid grassland community between 13,000 and 10,000 BP which later changed to a tree cover. South of this area, in the western Apure savannas, an extensive sand plain exists in an area that today is too moist to allow sand movement (Roa Morales 1979; Figure 2.1). Thermoluminescence dating of quartz sand from two fossilized dunes in this area has given dates of 11,600 and 36,000 years (Vaz and Miragaya 1989), while two palaeosols beneath the sand have yielded dates of 11,100 and 12,300 BP (Roa Morales 1979). Livingstone (1993) has concluded that African forests were subject to more severe drought than those of the Neotropics, and shrank to scattered local refugia in many areas. Bonnefille et al. (1990) have estimated a 30 per cent drop in rainfall in the Burundi region during the last glacial maximum.

Probably the most complete record of deforestation during glacial conditions comes from forested areas in north Queensland, Australia. Here Kershaw (1978) has provided a remarkably complete record of vegetation change over a period estimated to cover 120,000 years using a pollen core extracted from an extinct volcanic crater (Figure 2.6). This pollen record, carbon-dated over the past 40,000 years and dated by extrapolation beyond that, shows a complete cycle of vegetation change from the last interglacial period, through the last glacial period, to the present. During the last interglacial, tropical forest covered the site until about 79,000 BP, when drier conditions are indicated by increases in the proportion of *Eucalyptus* and *Casuarina* pollen, trees that dominate dry sclerophyll forest and savanna in the area today. Beginning at about 38,000 BP the latter pollen types increased, and during the period 26,000–10,000 tropical forest was completely displaced from the area around the site. At this time, a large increase in charcoal fragments was found in the core, indicating extensive burning in the dry forests and savannas that surrounded the site. This period of inferred aridity came to an end with a rapid re-establishment of tropical forest around the site, beginning about 10,000 BP. Recent dating of charcoal from soil beneath many present rain forest tracts in the region (Figure 2.5) has confirmed that extensive deforestation took place during the period 13,000–8,000 BP (Hopkins *et al.* 1993).

While the results presented above all indicate deforestation at the time of the last temperate glaciations, the data are all from regions that are in some way marginal for the sustenance of tropical forest at the present. The Atherton Tableland, where Kershaw's (1978) data come from, lies only some tens of kilometres from the western boundary of rain forests in north Queensland, where lowered rainfall results in a replacement of forest by *Eucalyptus* woodland and savanna (Figure 2.5). Africa generally is the driest of the three large tropical regions, and the severity of forest shrinkage in that continent during glacial periods is therefore not surprising (Livingstone 1993). Furthermore, low rainfall at present makes northern Venezuela a marginal environment for tropical forest. While the Peten area of Guatemala today receives heavy rainfall, it nevertheless lies close to the northernmost extent of tropical forest in the Neotropics and its forest cover may be especially vulnerable to climate change.

No unequivocal evidence for widespread deforestation exists in lowland areas that are today climatically less marginal for tropical forests. In South-East Asia, the available pollen record shows no sign of deforestation, although there is some vertical depression of vegetation zones at higher altitudes (Maloney 1980; Morley 1982; Newsome and Flenley 1988). In the Neotropics, some evidence exists for the intrusion of montane forest elements into the lowland rain forests of adjacent regions during glacial conditions (Bush and Colinvaux 1990; Bush *et al.* 1990; Hooghiemstra and Van der Hammen 1993). Bush (1994) and Colinvaux (1993) both conclude that floristic reorganization driven by lower temperatures rather than drought was the primary source of instability in these forests during glacial conditions. However, this reorganization appears to have stopped well short of regional deforestation, although there may have been partial deforestation in some areas (Ledru 1993). In a recent study in Brazil that used soil carbon isotopes to detect the earlier presence of savanna grasses, Martinelli *et al.* (1996) found only a few sites showing signs of Quaternary deforestation, primarily in the southern part of the country. There was little evidence for forest replacement in Amazonia. Partial deforestation has also been reported for the wet region of west Cameroon (Giresse *et al.* 1994).

In summary, current data indicate that during glacial periods sea levels in the tropics were approximately 100 m lower and sea surface temperatures were approximately 5°C cooler than at present. Drier conditions were widespread, but led to deforestation only in areas that are today marginal for the support of tropical forest. In these areas, tropical moist forest was replaced by savanna, dry forest, or other forms of drought-tolerant vegetation, and in some cases dune-fields developed in areas now too moist to permit these. In areas not subject to deforestation, competitive interactions and forest composition were probably modified considerably by lower temperatures and precipitation, and in those areas adjacent to mountains, considerable intrusion of montane species probably took place. The return to contemporary interglacial conditions in the tropics appears to have been especially rapid around 10,000 BP, but this usually did not involve a simple switch from cool, dry to warm, wet conditions. Detailed post-glacial pollen records (e.g. Ledru 1993; Giresse *et al.* 1994) show evidence of a great deal of short-term temperature and moisture fluctuation during the Holocene. Assuming that the events that have been documented during the last glacial period in low latitudes were repeated multiple times during the Pleistocene in response to periodic global climate cooling, this period clearly emerges as a time of considerable flux and instability in low-latitude environments.

Forest survival and recovery

While not all areas now covered by tropical moist forest were deforested during Pleistocene cold periods, clearly some were, raising questions both about where forest biotas survived during these periods, and how they were able to achieve reforestation of the landscape during interglacial periods. This question is particularly enigmatic given the widespread belief that tropical forests are in some way fragile and unable to withstand major disruptions.

The rapid forest recovery in many areas that were deforested during the last glacial period (e.g. Figure 2.6) implies that forest biotas survived in local rather than distant refugia, and present subhumid landscapes probably provide the best analogue of what these places were like during dry phases. In subhumid landscapes, forest survives primarily as diffuse systems of elongated gallery forests occupying riparian (stream-side) habitats and headwater basins. These are usually of very variable width, and are often highly interconnected (Figure 2.7). Recent research in these gallery forest systems in Central and South America has shown that they sustain a forest flora that is of comparable richness to that of continuous forests at local scales (Kellman *et al.* 1994; Meave and Kellman 1994; Tackaberry and Kellman 1996), and

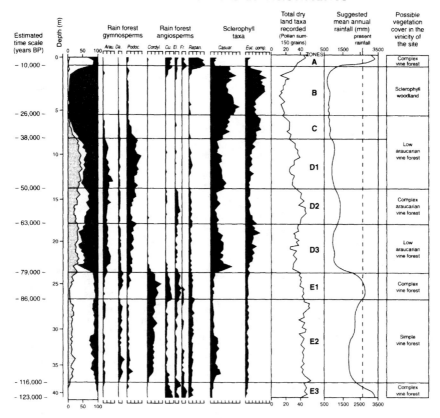

Figure 2.6 A pollen diagram from Lynch's Crater, Atherton Tableland, north-eastern Queensland (cf. Figure 2.5). The column at the left shows rain forest gymnosperms (light shading), rain forest angiosperms (white) and taxa of the dry sclerophyll forest (dark shading). The frequencies of individual taxa are shown as percentages of the pollen total, in 10 per cent increments. Abbreviations for taxa: *Arau.*, *Araucaria*; *Da.*, *Dacrydium*; *Podoc.*, *Podocarpus*; *Cordyl.*, *Cordyline* comp.; *Cu.*, *Cunoniaceae*; *El.*, *Elaeocarpus*; *Fr.*, *Freycinettia*; *Rapan.*, *Rapanea*; *Casuar.*, *Casuarina*; *Euc.* comp., *Eucalyptus* comp.

Source: Kershaw 1978: Figure 1

provide plausible temporary refugia for forest biotas during periods of Pleistocene drought. While in many areas these diffuse gallery forest systems may have been accompanied by some larger forest remnants, in others they are likely to have been the sole refuge site. In the Queensland sites studied by Kershaw (1978) (Figure 2.6), present forest cover is highly fragmented at a regional scale (Figure 2.5), and charcoal fragments of Late Glacial age exist in most of these (Hopkins *et al.* 1993). No viable forest refuge sites other than scattered riparian habitats appear to exist in this region, yet it has

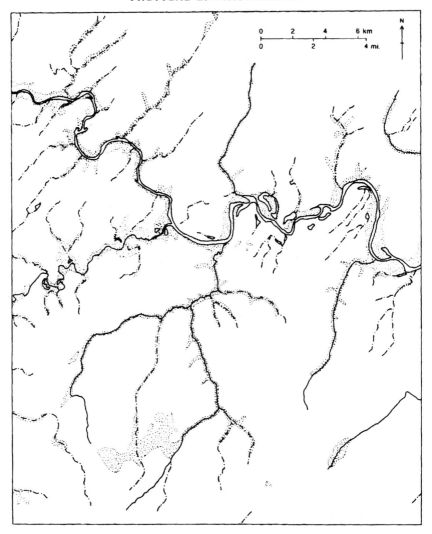

Figure 2.7 Distribution of gallery forests in the southern Apure savannas, Venezuela. These forests form a highly interconnected system of narrow forest corridors along most of the streams that permeate this savanna landscape. The parallel alignment of gallery forests in the northern part of the mapped area is related to the dunefields present in this zone (cf. Figure 2.1)

preserved a relatively rich forest biota through the Pleistocene disruptions (Adam 1992).

The potential for forest species to expand their ranges from scattered refugia of this sort has been examined in a savanna in Belize, where artificial fire suppression has been applied for fifty years (Kellman 1989). In this area, low soil fertility poses a special problem for invading tree seedlings. Fire cessation has been followed by an increase in the population densities of savanna shrubs and trees (Miyanishi and Kellman 1988), leading to the development of a closed woodland of these species. Comparable post-fire changes have been reported in the savannas of Venezuela by San José and Fariñas (1983), and Wijmstra and Van der Hammen (1966) have recorded palynological evidence for a similar vegetation type in the Rupunnuni savannas of Guyana in the past. A gradual accumulation of nutrients in surface soils beneath these trees and shrubs takes place (Kellman 1979), driven by increased nutrient cycling, and drawing upon gradual atmospheric accessions as the major nutrient source. Transplant experiments have shown that seedlings of a subset of slow-growing forest tree species are able to establish beneath these savanna trees and shrubs and continue the nutrient accumulation process (Kellman 1985a; Kellman and Miyanishi 1982). If nutrient capture is effective, it has been estimated that the nutrient capital of a tropical forest could be accumulated from atmospheric sources in as little as five hundred years (Kellman 1989).

It thus appears that in parts of the tropics where Pleistocene drought was sufficient to facilitate frequent fires, forests probably shrank to diffuse fragment systems concentrated primarily in riparian habitats. Here, temporary survival of forest species was achieved, and from here forest re-expanded relatively rapidly in fire-free intervals to cover the intervening uplands.

IMPLICATIONS OF ENVIRONMENTAL HISTORY FOR TROPICAL ECOSYSTEMS

Until recently, tropical ecosystems, and especially tropical forests, were thought of as the ultimate biological museum – sheltered from global climatic events, and developing characteristics such as diversity and 'fragility' that were intimately related to this assumed history of stability. Clearly, an accumulation of recent data shows a very different historical pattern. Tropical environments have been much more widespread in the distant past than at present, and more recently in the Quaternary have been buffeted by many more changes than was once assumed. While these recent changes have not extended to wholesale displacement by ice masses, as occurred in some high-latitude regions, they have nevertheless been substantial. What general implications does this history hold for understanding the form and functioning of tropical ecosystems? We suggest two possibilities:

1 Many characteristics of tropical biotas may have evolved under global biogeographic conditions very different from those which now prevail. As evolution is a gradual process, the evolution of a biota's characteristics cannot be adequately discussed in isolation from its geological history. As we have seen, humid tropical environments were far more widespread globally at the time when angiosperms were diversifying and beginning

their domination of the earth's vegetation. Conversely, tropical arid environments are geologically recent. In Chapter 5 we will discuss the degree to which the characteristic high and low diversities of humid and arid tropical biotas may be attributable to these differing histories.

2 The Quaternary history of the tropics presents a picture of ecosystems that are much more robust, and less fragile, than popular interpretations imply. This robustness has been manifested in the ability of some forest systems to undergo severe shrinkage and fragmentation during periods of drought, yet to preserve rich biotas which rapidly re-established a forest cover later. Even where deforestation did not occur, tropical forest ecosystems have also been able to survive the disruptions caused by reduced temperature and competition with montane taxa. This robustness gives some reason for cautious optimism about the prospects for conserving tropical forest biotas in the future. We will discuss this further in Chapter 5.

3

TROPICAL CLIMATES AND MOISTURE REGIMES

As we have seen in Chapter 2, the origin of the modern pattern of tropical climates dates from the early Miocene, when subtropical high-pressure zones and the arid climate conditions associated with these first appeared. The latitudinal zone lying between the subtropical high-pressure systems that occur at about 20°–30° now contains a variety of climates that lack the seasonally low temperatures of higher latitudes, but differ greatly in moisture regime (Figure 3.1). This low-latitude zone also comprises a special subsystem of the general circulation of the atmosphere, and most characteristics of the climate of any location within it can be traced ultimately to this circulation pattern. We therefore begin our consideration of tropical climates by describing the major characteristics of this atmospheric circulation.

ATMOSPHERIC CIRCULATION IN THE TROPICS

At a global scale, the atmosphere of low latitudes receives more solar radiation than it re-radiates back to space, while latitudes greater than approximately 38° re-radiate more than they receive. This energy imbalance serves to drive the general circulation of the atmosphere, which achieves the necessary latitudinal energy transfer and maintains the atmospheric temperatures at any place in a quasi-steady state. On a non-rotating planet, such an energy imbalance would set up a vertical convective circulation pattern involving atmospheric uplift in low latitudes, poleward movement at high altitudes, and a reverse equatorward airflow at low altitudes. However, on a rotating earth, a parcel of equatorial air at high altitude begins its poleward flow with a westerly velocity imparted by the rotation of the earth at the equator. As it moves poleward, it moves over a surface whose velocity is decreasing because of the earth's spherical shape and, as a result, the air parcel takes on an apparent westerly flow component. This apparent distortion of airflow direction has been called the Coriolis effect, and becomes more pronounced as latitude increases. It operates also in reverse; equatorward-moving air parcels take on an apparent easterly airflow. The net result of the Coriolis effect is a deflection of air movement to the right in the northern hemisphere, and

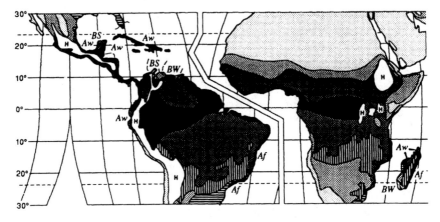

Figure 3.1 Distribution of the major Koppen–Geiger climate types in tropical latitudes. H, highland climates not classified

to the left in the southern hemisphere. Airflows that cross the equator in low latitudes undergo a partial reversal in flow direction.

At a global scale, the Coriolis effect distorts the poleward-moving high-altitude airflows into two large westerly circumpolar vortices in the mid-latitude zone of each hemisphere. Each contains one or more high-velocity jet streams at their core, beneath which lie major discontinuities between cold and warm air. Meandering patterns in these jet streams serve to inject cold and warm air masses equatorward or poleward, and in so doing achieve the major latitudinal energy transfer required to maintain thermal equilibrium in the atmosphere. In low latitudes, where the Coriolis effect remains relatively weak, some vestiges of the vertical convective circulation pattern persist, forming what have been termed Hadley circulation cells in each hemisphere. These involve vertical uplift at equatorial latitudes in a zone known as the Inter-Tropical Convergence Zone (ITCZ), or equatorial trough, followed by poleward flow to subtropical latitudes, where subsidence occurs forming persistent zones of high pressure at the surface. Finally, some of the subsiding air returns to equatorial latitudes at the surface. Because of the Coriolis effect, the poleward-moving air at high altitudes takes on a partially westerly flow path, while the return airflow at the surface assumes an easterly flow. These low-latitude easterly winds have been traditionally termed the trade winds; in the northern hemisphere they assume a north-easterly source direction, while in the southern hemisphere they assume a south-easterly direction.

The subtropical high-pressure belts comprise a series of semi-permanent discontinuous elongated high-pressure cells, centred over eastern ocean basins and more weakly developed over the western portions of these (Figure 3.2). Subsiding air in these subtropical cells assumes a westerly airflow when moving poleward, but an easterly airflow when moving equatorward, to

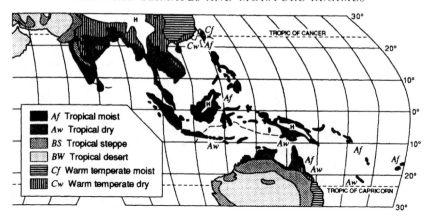

Figure 3.1(b)

comprise the trade winds. Although vertical circulation cells in low latitudes are predominantly latitudinal (the Hadley cells), there exists also a longitudinal component to vertical circulation (Figure 3.3); this was termed the Walker Circulation by Bjerknes (1969), in memory of Sir Gilbert Walker, who pioneered the investigation of the El Niño Southern Oscillation (see below). In the subtropical inter-cell zones of lower pressure that are located over western ocean basins (Figure 3.2), surface airflows tend to be more latitudinal, and it is in these areas that most tropical–temperate surface air mass transfers take place. For example, in the western Atlantic/Caribbean zone, northward injections of humid tropical air into eastern North America take place throughout much of the year, while in the northern winter, occasional cold air masses are injected southward to produce the 'Nortes' of Central America.

Seasonal variation in the general tropical circulation patterns described above takes two major forms. Positions of the ITCZ and subtropical high-pressure zones shift latitudinally in a seasonal cycle in response to seasonal changes in maximum solar angles and radiative heating (Figure 3.2). The greatest latitudinal shifts in the ITCZ take place over land surfaces, while over water bodies shifts are minor because the pattern of sea surface temperatures, which controls the ITCZ position, is very stable (Webster 1987a). As a consequence, only very low-latitude sites in the tropics are influenced directly by the ITCZ. The second seasonal change is felt much more widely. This involves longitudinal expansion of subtropical high-pressure cells to form semi-continuous belts during hemispheric winters, and their shrinkage during summers (Figure 3.2). Some inter-annual shifts in the Walker circulation cells also take place in association with the El Niño Southern Oscillation. This is discussed further below.

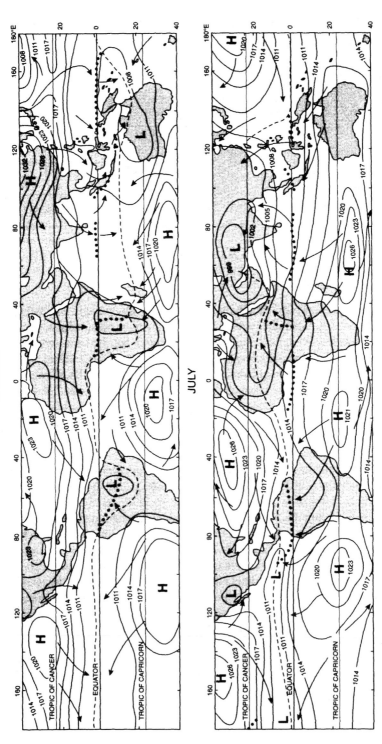

Figure 3.2 Mean sea-level surface pressure and predominant surface winds in tropical latitudes during January and July. The mean position of the Inter-Tropical Convergence Zone (ITCZ) is shown by a dashed line, and zones of secondary convergence are shown by dotted lines

Source: Nieuwolt 1977: Figure 4.2

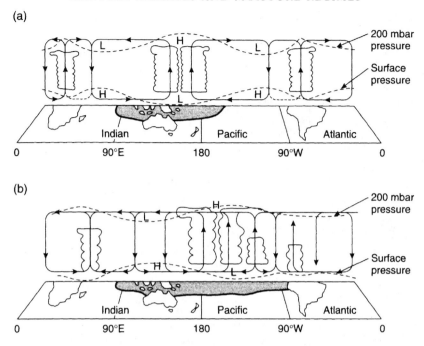

Figure 3.3 Schematic diagram of the Walker near-equatorial meridional circulation during years without (a) and with (b) an El Niño event. Shaded areas represent regions where sea surface temperatures exceed 27°C

Source: Webster 1987b: Figure 11.4

Trade wind airflows dominate the circulation patterns of most tropical sites not subject to monsoonal circulation (see below). The trade winds represent a highly persistent easterly surface airflow moving from the centres of subtropical high-pressure cells westward and equatorward to converge in the ITCZ. Embedded within the trades is a major vertical thermal anomaly known as the Trade Wind Inversion. Atmospheric temperature inversions represent layers in which temperature increases as one ascends, rather than decreases; their major effect is to suppress spontaneous vertical movement of air (convection), which depends upon a progressive decrease in temperature with height. In the trade winds, the inversion layer is lowest on the eastern sides of high-pressure cells over the eastern ocean basins, where it averages 500 m elevation but can occasionally be found at much lower altitude. The Trade Wind Inversion originates primarily from subsidence and warming of descending air in the high-pressure cells, with cold ocean water and advection of hot desert air playing a minor augmentative role on eastern ocean margins (Hastenrath 1991). As the airflow moves downstream over ocean waters, the inversion is gradually raised and degraded by convective activity,

31

and on the western sides of oceans and in the equatorial zone it exists only in a less intense form at 1,500–2,000 m (Hastenrath 1991). The inversion also shows a consistent seasonal pattern of variation, with the most intense development during hemispheric winters, and weakest development during summers. Because of its suppressive effect on convection, which is primarily responsible for tropical rainfall, the Trade Wind Inversion exercises a dominant influence on spatial and seasonal patterns of rainfall throughout much of the tropics. Its perennial intensity over central and eastern ocean basins results in these being persistently dry zones (Figure 3.4), while the western basins of oceans are subject to more convection and rainfall. Seasonal intensification of inversions accompanies the expansion of subtropical high-pressure cells during hemispheric winters (Figure 3.2) and results in a hemispheric dry season in the trade wind zone. Conversely, their weakening during hemispheric summers permits the increased convective activity and rainfall of the wet season.

Monsoonal circulations

The major anomaly in the global circulation pattern at low latitudes that has been presented above is in areas subject to monsoonal circulations. The term 'monsoon' refers to a seasonal reversal in airflow direction, which contrasts with the highly persistent patterns throughout most trade wind zones. Monsoons are driven primarily by the differing thermal characteristics of land and water bodies, and consequently are best-developed in areas on the margins of the large Asian land mass. Here, extreme surface temperature and pressure differences develop between summers and winters (Figure 3.2), resulting in summer inflows of air from surrounding water bodies, and a reverse winter outflow from the land mass. The extreme low pressure that develops over Asia during the northern summer precipitates inflows of moist, unstable air throughout South Asia at this time, constituting the rainy summer monsoon. Latent energy is transferred inland in these airflows and released during convectional activity, further intensifying the monsoonal circulation (Webster 1987a). In contrast, intense cooling and high pressure develop over Asia in winter, resulting in an outflow of cool, stable air over South Asia. This comprises the dry winter monsoon. This represents a highly simplified account of the Asian monsoon, which in detail is complex and much influenced by upper-atmosphere circulation patterns and the interaction of these with the Himalayan massif. A more elaborate summary is provided by Nieuwolt (1977), and the whole monsoon phenomenon is treated in great detail by Fein and Stephens (1987).

The Asian monsoon is coupled to a smaller monsoon in northern Australia. During the Asian winter monsoon, low pressure develops over Australia, which at that time is experiencing maximum radiative heating, and north-easterly airflows crossing the equator over Indonesia recurve to become north-westerly onshore airflows over northern Australia (Figure 3.2). This brings a wet monsoon season to the area. Conversely, high pressure develops over Australia during the Asian summer monsoon, and south-easterly offshore airflows from that continent pass over Indonesia and contribute to the Asian summer monsoon. Throughout much of the annual cycle of the Asian–Australian monsoon, the Malesian region, comprising Indonesia and peninsular Malaya, is subject to moist, unstable airflows, albeit from differing directions, and many stations in this region show reduced rainfall seasonality. This is also the tropical region which so far shows fewest signs of drought during the late Pleistocene (Chapter 2).

The climates of tropical Africa have also often been treated as monsoonal. While seasonal reversals of airflow affect much of sub-Saharan Africa, these can be interpreted as a consequence of large latitudinal movements in the ITCZ, and the reversals in airflow deflection that accompany a crossing of the equator, rather than the complete pressure reversal that drives the Asian monsoon. Thus, West Africa is subject to dry north-east trade winds from the Saharan high-pressure zone during the northern hemisphere winter. During the northern summer, the ITCZ moves well inland over the Sahel, and south-east trades from the southern hemisphere recurve into an onshore south-westerly airflow after crossing the equator, producing a rainy season in the region (Figure 3.2). In East Africa, the ITCZ undergoes a much larger seasonal latitudinal migration (Figure 3.2), but air masses converging on it throughout the year are generally of dry continental origin. Consequently, an abrupt oscillation between wet and dry season does not accompany the reversals of airflow that occur in this region (Nieuwolt 1977).

Ocean circulation

The general circulation of tropical oceans is closely linked to the pattern of surface winds, as these provide the primary stress that drive surface currents. Ocean circulation in the tropical Atlantic and Pacific oceans is dominated by subtropical gyres, closely linked to the subtropical high-pressure cells discussed earlier. The equatorial westward-flowing sectors of these gyres are driven by the trade winds and result in a pile-up of water in the western basins of both oceans; in the equatorial Pacific this results in a water surface approximately 40 cm higher than in the eastern Pacific (Hastenrath 1991). Accompanying this phenomenon is a deepening of the upper layer of water warmed by solar heating and wind mixing from 50 m in the eastern Pacific to 200 m in the west. The pile-up of water in the western oceans also drives an eastward-flowing equatorial countercurrent concentrated at about 100 m depth.

The circulation pattern of gyres also involves cooler currents of temperate origin moving towards the equator on the eastern ocean margins, as well as surface divergence and upwelling of deep ocean water in these zones. The low sea surface temperatures created by these processes contribute to the lack of convective activity and severe desert conditions that characterize these zones, but the upwelling of nutrient-rich water from great depths stimulates very high organic productivity in these zones. This provides the basis for important fishing industries, the most famous of which is the Peruvian anchovy fishery. As we shall see below, periodic suppression of this upwelling can occur, and has catastrophic consequences for the anchovy catch.

Ocean circulations in the Indian Ocean are considerably more complex as a consequence of the seasonal wind reversals of the Asian monsoon. The circulation in this ocean is treated at length by Hastenrath (1991), and will not be dealt with further here.

TROPICAL RAINFALL

In low latitudes, variations in precipitation rather than in temperature impose the basic temporal rhythms and spatial patterns on climate, and require that special attention be devoted to this phenomenon.

Mechanisms and sources

In the tropics, as elsewhere, rainfall is derived from uplift of parcels of air to a level at which condensation occurs and, later, cloud droplets coalesce to form raindrops. However, the primary sources of uplift are distinctive in the tropics. In temperate areas, most uplift occurs at the frontal systems associated with air mass boundaries; only under summer conditions do these areas receive much rainfall from spontaneous convection. In contrast, most tropical rainfall is convective, which has as a consequence certain distinctive properties. Chief among these is a tendency for rainstorms to be of short duration and high intensity, and to be patchily distributed at local scales. Hudson (1981) estimates that 40 per cent of all tropical rainstorms exceed 25 mm.h^{-1}, a rate considered to be the threshold intensity at which rainfall becomes erosive. In contrast, temperate areas are estimated to receive only 5 per cent of their rainfall in storms of this intensity.

The lowest levels of the atmosphere in tropical latitudes generally show steep lapse rates as a result of surface heating by high radiation loads; this leads to instability and a tendency for development of spontaneous convective cells daily when surface heating is greatest. Whether this convective activity continues to the point of cloud formation and later rainfall from thunderstorms depends upon the vertical structure at higher levels in the atmosphere. The most widespread source of suppression of convective activity is the Trade Wind Inversion, and during the hemispheric winter, when this tends to be best developed, small cumulus clouds are frequent but rarely develop further into thunderstorms. At a more regional scale, dynamic subsidence and divergence of air masses will also tend to suppress convective

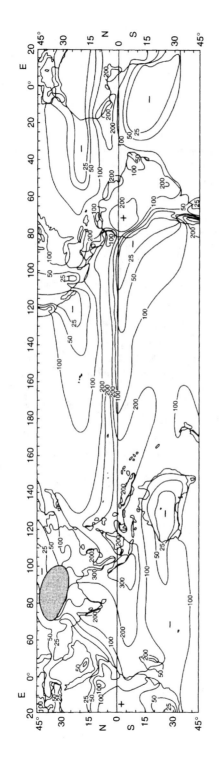

Figure 3.4 Mean annual precipitation (cm) in lower latitudes

Source: Riehl 1979: Figure 3.4

activity. For example, the many dry zones on the western sides of tropical mountain systems can be attributed to localized subsidence, while, at a larger scale, the southern Caribbean dry zone (Figure 3.1) has been attributed to the orientation of the coastal mountain ranges of northern South America, which leads to windfield divergence (Trewartha 1981). Less heating of the lower atmosphere would clearly also suppress convective activity, and within the tropics the coastal zones of cold oceanic currents on the eastern margins of oceans lead to some of the most intense desert conditions on earth (Figure 3.4). For similar reasons it is assumed that a 5°C lowering of sea surface temperatures in the tropics during glacial maxima (Chapter 2) must have reduced rainfall to some degree.

In contrast to these suppressive effects, any conditions encouraging convection to higher altitudes may be expected to increase rainfall. At a global scale, airflow convergence in the ITCZ leads to frequent thunderstorm activity, and it is therefore not surprising that a zone of high annual rainfall coincides with the mean position of this (Figures 3.2 and 3.4). Regionally, the weakening of the Trade Wind Inversion during the hemispheric summer is the principal stimulus to greater rainfall at this time. At other times, temporary convergence can be induced in trade winds by weak waves that travel slowly westward in these, giving brief periods of rainfall during their passage. At local scales, orographic uplift caused by major mountain barriers leads to some of the highest recorded rainfalls on earth, during the wet seasons of trade wind zones, or the onshore monsoons of Asia.

The diurnal pattern of rainfall at most tropical stations reflects its predominantly convective origin. Over land surfaces, maximum solar heating and convective activity take place between late morning and the late afternoon, during which time most thunderstorm development and rainfall occur. Figure 3.5, from Barro Colorado Island, Panama, shows this pattern, with higher probabilities of rainfall restricted to a distinct wet season, within which most rainfall is received between 1100 and 1800h. Particularly intense thunderstorms may persist into the evening hours in some terrestrial locations. However, nocturnal rainfall is more characteristic of marine and coastal locations, owing to the thermal properties of water, which lead to a steepened lapse rate in the lower atmosphere at night and result in more convection. Of course, tropical rainfall of atypical origin, such as that caused by hurricanes or cold outbreaks and frontal activity (see below), does not produce this characteristic diurnal signal.

Over ocean bodies, a readily available moisture source for tropical precipitation exists in water surface evaporation. The existence of extensive dry zones in subtropical latitudes over the oceans (Figure 3.4) is a clear indication of the control that general circulation patterns, rather than localized moisture supply effects, have on tropical rainfall patterns. Much less clear is the control of localized moisture supply effects on rainfall over tropical land masses, and the degree to which changes in vegetation cover may affect this. One would

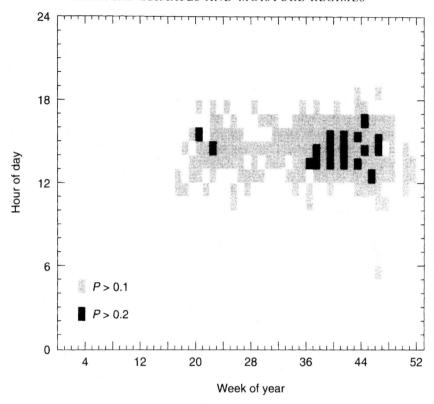

Figure 3.5 Distribution of the higher probabilities (> 0.1 and > 0.2) of receiving rainfall according to time of day and season at Barro Colorado Island, Panama, for the period 1929–84

Source: Windsor 1990: Figure 10

expect only minor effects over archipelagos, such as that of Malesia, which are embedded in maritime influences, and potentially major effects over the larger tropical land masses of Africa and South America. Recent work in Amazonia has provided some data on this phenomenon. Salati and Vose (1984) have used isotope signatures and mass budgeting to estimate the provenance of rainfall in the Amazon basin, and concluded that approximately half of the rainfall is internally generated, and half advected from the Atlantic Ocean. If deforestation were to lead to significant decreases in evapotranspiration from the terrestrial surface, there would clearly be cause for concern. Shuttleworth (1988) has estimated that a 10–20 per cent reduction in evapotranspiration would follow conversion of Amazonian forest to pasture as a result of increased albedo, decreased rainfall interception and reduced rooting depth and water uptake by pasture. Nepstad *et al.* (1994) have since

37

confirmed the last condition by comparing deep-soil moisture profiles in forests and pastures. Moreover, Shuttleworth's estimates assume no climate change, and more recent work by Shukla *et al.* (1990) suggests that this may be an unrealistic assumption. These authors use a coupled biosphere–atmospheric general circulation model to examine the effects of complete deforestation of Amazonia. The results of their simulation predict a regional reduction in precipitation of 26 per cent, which is larger than the regional reduction in evapotranspiration, and implies that dynamical moisture convergence into the basin may also decrease as a result of deforestation.

Spatial pattern in rainfall

The mechanisms of rainfall generation described above, coupled with the general atmospheric circulation patterns described earlier in the chapter, result in patterns of mean annual precipitation throughout the tropics that are generally interpretable (Figure 3.4).

Near-equatorial zones that lie close to the mean position of the ITCZ are generally high-rainfall areas, beyond which annual rainfall decreases as the subtropical high-pressure zones are approached. The central and eastern ocean basins that are most subject to the subtropical high-pressure cells are also distinctly drier zones than the western ocean basins. The latter areas represent discontinuities between these high-pressure cells and are much more subject to a variety of rainfall-inducing conditions. Chief among these are a well-elevated Trade Wind Inversion that allows convective activity throughout a large part of the year. However, these are also the areas subject to most hurricane activity (see below), and the very high rainfalls resulting from these can have a significant effect on long-term rainfall averages. Finally, these are the principal tropical areas subject to outbreaks of cooler air masses from higher latitudes and the frontal precipitation that this may bring. Because these latitudinal injections take place over an expanding surface area they thin rapidly and lose their temperate characteristics; however, they normally bring several days of rainfall in a winter season that is otherwise dry in this zone. In the Malesian region, the western ocean basin effects are augmented by pervasive maritime influences and persistently high sea surface temperatures, coupled with alternating monsoonal airflows. It is therefore not surprising that this area represents the most extensive zone of high precipitation within the tropics (Figure 3.4).

At a more local scale, orographic effects and prevailing airflow patterns result in the eastern seaboards of mountainous areas in the trade wind zone being high-rainfall areas, while adjacent areas in the lee of the mountain barrier are often very dry rainshadows (Figure 6.7). This effect is apparent in island systems, such as Hawaii, Fiji and most of the West Indies, where 'windward' and 'leeward' shores experience very different rainfall totals. It is also apparent throughout much of Central America, whose wet eastern seaboard was originally covered by evergreen forest, while the drier west coast supported mainly deciduous forest. Orographic effects are also apparent in monsoonal zones, where mountainous areas subject to onshore airflows may receive especially heavy rainfalls. The Western Ghats of the Indian subcontinent (Figure 3.9) and the western mountain ranges of Burma both experience exceptionally heavy rainfalls.

While much of the patterning of tropical rainfall within specific regions can be interpreted by reference to these broad generalizations, it is important also to emphasize that almost every region possesses local idiosyncrasies that are often much more difficult to explain. Moreover, larger-scale regional anomalies also exist, such as the southern Caribbean dry zone mentioned above. Many of these larger-scale anomalies are treated by Trewartha (1981), and the reader is referred to this author for a fuller discussion of these 'problem' climates.

Rainfall seasonality and variability

Most tropical areas experience at least some seasonality in rainfall, representing periods when convective activity is either suppressed or enhanced, and much of the variation in annual rainfall totals at tropical stations is attributable to variations in the relative lengths of these two seasons. In trade wind areas seasonality is driven by seasonal weakening and intensification of the Trade Wind Inversion, while in more equatorial latitudes the passage of the ITCZ brings with it increased rainfall. In monsoon areas, onshore-converging vs. offshore-diverging airstreams provide for a dramatic contrast in rainfall regimes. In detail, wet and dry seasons comprise extended periods of several months in which daily rainfall occurs with high or low frequencies (Figure 3.6). Shorter periods with the converse conditions may be embedded within the multi-month seasons; these may possess complex periodicities. For example, the Indian monsoon is subject to regular 'breaks' in precipitation at time-scales of 10–20 days and 40–50 days (Webster 1987b).

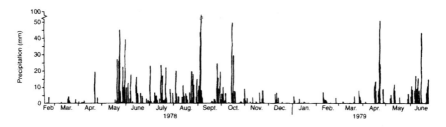

Figure 3.6 Daily rainfall recorded at Siguatepeque, central Honduras, during a 16-month period. This area experiences distinct wet and dry seasons, but each season can contain brief periods of the converse weather conditions

Data source: Kellman *et al.* 1982

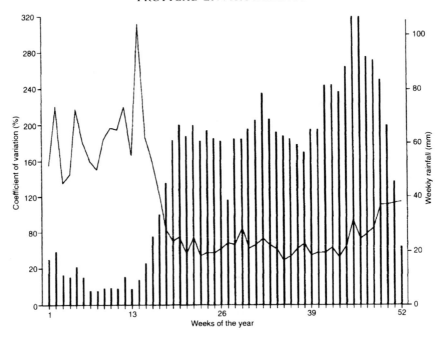

Figure 3.7 Seasonal distribution of weekly rainfall, and the variability in this, at Barro Colorado Island, Panama

Data source: Windsor 1990 (for the period 1929–86)

Some tropical climate records exhibit a distinct bi-phasic rainfall regime, with an abrupt transformation between phases. For example, Figure 3.7, from Barro Colorado Island, Panama, shows a 15-week dry season, during which less than 20 mm of rain per week falls, followed by an abrupt shift to a prolonged wet season of 32 weeks, during which weekly rainfall exceeds 50 mm. The two transition periods comprise only 5 weeks in total. More commonly, the seasonal change in rainfall is more gradual, with monthly rainfall building gradually to one or two peaks (e.g. Figure 3.8). Wet seasons with double peaks in equatorial latitudes are normally associated with a double passage of the ITCZ. At higher latitudes, some double peaking may also occur in response to latitudinal migration of the overhead sun and maximum convective activity or periodic hurricane activity. A monthly rainfall total of 100 mm is commonly used as an arbitrary definition of dry season conditions.

The inter-annual variability in rainfall totals and seasonality is of critical concern for the planning of agricultural activities in tropical areas. There is a well-established negative correlation between annual rainfall totals and the inter-annual variability of these that applies at all latitudes, including the

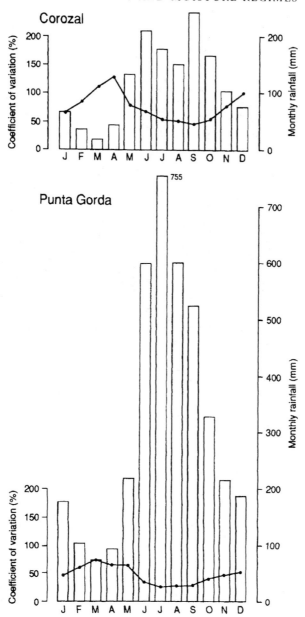

Figure 3.8 Monthly rainfall and its variability at two stations in Belize for the period 1935–70. Although both are located at sea-level and only 250 km apart, these two stations receive very different rainfall totals, but exhibit a common seasonal pattern of rainfall. Notice the greater proportional variability of rainfall that is characteristic of the drier station generally, and the dry seasons at both stations relative to wet seasons

Data source: Walker 1973

41

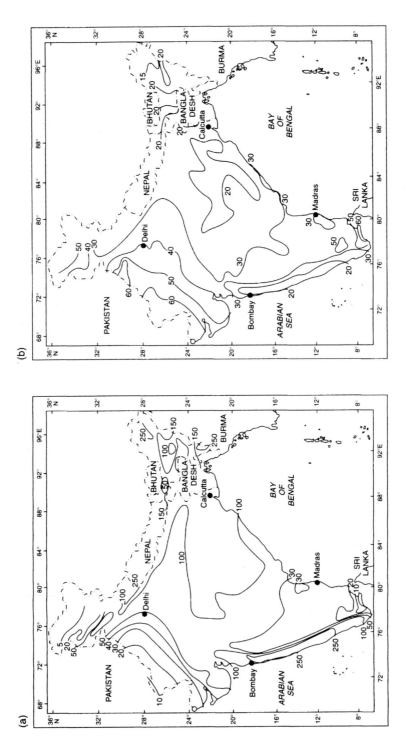

Figure 3.9 Distribution of (a) mean annual rainfall (cm), and (b) its variability (percentage coefficient of variability) in India

Source: Das 1987: Figures 7.4 and 7.5

tropics, and at both global and local scales (Figure 3.9). Thus drier parts of the tropics, such as north-eastern Brazil, north-west India and much of Africa, suffer not only from low rainfall, but also a high degree of variability in this. The droughts and accompanying human misery that have periodically afflicted parts of Africa in recent decades are often ascribed, in the popular literature, to human-induced 'desertification'. However, it should be remembered that most of these areas possess an inherently variable rainfall regime. As the carrying capacity of the habitats in these areas is usually moisture-limited, this too varies erratically. Under low population densities such variability can be accommodated by migration or storage of crops. However, where a large population has developed, a drought-induced fall in carrying capacity can have disastrous consequences. Rainfall variability creates special problems for animal husbandry in drier tropical areas; these will be discussed more fully in Chapter 10. Hodell *et al.* (1995) have recently documented a prolonged period of drought in the central Yucatán that was coincident with the collapse of the Classic Maya civilization in this region, and suggested a potential connection between the two events. This interpretation is strengthened by the work of Deevey *et al.* (1979) elsewhere in the Maya area, which suggests that the human population was approaching or had exceeded the environmental carrying capacity by the Late Classic.

Potential periodicities in inter-annual rainfall variability in drier parts of the tropics have been the subject of much research in recent decades. This work has been summarized by Hastenrath (1991), who concludes that there is evidence for a quasi-biennial periodicity in Indian rainfall, as well as that of the equatorial Atlantic and north-eastern Brazil. The latter two areas also show a periodicity at 13–14 years, while the outer tropics and subtropics of the Americas and Africa show periodicities at time-scales of 2–3 decades.

The negative correlation that exists between annual rainfall totals and inter-annual variability can also be found within seasonal patterns. The relative variability of rainfall (usually measured as the coefficient of variation) is characteristically higher during the dry season at most tropical stations (Figures 3.7 and 3.8). This clearly poses problems for dry-season agricultural activities that cannot draw upon irrigation. Greater variability is often present during the dry season–wet season transition, reflecting a large inter-annual variability in the timing of the onset of rains (or the summer monsoon). The date of onset of the summer monsoon in Kerala, southern India, has a standard deviation of 7.7 days about its normal onset date of 1 June (Das 1987). However, the absolute range of onset dates for the 1901–78 period has varied by more than a month (11 May–18 June).

The El Niño Southern Oscillation

Over the past two decades, an accumulation of evidence suggests that many of the inter-annual variabilities in rainfall throughout the tropics are to some degree synchronous and may be linked to a periodic oscillation in the atmosphere–ocean system of the tropical Pacific that has become known as the El Niño Southern Oscillation (ENSO). The event's name derives from a periodic warming of the Pacific waters off Peru and Ecuador that appears about Christmastime (the time of 'El Niño' in Spanish). While the mechanisms initiating an ENSO event remain obscure, the details of the oceanic and atmospheric changes associated with it have in recent decades become increasingly well-documented, especially since the particularly severe event of 1982–3 (Rasmusson and Wallace 1983).

ENSO events have been recorded since the eighteenth century and occur on average at 4-year intervals. However, intervals vary between 2 and 10 years (Cane 1983), and there has been a particularly persistent warming since 1990 (Kumar et al. 1994). Events begin with a weakening of the Walker circulation cell and the easterly trades in the central equatorial Pacific (Figure 3.3). The lessened wind stress results in a weakened westerly current and reduced pile-up of warm water in the western Pacific and the development of high sea surface temperatures throughout the equatorial Pacific, extending to the coasts of Peru and Ecuador (Horel and Wallace 1981; Hastenrath 1991). Coincident with this is an eastward shift of the zone of major convection and rainfall of the Walker cell, resulting in anomalously heavy rainfall in the normally dry eastern and central Pacific, and drier-than-average conditions in the normally wet Malesian and Indian Ocean region (Figure 3.4).

The most obvious biological consequence of an ENSO event is the suppression of cold upwelling in the eastern Pacific, leading to a collapse of marine productivity and the anchovy fishery (Barber and Chavez 1983). Other effects include the severe forest fires of Borneo that occurred during the 1982–3 event (Leighton and Wirawan 1986; Goldammer and Seibert 1990) and fires in the Galápagos Islands during 1984 fuelled by the death and drying of biomass that accumulated during the preceding wet event. While the most direct climatic effects of these ocean/atmosphere oscillations are felt in the Pacific basin, there is an accumulation of evidence pointing to links between them and weather conditions elsewhere in the tropics (e.g. Meehl 1994) and beyond the tropics, especially in North America (Graham et al. 1994).

TROPICAL STORMS

In the tropics, as elsewhere, thunderstorms may be accompanied by severe downbursts of wind that may cause localized damage. These have recently been implicated in areas of tropical forest damage in Amazonia observed on satellite imagery (Nelson et al. 1994). However, the tropics are home to one class of unique, and especially damaging, storms that have been called hurricanes in the Atlantic and eastern Pacific oceans, typhoons in the western Pacific, and cyclones in the Indian Ocean. For convenience, we will refer to these as hurricanes. These are rotational storms with high wind speeds (defined as >32 m.s^{-1}) and heavy rainfalls that cause severe wind damage and flooding in certain parts of the tropics.

Approximately 80 hurricanes are experienced worldwide per year (Hastenrath 1991), and they occur in well-defined regions and seasons (Figure 3.10). Hurricanes are essentially oceanic phenomena, and most damage is caused in coastal areas when they make landfall. Three conditions are known to be essential for their formation (Hastenrath 1991): (i) sea surface temperatures of at least 26.5°C; (ii) a Coriolis effect experienced only at latitudes greater than 5° N or S latitude; and (iii) small vertical wind shear. The requirement for high sea surface temperatures results in their being seasonal phenomena, characteristic of hemisphere summers and autumns, when sea surface temperatures are at their highest. Their origins can be traced primarily to minor disturbances in the trade winds, although the reasons why most disturbances meeting the above conditions do not metamorphose into hurricanes remain enigmatic (Simpson and Riehl 1981). Once formed, they generally drift westwards in the trades, with a lifespan of about a week, and often follow parabolic paths in ocean basins that recurve around the subtropical high-pressure anticyclonic cells (Figure 3.10). Most damage is caused on islands and eastern continental shorelines that lie in these movement corridors.

Fully developed hurricanes comprise an intense low-pressure core (the 'eye') into which high-velocity winds converge and spiral upward, producing copious quantities of rainfall and thereby releasing the large quantity of latent energy that sustains the system. Access to warm ocean water is essential to sustenance of the hurricane, and passage over land surfaces or cooler ocean water normally leads to rapid degradation of the storm. Maximum wind speeds are normally experienced within 30 km of the eye, and rainfall of 100 mm per day can be experienced in this zone. While the most spectacular forms of damage are caused by high wind speeds, much more widespread destruction, and most deaths, are caused by flooding induced by the heavy rainfall and, in coastal areas, the large tidal surges caused by the onshore winds.

The mean hurricane recurrence interval, even in hurricane-prone zones, is usually of the order of decades. For example, the country of Belize, which lies in such a zone and has an eastern shoreline 300 km in length, has experienced 12 hurricanes during the period 1871–1963, representing a mean recurrence interval of one hurricane every 7.7 years for the country as a whole (Cry 1965). However, if one assumes that major hurricane effects are experienced only within 50 km of the eye (Simpson and Riehl 1981), or along a 100-km stretch of shoreline, this translates to a recurrence interval of 23 years for any place on the country's eastern shore. Despite this apparent low frequency, hurricanes impose a large economic burden on potentially susceptible areas. Buildings must be constructed to withstand hurricane-force winds, or periodically be rebuilt. Large-scale crop losses can also be expected, together with flood damage to transportation networks. Tropical forest trees

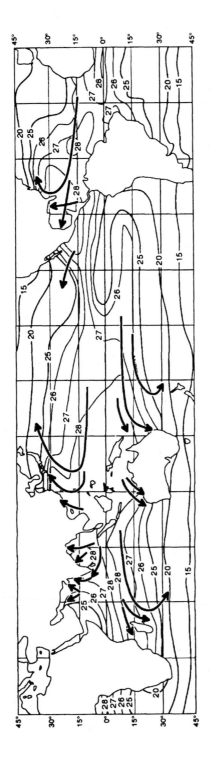

Figure 3.10 The principal tracks of tropical hurricanes and sea surface temperatures (°C) during the warmest season

Source: Palmen 1948: Figure 4

are especially susceptible to wind damage (Boucher *et al.* 1990), and slow recovery of these after hurricanes can leave the forests of some zones in a perennially damaged state (Webb 1958; Tanner *et al.* 1991). The dependence of hurricanes on high sea surface temperatures implies that hurricanes must have been virtually absent during full glacial conditions (Wendland 1977). Conversely, any future increase in sea surface temperatures as a result of global warming can be expected to result in an increased frequency, and possibly intensity, of these storms.

TROPICAL MOUNTAIN CLIMATES

Mountainous terrain occupies a significant proportion of the land surface area of the tropics and historically has been an important locus for human populations, especially in Central and South America. These mountain areas are often centres of volcanic activity and frequently contain more favourable soils than other tropical areas (Chapter 4), but they also possess specialized climatological conditions. Unfortunately, they rarely possess a set of weather stations of a sufficient density to capture the complexity of topographically induced climate variability, making generalizations about their climates rather tentative.

Environmental temperature lapse rates in the free atmosphere approximate 6°C per 1,000 m, but are subject to considerable variability by season and climate zone over land surfaces. Assuming a mean annual sea level temperature for the tropics of 26°C, this lapse rate places mountain areas with a mean annual temperature of 0°C at *c.* 4,300 m, which generally corresponds with the observed level of the lower snowline in these latitudes (Barry 1992). At these altitudes, seasonal variability in temperatures is usually as low as in the lowlands (<5°C), but diurnal temperature variations of 10–15°C may occur. While these specialized very-high-altitude tropical climates, and the equally specialized organisms that occupy them, have received considerable attention in the scientific literature (e.g. Rundel *et al.* 1994), it is mountain climates at lower altitudes (500–2,500 m) that are more widespread and affect most human populations in tropical mountain areas. At these altitudes, the climatic conditions and seasonality imposed by patterns in general circulation of the atmosphere persist, but are modified to varying degrees by lower temperatures, increased cloudiness and orographically enhanced precipitation.

Over much of the tropics, temperature decreases of *c.* 6°C per 1,000 m prevail throughout the lower atmosphere, but are reversed at the Trade Wind Inversion. Convective activity below the Trade Wind Inversion, augmented by orographic uplift, creates cloud masses that envelop most tropical mountain areas in a diurnal cycle that usually begins in late morning and persists into the evening hours. The cloud base is typically at 500–700 m in coastal areas and 600–1,000 m inland (Barry 1992). Associated with this convective activity is an increase in rainfall with altitude that generally peaks at

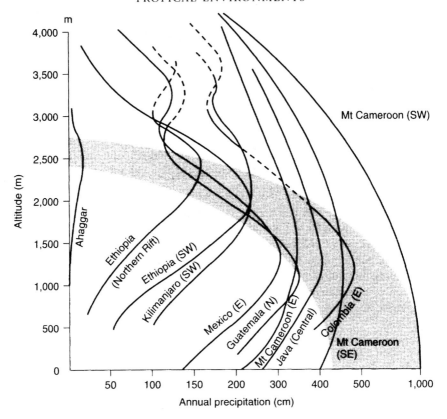

Figure 3.11 Generalized profiles of mean annual precipitation (cm) versus altitude for different tropical locations. The shaded area denotes the zone of maximum precipitation

Source: Barry 1992: Figure 4.14

intermediate elevations and then decreases above the Trade Wind Inversion. The zone of maximum rainfall lies generally between 1,000 and 1,500 m, but tends to be lower in high rainfall zones and higher in drier zones (Figure 3.11). Very-high-altitude sites in the trade wind zone, such as the highest peaks of the Hawaii volcanoes, are persistently above the Trade Wind Inversion, and experience virtual desert conditions. When coupled with high solar radiation loads and extreme diurnal temperature variations, this creates an especially harsh environment for plant life.

The persistent cloud cover at mid-elevation sites creates a comparably unique, but much less harsh, habitat for plant growth. These sites are not only subject to high rainfall, but also experience much reduced transpirational stresses, and augmentation of rainfall by the impaction of cloud droplets on plant surfaces. This phenomenon, commonly called 'fog drip', may double effective precipitation in drier areas, especially on windward slopes (Cavelier and Goldstein 1989). The 'cloud forests' that occupy this zone will be described in Chapter 6. Exposed peaks and ridge crests generally experience higher wind speeds as a result of reduced frictional drag on free air movement, but in the tropics, trade winds generally weaken with altitude, reducing this effect of exposure. However, such places may be especially vulnerable during tropical hurricanes (Tanner *et al.* 1991).

TROPICAL HYDROLOGY

The seasonal rhythm between wet and dry seasons that exists in some form in most tropical regions is reflected in the hydrological regimes of these places. With increased rainfall seasonality, both soil water regimes and the surface flow of rivers begin to diverge into characteristic wet- and dry-season modes. These differences have profound effects on the functioning of natural terrestrial and aquatic ecosystems, as well as upon agricultural activities.

Soil moisture regimes

Although exceptions occur, tropical soil profiles and their underlying weathered mantles are generally deep and contain large quantities of clay-sized soil particles, conditions which produce relatively large soil moisture storage reservoirs (see Chapter 4). The profiles may sometimes contain impediments to free vertical movement in soil moisture, such as a horizon of increased clay content, unweathered bedrock or an ironstone cuirass, which may also serve to inhibit root penetration to great depth.

In areas with little rainfall seasonality, the entire soil profile remains at close to field capacity throughout most of the year. Figure 3.12 provides data from San Carlos de Rio Negro in southern Venezuela, which illustrates such a condition. The site receives 3,565 mm of rainfall per year and does not have a distinct dry season, but experiences occasional dry spells (Franco and Dezzeo 1994). Soil water suctions measured in the upper 100 cm of a well-drained sandy soil throughout a 15-month period showed suctions persistently above field capacity ($c.$–100 mb). Minor decreases in suction in upper soil layers were halted abruptly by frequent daily rainfall (Figure 3.12). The only exception to this pattern was a period with little rain in January–February 1983, when soil suctions fell below field capacity throughout much of the profile, and moisture conditions at the wilting point are estimated to have prevailed for about a week (Franco and Dezzeo 1994). Under a soil moisture regime of this sort, relatively frequent drainage of soil water from the profile can be expected. In sandy soils beneath forest in Veracruz, Mexico, Kellman and Roulet (1990) found that drainage output from the profile tended to occur on any day when rainfall exceeded 10 mm, once the soil profile had become completely wetted. Working in deciduous forest in central

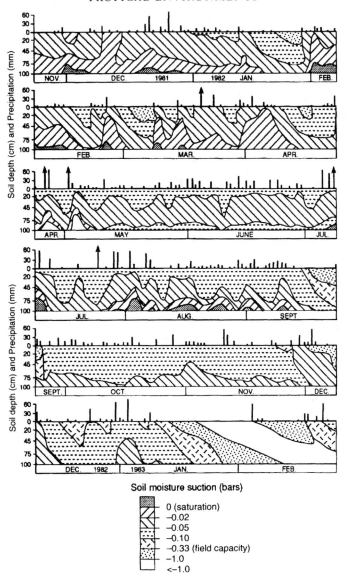

Figure 3.12 Variations in soil moisture suction in the upper metre of an oxisol beneath evergreen tropical forest at San Carlos de Rio Negro, Venezuela, during a 15-month period

Source: Franco and Dezzeo 1994: Figure 2

Venezuela, San José *et al.* (1995) found soil moisture drainage to occur during the wet season on any day with rainfall exceeding 30 mm; the difference as compared with the Veracruz data presumably reflects the larger moisture storage capacity of the finer-textured soils in Venezuela. Where a horizon with lower permeability exists within the profile, vertical movement of drainage water may be accompanied by periodic saturation, and considerable lateral downslope flow above the horizon. In high-intensity rainstorms this may lead to saturation overland flow, especially in downslope locations, which will contribute to storm flow in rivers.

Soils in regions experiencing a prolonged dry season undergo a progressive drying of the profile as the season advances as a result of moisture use in plant transpiration that exceeds rainfall input. For example, potential evapo-transpiration rates (Penman formula) for Barro Colorado Island average 139 mm per month during the dry season (Dietrich *et al.* 1982), when rainfall generally averages less than 50 mm per month (Figure 3.7). Soil moisture suctions in the upper soil layers may fall well below the wilting point by the late dry season, and plants rooted exclusively in this zone undergo a period of deciduousness (Jackson *et al.* 1995). It has usually been assumed that few plants have access to moisture in deeper soil layers, but the recent work by Nepstad *et al.* (1994) in Amazonia shows that forest trees continue to draw down soil moisture reserves from depths exceeding 8 m. The early rains of an ensuing wet season arrive on soils that have a large, and depleted, storage capacity. Sequential wetting by early rains is gradual and stimulates a rapid proliferation of fine roots and, presumably, increased moisture absorption (Kavanagh and Kellman 1992; Kummerow *et al.* 1990). As a result, soil moisture does not begin to drain from the soil profile until well after the rains have begun. In the sandy soils of low moisture storage capacity studied in Veracruz, drainage during three wet seasons did not begin until 33, 45 and 48 days after the start of rains (Kavanagh and Kellman 1992; Kellman and Roulet 1990). The net result of these processes is that the frequency of soil water drainage, and the leaching that this induces, is considerably reduced in areas with a dry season. For example, San José *et al.* (1995) found only 11 and 12 days with soil water drainage at 50 cm in two 7-month wet seasons beneath deciduous forest in Venezuela. At this site, only 14 per cent of the annual rainfall of 1,133 mm is estimated to have percolated to ground water. This reduced leaching regime in drier parts of the tropics has some distinctive consequences for soil formation that will be discussed in Chapter 4.

River flow regimes

The drainage basins of most tropical rivers are subject to some seasonality in rainfall inputs, and the river flow regimes normally reflect this. For example, Figure 3.13 shows the mean monthly discharge at four locations on the Belize River in Central America. Although this area is subject to only

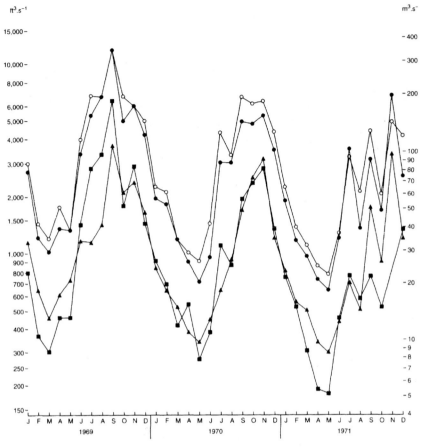

o   Mean monthly discharge at Bermudian Landing
•   Mean monthly discharge at Banana Bank
▲   Mean monthly discharge at Benque Viejo del Carmen
■   Mean monthly discharge at San Ignacio

Figure 3.13 Mean monthly river discharge at four stations on the Belize River during three consecutive years

Source: Jenkin *et al.* 1976: Figure 7

a moderate 4-month dry season, mean discharges vary by as much as an order of magnitude between minimum and maximum flows. On very large tropical rivers, such as the Amazon and Orinoco, the high wet-season discharges are accompanied by extensive flooding of the surrounding floodplain for periods of up to 5 months. On these two rivers, a highly specialized floodplain ecosystem exists, comprising forests capable of withstanding extensive deep flooding (Figure 3.14), a fish fauna that forages in these forests

Figure 3.14 Flooded forests on the floodplain of the middle Orinoco River near Mapire, Venezuela, under dry-season conditions. During the wet season, this forest is flooded for up to 5 months and to depths up to 9 m

during the wet season (Forsberg *et al.* 1993), and lagoon systems in which much of the aquatic biota becomes concentrated during the dry season.

Superimposed on the annual cycle of river discharge are brief periods of storm flow. These are usually most apparent in small headwater watersheds, most parts of which will experience rain from individual thunderstorms (Figure 3.15). On larger rivers, such flooding peaks will normally be much dampened, except in instances where basin-wide rainfall occurs, as may accompany the passage of a hurricane. Despite the very high discharges that may occur in these storm flows, these are normally of brief duration and, at least in forest-covered watersheds, their proportional contribution to total annual river discharge may be small. For example, Lesack (1993) has estimated that storm flows in a 23.4-ha forested headwater watershed in Amazonia (Figure 3.15) contribute only about 5 per cent of the total annual stream flow. Decreased flood peaks may be expected in watersheds with greater infiltration and moisture storage capacities, such as those covered by deep sands, or on limestone terrain. In contrast, increased flood peaks may be expected whenever infiltration capacities are decreased, especially as a result of deforestation. For example, data from a series of 1 × 7 m plots established on similar slopes beneath forest and various agricultural uses in the Philippines showed runoff beneath primary and secondary forest to be approximately

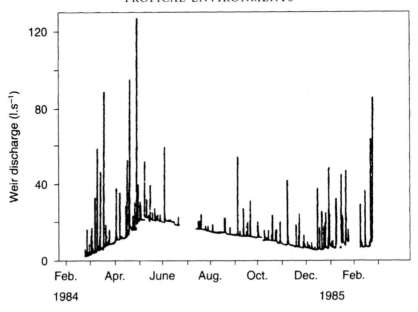

Figure 3.15 Stream discharge from a small (23.4-ha) forest-covered watershed in central Amazonia during 1 year. Brief high-discharge episodes produced by thunderstorms are superimposed on a dampened cycle linked to the seasonal rainfall pattern.

Source: Lesack 1993: Figure 4

0.26 per cent over a 227-day sample period (Kellman 1969). However, once cleared for agriculture, runoff increased to 1–2 per cent during the first two years of usage, and to greater than 12 per cent on a 12-year-old field.

4

SOILS AND NUTRIENT
LIMITATION IN TROPICAL
ECOSYSTEMS

Primary productivity, or the production of organic compounds by plants through the photosynthetic process, is fundamental to the functioning of ecosystems. Populations of plants have the potential continually to increase the level of productivity at any site, through increasing their individual size (vegetative growth) or the size of the population, provided that environmental conditions are not limiting. Controls on the level of productivity are normally set by either the availability of resources needed for growth, or the existence of other limiting conditions (such as temperature) that impinge upon the physiology of the growth processes. In the presence of abundant resources and the absence of other constraints on growth, short-term rates of productivity among plants are set by the energy conversion efficiency of the photosynthetic process and the architectural efficiency of plants in establishing a light-intercepting canopy. However, over annual time periods, other limits to productivity normally appear. In temperate areas, climate generally sets the limits to annual productivity by way of low temperatures during winter months. In the tropics, the absence of a comparable temperature limitation to annual productivity means that this should exceed that of temperate areas by a factor comparable to the relative length of the growing period (i.e. ~×3). However, estimates of net primary productivity for tropical and temperate biomes suggest considerable overlap in values. Whittaker and Likens (1975) estimate net primary productivity in tropical forests to range from 1,000 to 3,500 $g.m^{-2}.yr^{-1}$, while those of temperate forests range from 600 to 2,500 $g.m^{-2}.yr^{-1}$. For tropical and temperate grasslands, the same authors estimate respective ranges of 200–2,000 and 200–1,500 $g.m^{-2}.yr^{-1}$.

This degree of overlap between temperate and tropical productivities indicates that factors other than temperature must limit annual primary productivity in the tropics. In tropical areas with a prolonged dry season, moisture becomes the most obvious limiting factor, while in moister areas other factors must be involved. In Chapter 1, two potentially important factors interfering with productivity were identified: intense biotic interactions and soil nutrient limitations. In this chapter we examine nutrient limitation to plant productivity generally, and in the tropics in particular, and the soil processes that contribute to this limitation.

PLANT NUTRIENTS

Primary productivity comprises the synthesis of carbon compounds from carbon, hydrogen and oxygen derived from CO_2 and H_2O in the photosynthetic tissues of plants (normally the leaves). In addition to these three elements, plants require at least 13 other elements for tissue construction or biochemical processes (Table 4.1); these are commonly referred to as 'nutrients' or 'mineral nutrients', and are derived primarily from the soil. Plant nutrients are commonly classified into 'micro-nutrients' and 'macro-nutrients' (Table 4.1). The former comprise elements that are required in very small quantities and, as such, are sufficiently available in most soils. The latter are required in larger quantities, and one or more is therefore more commonly limiting to plant growth. Consequently, most attention has been focused on these macro-nutrients. Nutrients are absorbed in ionic form from soil solutions by plant roots or the fungal symbionts that are commonly associated with the roots. With the exception of nitrogen, geological materials form the ultimate source of these plant nutrients, which are made available to plants primarily by ion release during the weathering (or decomposition) of geological materials in the soil. For nitrogen, N_2 in the atmosphere is the ultimate source, from which it must be 'fixed' to available forms, primarily by specialized micro-organisms. Small quantities of all nutrients are also deposited in ecosystems from atmospheric sources. In temperate areas these have often been treated as an insignificant input relative to that by weathering,

Table 4.1 Mineral nutrients required for plant growth

Element	Abbreviation	Common ionic form absorbed
Macro-nutrients		
Nitrogen	N	NH_4^+ (ammonium), NO_3^- (nitrate)
Phosphorus	P	PO_4^{3-}
Potassium	K	K^+
Calcium	Ca	Ca^{2+}
Magnesium	Mg	Mg^{2+}
Sulphur	S	SO_4^{2-}
Micro-nutrients		
Iron	Fe	Fe^{2+}, Fe^{3+}, Fe-chelates
Manganese	Mn	Mn^{2+}
Copper	Cu	Cu^{2+}
Zinc	Zn	Zn^{2+}
Molybdenum	Mo	MoO_4^{2-}
Boron	B	$B(OH)_4^-$
Chlorine	Cl	Cl^-

but, as we shall see, these can become significant sources to tropical ecosystems with very infertile soils.

Within soils, plant nutrients exist in four forms: in organic matter; in geological materials; as ions in the soil solution; and 'adsorbed' to varying degrees on to the surfaces of mineral and organic particles. Nutrients in organic matter will be discussed more fully below in the section on nutrient cycling; here we will examine the latter three conditions. In many young soils, elements in geological materials form the largest reservoir of nutrient elements in the soil. However, in this form the elements are locked in crystalline structures of either 'primary' minerals, which are derived from volcanic activity, or various resynthesized 'secondary' minerals, which have been formed during weathering. In this form, these elements remain essentially unavailable to plants, as they are released only very gradually over geological time periods during the decomposition of the crystalline structures in weathering. In contrast, nutrient ions in soil solution are readily available to plants, but in this state are also vulnerable to being lost in percolating water ('leaching'). However, the quantities of nutrients existing in soil solution usually form only a small proportion of the plant-accessible nutrients existing in the soil; the major proportion of the latter exist in an adsorbed state. Because of the importance of this reservoir of adsorbed nutrients to plants, we will pay particular attention to it in the section that follows. As we shall see, the nutrient adsorption system of many tropical soils has some unique properties, the basic features of which must be appreciated if one is to understand the processes governing plant nutrition in these soils, and the ways in which this can be manipulated in agricultural systems.

Nutrient ion adsorption

The surfaces of both the organic and the mineral fragments that comprise the matrix of soil contain unsatisfied positive and negative electrical charges of varying density. These unsatisfied charges tend to attract ions of opposite charge from the surrounding soil solution, each of which possesses one or more positive or negative charges. Ions that possess positive charges are termed 'cations', and those with negative charges are termed 'anions'. The ionic forms of the various nutrient elements used by plants are indicated in Table 4.1. Ions that are held on the surfaces of soil particles by these electrical charges are said to be in an 'adsorbed' state, and are often referred to as 'exchangeable' because they can be exchanged with one or more other ions in solution that possess similar charge. Most surface area in a soil is contributed by particles of small size (usually defined as 'clay-sized', with a diameter <0.002 mm), and because most mineral particles of this size are secondary minerals, the mineral component of the soil's charge system is dominated by the so-called 'secondary clay minerals'.

Two forms of electrical surface charge exist: permanent and variable (Sollins et al. 1988). In permanent-charge clay minerals, the charge is a product of the lattice structure present at particle surfaces, and is almost always negative (Uehara and Gillman 1981). As a consequence, these types of clay minerals attract cations to their surface, and are said to possess 'cation exchange capacity', or CEC. The exchange capacity of most temperate-area soils is dominated by permanent charges, and most introductory treatments of soil chemistry deal only with these relatively simple systems. Unfortunately, the exchange capacity of the most common tropical soils is dominated by variable charges, which requires that we sketch the basic features of these somewhat more complex ion adsorption systems.

In variable-charge systems, which include soil organic matter (SOM) plus some secondary clay minerals, the quantity and sign of the surface charge depend on the degree to which hydroxyls (OH^-) or protons (H^+) are complexed at the surface (Uehara and Gillman 1981). This, in turn, depends on the H^+ concentration in the soil solution (its pH), with lower pH values and greater H^+ concentrations leading to more protonation of the surface. As this occurs, the surface takes on an increased positive charge and hence an anion exchange capacity (AEC), and the point at which equal numbers of positive and negative charges coexist has been called the point of zero charge or PZC (Sollins *et al.* 1988). At pH levels below this point the material takes on an increasing net AEC, while at points above, increasing net CEC prevails. At the PZC, the variable-charge component of the soil provides no adsorptive capacity, although some CEC may be provided by any permanent-charge clay minerals present.

The PZC of soil organic matter is at a pH level much lower (<2.5) than that found naturally in soils (Sollins *et al.* 1988), so SOM always imparts CEC to soils, especially in upper soil horizons, where it is most abundant. In contrast, the PZC of variable-charge clay minerals is usually much higher, which may lead to a net AEC, rather than CEC, in soils dominated by these materials. For example, the PZC of iron and aluminium oxides generally lies between pH 7 and 9, which is far higher than that typically found in most tropical soils (Uehara and Gillman 1981). As these oxides are particularly abundant in many highly weathered tropical soils, these soils may contain little CEC in deeper soil layers where little SOM exists, but rather a net AEC; this leads to adsorption of anions such as NO_3^-, which tends to be weakly retained in soils dominated by permanent-charge clays. Bulk samples of most tropical soils contain both organic matter and a mixture of clay minerals of varying properties. The PZCs of these soils tend to range between pH 3.5 and 5.0 (Uehara and Gillman 1981), which lies within the range of natural soil pH variability. Consequently, these soils may take on either CEC or AEC characteristics, depending on local circumstances and type of management. For example, the PZC of agricultural soils in north Queensland was found to be higher at all soil depths than in similar soils beneath tropical rain forest, indicating a reduction of CEC (Gillman 1985). This was attributed primarily to large reductions in the SOM after clearing, although the effect had been partially offset by the addition of phosphorus fertilizer.

Because of the predominance of permanent negative charges in the soils of temperate areas, where most soil analytical techniques originated, and the importance of these in retaining such nutrient cations as K^+, Ca^{2+} and Mg^{2+}, exchange capacities have traditionally been measured only as CEC. Measurement has normally involved saturation of the soil with NH_4OAc at pH 7, followed by displacement and measurement of the NH_4^+ ions adsorbed, as milliequivalents per 100 g of soil. While this method gives reliable estimates of relative CEC in soils whose mineral fractions are dominated by permanent-charge clays, its reliability in many tropical soils is being increasingly questioned, as the artificially high pH of the exchange solution results in an overestimate of the variable-charge CEC. Unfortunately, most standard chemical analyses of tropical soils are still based on techniques developed for temperate-area soils, and their results must be interpreted cautiously.

Ions held in the exchange complexes of soil usually represent a reservoir of soil nutrients that is far larger than that existing in the soil solution. In this state they are protected from leaching but also unavailable for plant uptake until displaced into soil solution. Displacement involves an exchange reaction, in which the ion displaced from the exchange complex is replaced by one or more ions of similar total valency. Ions of individual species exist in a dynamic, and highly buffered, equilibrium between their concentration in solution and the exchange complex. Thus any reduction in concentration in solution as a result of, for example, root uptake is compensated for by displacement into solution of similar ions from the exchange complex, and their replacement by ions of other species. Similarly, in the absence of mechanisms achieving a continual return of nutrient ions to exchange complexes from solution, these will gradually be displaced by H^+ ions in incoming rainwater and be lost from the soil, in a form of progressive acidification. Ions in the exchange complex can also be manipulated in managed ecosystems. One of the commonest forms of management is to apply lime (a soluble calcium compound) to correct excessive soil acidity. This serves to saturate the soil solution with Ca^{2+} ions, which subsequently displace H^+ ions in the exchange complex.

Nutrient absorption by roots

Ion absorption by roots is a complex process of which not all the details are as yet fully understood (Marschner 1986). Differences between the ion concentration in root sap and external soil solutions indicate that uptake is selective, and the need to maintain an electrochemical balance in both root cells and external solution requires net excretion of ions by roots to balance uptake. Excreted ions are predominantly H^+ and HCO_3^-. Nutrient uptake rates are usually greatest in young roots, in the zone immediately behind actively growing root tips where a large concentration of root hairs is found. Absorption normally creates a zone of depletion surrounding the plant root, into which further nutrients must move by mass flow of the soil solution, or diffusion within it, if uptake rates are to be maintained. The rates of movement in soils of the major nutrient ions (PO_4^{3-}, K^+, NO_3^-, NH_4^+) are usually much slower than the absorption rates needed for optimal growth, which results in this being a rate-limiting step in plant nutrition over which plants have little control (Chapin 1980).

Plant response to such limitation can take one or more of several forms: (i) reduced growth rate; (ii) increased root proliferation to explore new areas of soil; and (iii) the development of specialized structures facilitating access to nutrients beyond the soil solution depletion zone. Plants adapted to low-fertility soils are often obligate slow growers, which enhances their survivorship, but they may be incapable of switching to a rapid growth rate in the presence of high soil nutrient concentrations (Chapin 1980). High root proliferation rates imply a large investment in root tissue, and it is therefore not surprising that root : shoot ratios tend to increase on soils of lower fertility (Chapin 1980). The most widespread specialized structures enhancing nutrient uptake by plants are root–fungus symbioses termed 'mycorrhizas' that are increasingly being found on a wide range of higher plants.

Mycorrhizas represent fungal infections of plant roots with varying degrees of intrusion of the root tissues by fungal hyphae, together with considerable fungal tissue external to the root. Two broad classes of mycorrhizas exist: ectomycorrhizas and endomycorrhizas. Ectomycorrhizas (EM) are found most commonly in the trees of temperate areas. They are also found sporadically in tropical forests (Janos 1983) and are common among trees of moist savannas (Hogberg 1989), including the economically important tropical plantation tree *Pinus caribaea*. Ectomycorrhizas are structurally distinct and tend to be host-specific, which can limit host inoculation. In contrast, endomycorrhizas, among which the most common group are vesicular-arbuscular mycorrhizas (VAM), are widespread in both temperate and tropical plants, and show little host specificity, which greatly facilitates inoculation of new roots. Because mycorrhizal fungi impose a respiratory load upon their host plants, one would anticipate strong selection for resistance to infection unless this provided some net benefit to the host. An accumulation of experimental evidence over recent decades has shown convincingly that enhanced nutrient absorption, especially of phosphorus, represents the major benefit of infection (Janos 1983; Alexander 1989), and increased soil phosphorus deficiency promotes greater infection (Marschner 1986). Mycorrhizas improve the absorption of all nutrients because their fungal hyphae can exploit soil volumes far exceeding those exploited by root hairs. There is also increasing evidence that mycorrhizas are capable of gaining access to phosphorus and other nutrients that are complexed with iron and aluminium oxides (Jehne and Thompson 1981; Tiessen *et al.* 1991). Because of the abundance of these oxides in many tropical soils, and the tendency for phosphorus to be 'fixed' by these (see below), this capacity may be especially important to plant nutrition in the tropics. Connell and Lowman (1989) have suggested that EM association imparts greater competitive benefits to host plants than VAM association, and that occasional low-diversity tropical forests may be the product of gradual displacement of diverse VAM-associated species assemblages with low-diversity EM-associated assemblages.

NUTRIENT CYCLING

On an annual basis, the quantities of nutrients absorbed by plant roots and used in new tissue construction and biochemical processes are usually much larger than the quantities of nutrients released by chemical weathering. For example, Table 4.2 provides estimates of the quantities of nutrients released by chemical weathering in the forest-covered Caura Basin in the Guiana Shield region of Venezuela, together with estimates of the quantities of nutrients cycled in litterfall or used in annual root production. Plant nutrient usage is underestimated in these data as they do not include intra-annual root turnover and exclude nutrient accumulation in tree boles; yet values are clearly far larger than the weathering inputs. Moreover, in the tropics, nutrients are often released by chemical weathering at a depth that makes them inaccessible to plant roots. With the exception of phosphorus, Lewis *et al.* (1987) found few signs of retention of nutrients released by weathering in the Caura basin, and Kellman and Sanmugadas (1985) have concluded that savanna ecosystems on deeply weathered soil mantles in Belize are effectively decoupled from the geochemical cycle driven by weathering. These considerations lead to the conclusion that most nutrients used by plants in these situations must be recycled, rather than derived from weathering. Consequently,

Table 4.2 Estimated quantities (kg.ha^{-1}.yr^{-1}) of four macro-nutrients released annually by weathering in a Guiana Shield watershed in Venezuela, and the quantities of these nutrients cycled annually in litterfall, and used in annual root production, by forests in a nearby watershed

Nutrient	Weathering input*	Forest nutrient fluxes		
		Litterfall†	Root production‡	Total
Ca	7.6	13.0	4.2	17.2
Mg	3.4	5.2	5.2	10.4
K	7.6	15.4	13.5	28.9
P	0.22	2.12	2.27	4.39

Sources:
* From Lewis *et al.* 1987
† From Cuevas and Medina 1986
‡ Combines root production data from Jordan and Escalante 1980 and root nutrient concentrations from Cuevas and Medina 1988

the processes responsible for recycling these nutrients, and minimizing leakage in percolating water, are of critical concern to ecosystem functioning.

Decomposition and nutrient turnover

Nutrients used by plants are ultimately returned to the soil as dead organic material, either directly as dead plant tissues, or indirectly as dead animal tissue. The vast proportion of dead organic material is provided by plant tissues, so measurement of this and the nutrients that it contains should provide a means of assessing rates of nutrient cycling in any community. In practice, this is difficult to do accurately; turnover of fine litter comprising leaves, twigs and small branches can be accurately measured, but accurate assessments of the rate of tree bole turnover are much more difficult to achieve, and accurate assessments of root turnover rates have rarely been achieved in either temperate or tropical ecosystems (but see Vogt *et al.* 1982 for an exception). As a consequence, rates of nutrient turnover in fine litter have been widely used as a means of comparing nutrient cycling rates in different ecosystems, but it must be recognized that these underestimate absolute turnover rates to an unknown, and possibly quite variable, degree. A ratio k, of the annual fine litterfall divided by the standing stock of litter on the soil surface, has commonly been used as a means to compare decomposition rates between communities that are assumed to be in a steady state. A ratio of 1 indicates litter turnover in approximately 1 year, while values >1 indicate more rapid turnover and those <1 slower turnover. Ratios for forests summarized by Anderson and Swift (1983) indicate tropical forest values in the range 1.1–3.3 and temperate forest values between 0.4 and 1.4. While there is much variability in these data, and some overlap between temperate and tropical values, these figures indicate the general tendency for organic materials to be decomposed more rapidly under tropical conditions.

Decomposition of dead organic material is a complex biological process involving decomposer communities composed of invertebrate animals such as earthworms and termites, fungi and bacteria. In general, macro-fauna are more prominent in tropical decomposer communities, and micro-fauna more prominent at high latitudes (Swift *et al.* 1979). The decomposition process may be envisaged as a cascade, in which different components of the biota engage in decomposition activities, and are themselves later decomposed by other components of the biota, until final mineralization of contained nutrients is achieved. Within litter parcels, rates of decomposition vary widely, depending on the type of material, with woody and other lignified tissues, and those low in nutrients or high in secondary metabolites, disappearing most slowly, and leaves usually the most rapidly. Slowly-decomposing organic material is sometimes termed 'low-quality', and rapidly-decomposing material termed 'high-quality'.

Figure 4.1 A thick superficial root mat from forests occupying infertile quartzitic sands high in aluminium, southern Venezuela. The mat consists of plant litter thoroughly impregnated with living and dead roots

Overall rates of litter decomposition should reflect the overriding influences of climatic conditions and quality of litter substrate. However, while the favourable temperature and moisture conditions for decomposer activity in the tropics are reflected in generally higher decomposition rates at low latitudes, there exists considerable localized idiosyncratic variability that is not always readily explained (Anderson and Swift 1983). Deeper litter layers and more organic matter accumulation is characteristic of many tropical mountain ecosystems and is presumably climate-induced. However, the degree to which this reflects only lower temperatures, or is compounded by more frequent waterlogging of soils in this area, is difficult to evaluate. Within local tropical regions, there is a pronounced tendency for the forests on poorer soils to support a thick superficial litter layer, impregnated with roots to form a 'root mat' (Figure 4.1). This suggests depressed litter decomposition rates due to litter of poorer quality, but Anderson *et al.* (1983) have found no evidence for reduced decomposition rates in a heath forest in Sarawak that possesses such a litter layer. Kingsbury and Kellman (1997) have recently compared root mat depths in southern Venezuela with various surface soil properties, and found that deeper root mats are most closely correlated with high levels of aluminium in surface soils. As this element is known to be toxic to the roots of many plants (see below), these results suggest that the root mats

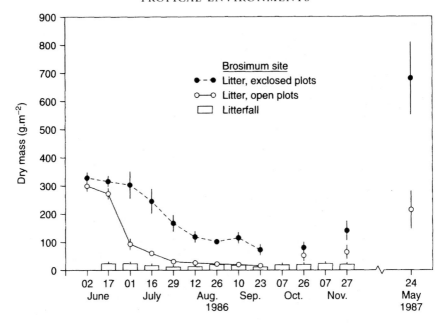

Figure 4.2 The effect of crab detritivory on the litter layer of a dry coastal forest in Veracruz, Mexico. Under normal conditions, the soil remains bare of litter throughout most of the wet season, but when crabs are excluded artificially, a normal litter layer develops

Source: Kellman and Delfosse 1993: Figure 2(a)

may be primarily a product of dead root material being deposited upon, rather than within, soils. At the other extreme, some very rapid decomposition rates have been demonstrated to be the product of a specialized decomposer community. Land crabs are active detritivores in many tropical coastal forests, and one exclosure experiment has demonstrated a dramatic increase in litter accumulation in their absence (Figure 4.2).

The rates of litter turnover and nutrient release in tropical savanna communities have received much less attention than those in forests. This reflects both the greater difficulties of measuring above-ground litter production and decomposition in graminoid communities, and the fact that in these communities 'decomposition' and mineralization of nutrients in above-ground tissues is normally achieved by burning at frequent intervals (see Chapter 7). As a result, above-ground nutrient turnover in these communities involves accumulations of small quantities over one to a few years, followed by their instantaneous release at the time of burning. The quantities of nutrients released by such events have been estimated for a Belizean savanna by Kellman *et al.* (1987), 1 and 5 years after a fire (Table 4.3). While the quantities released are smaller than those cycled annually by forests (Table 4.2), they neverthe-

Table 4.3 Nutrients (kg.ha^{-1}) accumulated in the herbaceous layer of a tropical savanna in Belize, 1 and 5 years after a fire

Nutrient	1 Year	5 Years
Ca	1.39	3.09
Mg	1.79	3.21
K	6.52	9.05
P	0.42	1.38

Source: Kellman *et al.* 1987

less represent a large instantaneous loading of the soil, raising the potential for nutrient 'leakage' from the system. Between fires, and below the soil surface, nutrient turnover in these communities is probably much more muted. Menault *et al.* (1985) have suggested that a functional dichotomy exists between humid and arid versions of African savannas. In the former, almost continual mineralization of nutrients takes place, while in the latter most decomposer activity is moisture-limited and concentrated in one brief period of rapid mineralization during the short wet season. Abbadie *et al.* (1992) have shown that nitrogen recycling by grasses in a moist savanna in Africa is extremely effective; nitrogen in dead roots is rapidly mineralized and reabsorbed, freeing the grasses from dependence on other sources of this nutrient.

Nutrient retention and leakage

Nutrients mineralized during the decomposition process enter the soil solution where they either are reabsorbed by plant roots, are adsorbed to soil particles or remain in solution, where they are susceptible to loss in percolating water. Nutrients in soil solution are thus at their most vulnerable stage in the nutrient cycle, a vulnerability that is exacerbated in the tropics by the prevalence of high-intensity rainfall. The mechanisms by which leakage from this stage of the cycle is avoided have been the subject of much speculation and, more recently, some detailed investigations. Much of this recent investigation has been focused upon some of the most infertile soils in the tropics (e.g. Stark and Jordan 1978; Walker *et al.* 1981), and there has been some tendency to extrapolate results achieved in these extreme systems to tropical soils generally, about which much more remains to be learned.

Although much attention has been devoted to potential biological mechanisms promoting nutrient conservation (e.g. Jordan *et al.* 1980), nutrient diffusion rates in the soil provide the rate-limiting step in nutrient reabsorption by plants, which means that plants are limited in the conservation options available to them once nutrients have been shed in dead tissue.

Within the soil, these options are limited to continual expansion of nutrient-absorbing surfaces, by way of fine root proliferation, mycorrhizal expansion, or both, but these processes may be metabolically and nutritionally costly (Vogt *et al.* 1982). In recent years, there has been increased interest in non-biological mechanisms that may help to suppress nutrient leaching from tropical soils. Here we will discuss both biological and non-biological processes that may help prevent leakage of mineralized nutrients from soil solutions, focusing upon three that seem to be of special significance: specialized percolation patterns by soil moisture; special adsorption processes in tropical soil; and patterns of root proliferation. Other mechanisms that are internal to plants will be treated later, when we discuss natural selection in plants occupying tropical soils of very low fertility.

Soil water percolation

Leakage of nutrients in soil solution depends both on the volume of water percolating through the soil, and the pathways that it follows. The volume of water moving through tropical soils is usually large under wet-season conditions, once the profile becomes fully wetted, as much of the rainfall is received as high-intensity thunderstorms. Traditional conceptions of the percolation process have envisaged a wetting front moving uniformly through the pore space of the soil profile. However, this model assumes a relatively uniform pore size structure, and research has shown that this rarely exists in most soils, with the exception of recent sands (Beven and Germann 1982). Instead, soil particles tend to aggregate to varying degrees to form structural units or 'peds', between which voids exist, especially under drier conditions when some ped shrinkage may occur. Aggregation is promoted by soil organic matter, coatings of oxides, or an abundance of polyvalent cations, such as Ca^{2+}, in the exchange complex. Among variable-charge clay minerals, a decrease in net surface charge also leads to less repulsion between particle surfaces, and more stable aggregates (Sollins *et al.* 1988). Other voids within the soil matrix are provided by the growth and death of plant roots, and the burrowing activity of soil macro-fauna. As a result, the pore structure of most soils is highly heterogeneous, varying from the 'micro-pores' surrounding individual soil particles within a ped, to a variety of larger pore sizes, usually referred to collectively as 'macro-pores'. These are usually defined as pores that are too large to retain water by capillarity at field capacity (Sollins and Radulovich 1988).

Because permeability of a pore to water increases with the fourth power of the pore radius, macro-pores have an enormous capacity to transmit large volumes of water rapidly, essentially bypassing the micro-pores of the soil aggregates and the soil solutions contained by these. In theory, macro-pore flow requires soil saturation and ponding of water at the soil surface, but Radulovich *et al.* (1992) have shown experimentally that outflow from tropical soil cores begins well below saturation levels. These authors suggest that macro-pore flow is widespread in tropical soils, beginning whenever the hydraulic conductivity of soil micro-aggregates is exceeded, and proceeding initially as films along the walls of macro-pores that may later thicken to fill the pore completely. The implication of this process for soil leaching is clearly profound, as it suggests that much of the soil solution may remain effectively isolated from percolating water during rainstorms, even when these are of high intensity. While surface-applied materials such as fertilizers may be especially vulnerable to leaching by macro-pore flow, nutrient ions released gradually into soil solutions during organic matter decomposition would be little affected.

Ion adsorption

The vulnerability of an ion to leaching depends also upon the degree to which it is adsorbed on to a soil's exchange complex. In temperate areas whose soils are dominated by permanent-charge clays with net CEC, anions such as nitrate (NO_3^-) may exist almost entirely in solution and so be vulnerable to leaching. In contrast, most cations are less vulnerable under these conditions. The low CEC of many tropical soils has led to the assumption that there should be little cation adsorption in tropical soils, leaving these vulnerable to leaching. Conversely, nutrient anions, such as NO_3^-, should be retained by the net AEC that develops in tropical soils, and pH-dependent anion adsorption has indeed been demonstrated (Singh and Kanehiro 1969; Kinjo and Pratt 1971). Iron and aluminium oxide coatings are especially active in this role (Chao *et al.* 1964), and the subsoils of acid tropical soils appear to be sites of preferential nitrate adsorption (Cahn *et al.* 1992). A further important extension of the anion adsorption process is that when polyvalent anions, such as SO_4^{2-} and PO_4^{3-}, are adsorbed, their ionization charge may be only partially neutralized, leaving one or more unsatisfied negative charges at their site of adsorption (Wiklander 1980). As a consequence, temporary CEC is created, which can serve to adsorb cations. Phosphate adsorption is especially effective in this capacity (Figure 4.3). Thus, a flush of mineralized nutrients entering soil solution, for example after a savanna fire, does not remain in solution and vulnerable to leaching, provided that sufficient polyvalent anions are involved. This process can effectively suppress 'acute' leaching losses from tropical soils, providing the time delay necessary for root proliferation and absorption to achieve more complete immobilization of nutrients in plant biomass.

The effectiveness of this process has been demonstrated in a savanna soil from Belize, where flushes of nutrients created by savanna fire, or artificial nutrient loading, created temporarily increased nutrient concentrations in surface soil solutions, but no change at depth (Kellman 1985b; Kellman *et al.* 1985). Increased concentrations returned to ambient levels over the

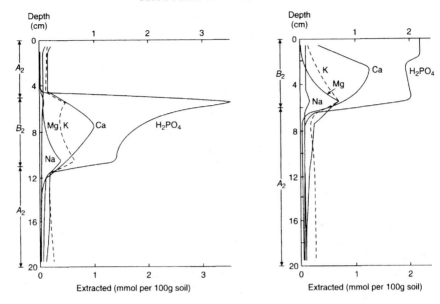

Figure 4.3 Distribution of H_2PO_4, Ca, Mg, K and Na retained in two artificial soil leaching columns built from podzol horizons that were high (B_2) or low (A_2) in phosphate adsorption capacity

Source: Wiklander 1980: Figures 11 and 12

subsequent week in a pattern that could be replicated experimentally with sterilized soil, indicating non-biological immobilization (Kellman and Sanmugadas 1985). The longer-term effectiveness of this nutrient retention mechanism, when combined with plant uptake and macro-pore flow, is indicated by changes in mean nutrient concentration in solutions as these move vertically through the system (Figure 4.4). Solutions entering the surface soil as throughfall had concentrations persistently above those leaving the soil at 90 cm, despite having traversed surface soils of much higher soil solution concentration. Moreover, nutrient concentrations in river water were many times higher than those leaving the soil, indicating the effective isolation of the soil–vegetation system from weathering inputs at the base of the regolith.

Root proliferation

Immobilization of nutrient ions by adsorption can provide only temporary suppression of leaching, as in the absence of reabsorption by plants these remain vulnerable to gradual 'chronic' leaching from the soil. Deployment of roots to sites of nutrient mineralization would facilitate immediate access to nutrients released, and there is experimental evidence showing that tropical forest root proliferation is stimulated in the presence of dead organic matter

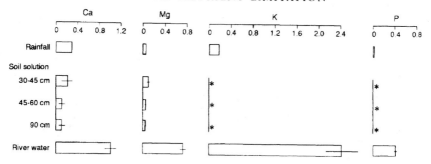

Figure 4.4 Variations in solution nutrient concentration (mg.l⁻¹) as water moves through a savanna ecosystem in Belize during 1 month. Rainfall is shown as volume-weighted means; soil solutions and river water are mean ± 1 SD of daily measurements.*, concentrations below instrument detection limits

Source: Kellman and Sanmugadas 1985

(St John 1983). Such a mechanism would ensure location of live roots at sites of decomposing root tissues within the soil, and would also account for the propensity of fine roots in tropical forests to spread upwards into decomposing litter layers. This phenomenon is widespread in the tropics, especially on soils of very low fertility (Figure 4.1), and the root mats formed at these sites have been shown to be exceptionally effective at retaining nutrients. Solutions containing radioactive isotopes of Ca and P sprayed on to root mats by Stark and Jordan (1978) showed almost complete disappearance of the isotope during passage through the mat, presumably reflecting effective adsorption by organic matter, followed by root absorption.

Research on temperate-area plants has shown that locally increased nutrient levels, especially of nitrate and phosphate, can also stimulate fine-root proliferation (Drew 1975; Hackett 1972; Wiersum 1958), and a rapid increase in nutrient uptake capacity by roots located in such enriched soil micro-sites has been demonstrated (Jackson *et al.* 1990). Less complete information is available on the response of tropical plant roots to increased nutrient concentrations. Cuevas and Medina (1988) found increased root production in ingrowth cylinders to which nutrients had been added selectively, but only among cylinders placed in root mats, rather than in topsoil. Similarly, Sanford (1987) found upward growth of roots from root mats to be stimulated by an artificial nutrient gradient. In contrast, only moisture, rather than nutrients, was found to be responsible for the proliferation of fine roots in a dry tropical forest at the start of the wet season (Kavanagh and Kellman 1992). However, in the same forest, roots were found to be more concentrated in a superficial layer in the topsoil of more weathered sands, and a bioassay showed increased root proliferation in this material (Kellman 1990). Weathered sand was also found to retain more nitrate than recent sand in a

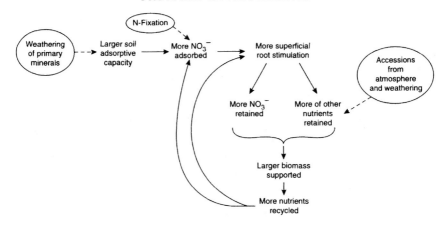

Figure 4.5 A postulated nutrient retention feedback loop that may develop in tropical soils as they weather

Source: Kellman 1990

leaching experiment. This led to the hypothesis that as soils weather in the tropics they develop a greater nitrate adsorption capacity which stimulates fine-root proliferation in topsoils; this, in turn, provides a better nutrient absorption system for other nutrients, which become entrained in an effective recycling process (Figure 4.5).

Considerable debate has focused on whether mycorrhizal fungi can engage directly in the decomposition of dead organic matter, and thus obviate the need for nutrients contained in this to enter soil solution. A widely cited hypothesis by Went and Stark (1968) proposed such a mechanism of 'direct' nutrient cycling, but has so far not been confirmed experimentally. Using a radioisotope of phosphorus, Herrera *et al.* (1978) were able to show transfer of phosphorus from plant litter to mycorrhizal hyphae and plant roots, but a decompositional role for the hyphae was not demonstrated (Janos 1983). Janos (1983) and Alexander (1989) consider that organic matter decomposition by the fungal hyphae of mycorrhizas remains unproven but is more likely to exist in EM than in VAM species. However, even in the absence of a saprophytic ability, the fungal hyphae of mycorrhizas enable close physical contact to be maintained between decomposing organic matter and absorption surfaces. The hyphae of both EM and VAM mycorrhizas are often found in close association with organic material (Alexander 1989), and Ca and Mg release from litter placed in contact with a tropical forest root mat was found to be faster than from litter without such contact (Cuevas and Medina 1988), although the mechanisms responsible for these differences were not identified.

The processes described above involving roots and mycorrhizas, when combined with the soil adsorption processes and specialized percolation patterns described earlier, appear to be sufficient to explain the enigma of nutrient retention by natural tropical ecosystems, without the need to invoke the existence of deeply rooted plants capable of returning lost nutrients to the surface soil. The existence of such 'nutrient pumps' has been frequently postulated but never satisfactorily demonstrated. Most data on rooting patterns in moist tropical plant communities show an overwhelming tendency for fine roots to be concentrated in superficial horizons (e.g. Stark and Spratt 1977; Kellman 1979; Kellman and Sanmugadas 1985). While Nepstad *et al.* (1994) have confirmed the existence of deep roots beneath an Amazonian forest, and provided convincing evidence for their use of soil water, their contribution to nutrient absorption remains unknown. A root-cutting experiment on the large tap roots of *Pinus caribaea* in a Central American savanna showed no adverse effects on tree growth rates or foliar nutrient levels over 5 years (Kellman 1986), indicating that the shallow lateral roots that this species develops in surface soils provide adequate nutrient supplies. Of critical importance for managed ecosystems is whether similarly effective nutrient retention processes can be retained or redeveloped within these.

TROPICAL SOILS AND SOIL FORMATION

Soils provide the rooting medium for almost all plants, including crop plants. As such, they provide physical support and supply water and soil nutrients to the plant. Earlier in this chapter we argued that the year-long growing season in the tropics makes these soil resources probable rate-limiting factors on primary productivity by plants in the tropics, with nutrients being especially critical in moister areas. In this section we will see that a variety of processes have led to many (although by no means all) tropical soils being relatively poor nutrient-supplying systems for plants. This makes it even more likely that productivity will be limited by nutrients in many tropical ecosystems.

Until the past half-century, knowledge of tropical soils was very limited and interpretations of what knowledge did exist were usually over-generalized and often contradictory (see Richter and Babbar 1991). Today, we have a much more complete empirical base with which to work; this shows much more localized complexity in tropical soils than was once assumed, but it also shows certain soil types to be more common than others, especially those that are now referred to as oxisols and ultisols in the United States Department of Agriculture classification (Soil Survey Staff 1975). We will pay special attention to their formation and properties. We will then go on to discuss some less widespread tropical soil types, especially those of greater fertility, which are especially important to agriculture.

Soils are composed primarily of mineral fragments of varying size, representing a stage in the physical and chemical disintegration of geological materials that are commonly referred to as 'parent material'. Mixed into these mineral fragments are varying amounts of dead organic material, water, gases and a soil biota. Soils normally exhibit a characteristic 'profile' comprising a vertical sequence of superimposed horizontal layers or 'horizons' of varying degrees of distinctness, which are the product of transformations and movements of materials within the soil profile. Almost a century ago, Dukachayev, the father of modern soil science, identified five 'soil-forming factors' that together produce chacteristic soil profiles: parent material, climate, organisms, local relief and time. These are still used as a conceptual framework in soil science. Interactions between these factors during the course of soil formation involve a large number of specific processes. However, two groups of processes are dominant: the chemical weathering of parent materials; and the vertical movement of materials within the profile, normally in a downward direction.

Weathering under tropical conditions

Weathering involves the decomposition of geological materials that may be consolidated and massive (e.g. granite), or unconsolidated (e.g. river alluvium); in the latter case rates are usually much more rapid as initially larger surface areas are available for chemical reactions. Although driven by chemical reactions, the process also involves physical disintegration of massive structures into smaller-sized particles, which may be microscopic in size. Virtually all of the chemical reactions require water and operate more rapidly at higher temperatures. Consequently, where conditions are both wet and warm continuously, as in parts of the tropics, rates of weathering will be especially rapid. Moreover, where parent materials have been exposed to weathering for especially long time periods, one would expect deeper weathered zones, unless rates of erosion are sufficient to preserve unweathered geological material close to the surface. Large parts of the tropics are made up of the old continental plates of Gondwana (the Guiana and Brazilian Shields in South America, and large parts of Africa and Australia), and we would therefore expect to find especially deep and intensely weathered soils in these places.

'Primary' minerals represent materials that have crystallized from cooling magma or been recrystallized during the metamorphosis of sedimentary rocks, and comprise the ultimate source of most mineral constituents of soils, with the exception of biogenic material, such as limestone ($CaCO_3$). Most primary minerals are dominated quantitatively by silica, and have as their basic structural unit the silica tetrahedron, comprising a silicon atom surrounded by four oxygen atoms (Figure 4.6). In silicate minerals, adjacent silica tetrahedra share other bonding atoms to form rigid crystalline structures. In their simplest structure, adjacent oxygens are shared to form the mineral quartz (SiO_2), which is extremely stable under most soil conditions. Silicate minerals crystallized under conditions radically different from those to which they are exposed at the earth's surface, become unstable there, especially in the presence of water and oxygen. The progressive decomposition that follows takes

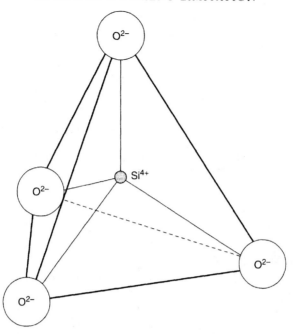

Figure 4.6 A silica tetrahedron, the basic building block of silicate minerals. The tetrahedron consists of a quadrivalent silica cation combined with four divalent oxygen anions; in quartz (SiO_2), adjacent oxygens are shared, to produce an extremely stable mineral that resists chemical weathering (cf. Figure 4.7)

place primarily at the surfaces of these crystalline minerals and involves a progressive removal of atoms critical to the crystalline structure, or their replacement with other ions. The process is complex (Loughnan 1969), and here we can only sketch its essential features.

Hydrolysis represents the primary weathering process and is driven by the presence of partially dissociated water molecules (H^+, OH^-) in the weathering medium. Hydrogen ions penetrate the surfaces of crystal lattices, creating a charge imbalance and leading to the dissociation of other cations into solution (White 1987). Any increase in H^+ ion activity in solution may be expected to increase weathering rates. Two processes are primarily responsible for this in non-flooded soils: (i) the dissolution of CO_2 in soil solution to form weak carbonic acid; and (ii) the presence of humic substances. As both processes are promoted by the decomposition of organic matter, the presence of organisms in a soil greatly increases weathering rates. Continuation of hydrolysis also requires the removal of dissociated cations and the replenishment of the weathering medium with new H^+ ions. Consequently, where frequent percolation occurs, as in perennial wet climates

Table 4.4 Mobilities of the common cations

Relative mobility	Cation	Conditions of mobility
1 (most mobile)	Ca^{2+},Mg^{2+},Na^+	Readily lost during leaching
2	K^+	Readily lost during leaching, but loss may be slowed by fixation in illite
3	Fe^{2+}	Mobile under anoxic conditions
4	Si^{4+}	Slowly lost during leaching
5	Ti^{4+}	Normally of very limited mobility
6	Fe^{3+}	Immobile under oxidizing conditions
7 (least mobile)	Al^{3+}	Immobile in the pH range 5.5–9.5

Source: Loughnan 1969, modified

and on coarser-textured parent materials, weathering will be most rapid. In contrast, where percolation is reduced, by virtue of less rainfall or finer-textured parent materials, or where ground water imports cations flushed from higher topographic positions, dissociated cations will remain in the local soil solutions and decrease the rates of hydrolysis.

The vulnerability of different minerals varies widely, as does the mobility of the cations comprising these (Table 4.4). Consequently, during the course of weathering, the relative proportions of different mineral constituents change, with the least vulnerable minerals and least mobile cations tending to increase with time. For example, at pH <9 quartz is only weakly soluble, and therefore tends to persist in most acid soils (Figure 4.7), while limestone dissolves readily, leaving only impurities to form the mineral fraction of soil. While materials in solution and resistant primary minerals form important weathering products, there exists a futher class of products that is of extreme importance in soils: the secondary clay minerals.

Secondary clay minerals

Secondary clay minerals comprise mineral material synthesized during the course of chemical weathering from primary mineral constituents. While more stable than the primary minerals from which they derive, they are subject to further transformation to form other clay minerals or soluble ions. Because they exist as clay-sized soil particles, or as coatings on the surfaces of other particles, they provide much of the contact surface between soil particles and solution, and are extremely important in the adsorptive processes in soils, especially those of the tropics, where most primary minerals may have been transformed to secondary clay minerals. Secondary clay minerals occur as both crystalline silicate minerals and non-crystalline or amorphous material. While a large number of different secondary clay minerals, and intergrades between these, have been identified using X-ray diffractometry, here we will concentrate on the most common forms. These include

Figure 4.7 An unweathered vein of quartz persisting in otherwise deeply weathered granite saprolite, Belize. Quartz (SiO_2) is relatively stable at pH < 9.0, and therefore tends to persist in acid soils, such as these, whose pH range is 4.5–5.2

crystalline illite, montmorillonite and kaolinite, and the amorphous materials composed of iron and aluminium oxides, and allophane.

Crystalline materials consist of microscopic platelets, whose basic constituents are sheets of silica tetrahedra and aluminium octahedra. These are combined in lamellae consisting of matched pairs in kaolinite, which is referred to as a 1 : 1 clay mineral, or as 2 : 1 clay minerals such as illite and montmorillonite, in which silica tetrahedra sheets are sandwiched between sheets composed of aluminium octahedra. In illite, adjacent lamellae are partially bonded together by cations, principally K^+, while in montmorillonite these have been lost by further weathering to provide larger and more variable spacing between adjacent lamellae. This creates a much larger surface area and imparts important physical and chemical properties to this mineral, which will be discussed below. Allophane is a weakly crystalline clay mineral with a hollow spherical structure (Uehara and Gillman 1981) that develops at an early stage in the weathering of many volcanic parent materials, especially volcanic ash. It often forms stable complexes with soil organic matter and has a low bulk density. Various chemical forms of iron and aluminium oxides exist in soils as weakly crystalline particles or coatings on the surfaces of other particles, including other clay minerals. Aluminium oxides are grey and their presence is normally masked by the brighter red and yellow colours of the iron oxides with which they usually co-occur.

Specific surface areas and adsorptive characteristics of the major secondary clay minerals differ radically (Table 4.5), and tend to fall into two groups. Allophane and montmorillonite possess large specific surface areas and a very large moisture retention capacity; in montmorillonite moisture is retained between lamellae, which causes this clay to undergo swelling and shrinkage during wetting and drying cycles. Both clay minerals also possess large adsorptive capacities, although of different form. In montmorillonite this is overwhelmingly permanent-charge CEC, while in allophane it is dominated by variable-charge CEC but with significant permanent-charge CEC and AEC (Table 4.5). At the other extreme, kaolinite and the oxides of iron and aluminium are of low specific surface area and possess almost no permanent-charge CEC; their small adsorptive capacities are dominated by variable-

Table 4.5 Specific surface area and adsorptive properties of common secondary clay minerals. Adsorptive properties of soil organic matter also shown for comparison

Material	Specific surface area $(m^2.g^{-1})$ *	CEC[†]			AEC[†] (meq per 100 g of clay)
		Perm.	Var.	Tot.	
Illites	90–130	11	8	19	3
Montmorillonites	750–800	112	6	118	1
Kaolinites	10–20	1	3	4	2
Hydrous Fe and Al oxides	25–42	0	4–5	4–5	4–5
Allophanes	500–700	10	41	51	17
Peat		38	98	136	6

Source:
*   From Sollins *et al.* 1988
†   From Sanchez 1976. Total CEC determined at pH 8.2

Notes: CEC, cation exchange capacity; AEC, anion exchange capacity

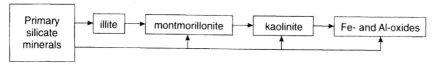

Figure 4.8 The generalized sequence of weathering products that may be expected to appear during the decomposition of silicate minerals

charge CEC and AEC. The characteristics of illite are intermediate between these two groups.

During the weathering of primary silicate minerals, secondary clay minerals tend to form in a predictable sequence, with a dominant pathway, plus some direct transformations (Figure 4.8). On volcanic ash, allophane typically degrades to a form of kaolinite (Sollins *et al.* 1988). Clay minerals found later in the sequence are progressively more stable and resist further weathering. Thus, while there is a general tendency for Fe and Al oxides to form as an ultimate weathering product, the process can be arrested by local conditions. For example, montmorillonite can form and persist in drier tropical climates that are subject to less leaching, and this clay mineral tends to be especially abundant in soils formed on parent materials high in Ca and Mg. Montmorillonite can also persist in tropical soils occupying low topographic positions, where ground water leads to an influx of cations that have been leached from elsewhere. As a result, the mix of clay minerals present in any soil reflects both the time that the parent materials have been exposed to weathering, and the local state of the other soil-forming factors that may ameliorate its intensity. Soils on older tropical land surfaces may be expected to be dominated by kaolinites and the oxides of iron and aluminium, and have low adsorptive capacities. However, under lower rainfall more montmorillonite may be expected (Figure 4.9), and in areas of recent vulcanism, significant quantities of allophane and montmorillonite may be expected in the soils. Uehara and Gillman (1981) estimate that in the tropics soils with variable-charge characteristics make up 60 per cent of the total, permanent-charge 10 per cent and mixed charge 30 per cent. This contasts with estimates for temperate areas of 10 per cent, 45 per cent and 45 per cent for variable-, permanent- and mixed-charge soils, respectively.

Vertical movement of materials and profile development

Materials that are moved vertically in soil profiles, and later deposited to form distinctive horizons, include organic matter in upper soil layers, soluble compounds such as $CaCO_3$ which are precipitated at depth where wetting fronts stop, and a variety of materials such as clay and iron compounds that are moved short distances within the profile and deposited as a 'B' horizon.

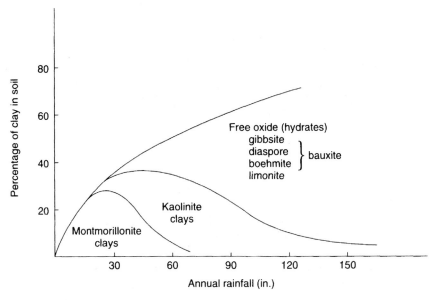

Figure 4.9 Progressive types of clay development in Hawaiian soils under an aseasonal rainfall regime

Source: Loughnan 1969: Figure 39

'Ah' horizons of organic matter accumulation are characteristic of all tropical soils, although they may be attenuated owing to high litter decomposition rates in the lowland tropics. Similarly, zones of carbonate and salt accumulation can be found at depth in the soils of tropical arid and semi-arid climates. However, well-developed B horizons are often less common in tropical soils than they are in the soils of temperate areas. This is probably attributable to the rapid weathering of parent materials to a clay-rich matrix in moister tropical areas, through which particulate material is less readily moved. As a result, clay eluviation tends to occur only on coarser-textured parent materials, and the organic colloids that act as chelating agents, moving Fe and Al to B horizons in temperate areas, are effectively filtered from solutions passing through clay-rich profiles (St John and Anderson 1982). Consequently, it is only on quartzitic sands that possess few readily weatherable minerals that the Fe- and Al-enriched B horizons characteristic of temperate-area podzols (spodosols) are found. However, where appropriate parent materials do occur, spectacular giant podzols 15–20 m in depth can form in the tropics (Thompson 1981).

Plinthite (laterite) formation

One of the most enduring myths about tropical soils is that after forest clearance, many profiles are subject to an irreversible hardening of superficial horizons to form an indurated mass termed 'laterite'. The term was coined by Buchanan (1807) to describe soft weathered material that hardened irreversibly after excavation and was being used as a source of building bricks in southern India. The fact that many compounds such as iron and aluminium oxides that were prominent in this material were also abundant in tropical soils generally, led to the assumption that this was a soil deposit (Richter and Babbar 1991), and the emphasis placed on genesis in the soil science of the early twentieth century led to its being designated a 'post-senile' stage of soil formation (e.g. Mohr and Van Baren 1954). These generalizations have persisted in recent decades, despite the absence of evidence for the process being widespread in the tropics, and a great deal of contradictory evidence. For example, parts of the tropics, such as West Africa and India, that have supported dense human populations for many millennia and been subject to repeated cycles of forest clearance have not been transformed into indurated ironstone pavements. Using recent survey data for Amazonia, Sanchez et al. (1982) estimate that only 4 per cent of that region's soils have soft plinthite in their subsoils which could be subject to hardening if exposed by severe erosion of overlying topsoils.

Hardened ironstone is now recognized to be a material that develops gradually over exceptionally long time periods, in places where water table fluctuations lead to gradual accumulation of iron oxides. Under anaerobic conditions, iron exists in the ferrous (Fe^{2+}) state and is mobile. When water tables fluctuate regularly, iron imported in ground water in the ferrous state can be partially oxidized to the ferric (Fe^{3+}) state as $Fe(OH)_3$. Oxidation normally takes place selectively at nodes, producing isolated iron mottles within a reduced matrix of white or grey colour (Figure 4.10). With time such mottles can grow to become hardened nodules which, over geological time, can coalesce to form an indurated lag deposit (Figure 4.11). Once formed, this material is extremely resistant to destruction and, over very long time periods, it may be exposed as soil material overlying it is eroded away. In this condition, it acts as an erosion armour on the land surface and is destroyed slowly only by undercutting and collapse at edges. Materials of this sort can be found scattered throughout the tropics on older land surfaces which, at an earlier erosion phase, were subject to the hydromorphic conditions necessary for its formation. To avoid the confused history of the term 'laterite', this material, in massive or nodule form, is now referred to as 'plinthite' (Soil Survey Staff 1975).

Figure 4.10 Iron concretions developing in an ultisol subject to seasonal waterlogging, coastal savannas, Belize

Oxisols and ultisols

Oxisols and ultisols are highly weathered soil types that may be regarded as the 'modal' soil types of the tropics. Together, they cover over 40 per cent of the non-desert areas of the tropics, and are especially abundant in South America (Table 4.6). Their primary equivalents in the FAO–UNESCO world map are, respectively, ferralsols and acrisols (FAO–UNESCO 1988). Both soil types are dominated by kaolinites and iron and aluminium oxides which impart low CEC but some AEC. In contrast to these poor chemical properties, these soils generally possess favourable soil structural properties as a result of strong clay aggregation and cementation by oxides, which impart some of the soil physical properties of sand, such as high infiltration capacities. Sample profiles of these two soil types are provided in Tables 4.7 and 4.8.

The major property differentiating the two soil types is the presence of a clay-enriched B horizon in ultisols (Table 4.8). Because such enrichment normally requires some vertical clay movement, ultisols tend to be more common on parent materials that contain slowly-weathering coarse-textured materials, such as quartz, through which some clay movement can occur. They are also most common on flatter land surfaces, where water movements are predominantly downward. The presence of a clay-enriched B horizon can

80

Figure 4.11 Massive indurated plinthite formed by the coalescence and hardening of iron concretions, Apure savannas, Venezuela. This deposit, which is probably of considerable geological age, is overlain by the sands of the Apure sandplain

lead to impeded drainage and lateral soil water movement in the wet season, as well as impeded root development. As a consequence, these soils are generally considered to have less agricultural potential than well-drained oxisols, except in circumstances such as irrigated farming, where impeded drainage may be desirable.

81

Table 4.6 Proportionate distribution of major tropical soil types (excluding desert soils) between the tropics of Cancer and Capricorn

Soil type	C. America	S. America	Africa	Asia	World
Oxisols	0.3	39.3	22.3	1.9	23.4
Ultisols	14.2	16.0	11.1	38.1	17.9
Alfisols	14.9	7.6	11.6	12.4	10.5
Vertisols	6.9	3.7	2.5	4.0	3.4
Andosols	8.2	0.7	0.3	1.1	1.0
Mollisols	21.3	1.8	0.1	2.4	2.2
Recent soils	16.1	11.4	30.8	25.4	22.4
Mountain soils	16.1	19.2	21.0	11.3	18.4
Organic soils	1.0	0.2	0.1	3.0	0.7
Podzols	0	0	0.1	0.4	0.1

Source: Data compiled by Richter and Babbar 1991 from the FAO–UNESCO *Soil Map of the World* (FAO–UNESCO 1988)

Note: All data expressed as % of the area specified

Table 4.7 Field profile description and selected physical and chemical properties of an oxisol from Puerto Rico

Location:	Cibuco, Puerto Rico
Vegetation:	Unimproved pasture
Parent material:	Residuum from mixed andesitic and basaltic flows
Topography:	Convex slope in mountainous terrain at 480 m elevation

Profile description:

Depth (cm)

0–15   Dark reddish-brown (5YR 3/4) clay; strong fine to very fine, granular structure; firm, slightly sticky, slightly plastic; abrupt, smooth boundary

15–35   Dark-red (2.5YR 3/6) clay; few mottles of yellowish-brown (10YR 5/4); weak, medium subangular blocky, breaking to moderate, very fine blocky structure; firm, slightly sticky, slightly plastic; thin, patchy films of clay; few, fine pores; gradual smooth boundary

35–65   Dark-red (10R 3/6) clay; fine, distinct mottles of reddish brown (5YR 4/4); moderate, very fine, blocky structure; firm, slightly sticky, slightly plastic; patchy films of clay; few, fine pores; clear, smooth boundary

65–92   Weak-red (10R 4/4) to red (10YR 4/6) clay; weak, medium, subangular blocky, breaking to strong, very fine, angular blocky structure; firm, slightly sticky, slightly plastic; few, thin patchy films of clay; gradual smooth boundary

92–120   Red (10R 4/6) clay; other features similar to above horizon

120–150   Dark-red (2.5YR 3/6) clay; weak, medium subangular blocky, breaking to moderate, very fine angular blocky structure; firm, slightly sticky, slightly plastic; few, thin patchy films of clay; clear wavy boundary

150–178   Dominantly weak-red (10R 4/4) to red (10R 4/6) clay loam to silty clay loam; colour is variable and ranges to strong brown (7.5YR 5/6); massive structure; friable, non-sticky, non-plastic

178–195   Dominantly yellowish-red (5YR 4/6) clay loam or silty clay loam; colour is variable and ranges from red (10R 4/6) to strong brown (7.5YR 5/6); massive structure; firm, fine pores.

195–445   Decomposed parent material, similar to the above horizon

Physical and chemical analysis:

Depth	% sand	% silt	% clay	% C	CEC*	Ca	Mg	K	Na	pH	% free Fe oxides
						Exchangeable cations (meq.100g⁻¹ soil)					
0–15	8.5	19.1	72.4	2.72	16.7	2.1	0.6	0.2	0.1	4.6	14.7
15–35	3.4	20.5	76.1	1.42	13.0	3.2	1.5	0.1	0.1	5.0	16.4
35–65	4.6	27.6	67.8	0.66	12.8	1.0	0.8	0.1	<0.1	5.1	16.2
65–92	5.4	35.6	59.0	0.24	11.9	0.5	0.6	0.1	<0.1	4.9	12.2
92–120	6.2	34.1	59.7	0.22	11.7	0.5	0.9	0.1	<0.1	4.9	13.5
120–150	7.7	34.7	57.6	0.26	11.4	0.6	0.9	0.1	<0.1	4.9	12.5
150–178	22.3	39.6	38.1	0.16	10.4	0.5	0.8	0.1	<0.1	4.8	11.8
178–195	23.3	41.8	34.9	0.21	11.1	0.5	0.6	0.1	<0.1	4.8	13.6
195–320	23.3	44.3	32.4	0.06	12.0	1.2	0.8	0.1	<0.1	4.8	10.9

Source: Soil Survey Staff 1960

Notes: Soil colours determined by the use of Munsell colour charts
* Cation exchange capacity (CEC) determined by NH_4OAc extraction, making these values probable overestimates for a soil that probably has a large variable-charge component

Table 4.8 Field profile description and selected physical and chemical properties of an ultisol from Trinidad. This soil shares many of the characteristics of the ultisols associated with the east-coast savannas of Central America (Figure 7.3), including the presence of iron concretions that harden to plinthite when dried (Figure 4.10)

Location:	Cumoto Village, Trinidad
Vegetation:	Natural savanna
Parent material:	Alluvium
Topography:	Flat with poor drainage during the wet season

Profile description:
Depth (cm)

0–18   Grey (5Y 6/1) fine sand; numerous fine roots and earthworm casts; single grain structure; clear smooth boundary

18–25   White (5Y 8/1) fine sand with frequent grey (5Y 6/1) earthworm casts; old root channels coated with strong brown (7.5YR 5/8) mottles; single grain structure; abrupt smooth boundary

25–35   White (5Y 8/1) silty clay (gleyed); intensely mottled; reddish-yellow (7.5YR 7/8) to strong brown (7.5YR 5/8) coatings along old root channels; weak platy structure; clear smooth boundary

35–60   White (5Y 8/1) silty clay; intensely mottled yellow (10Y 8/6) to dark red (10YR 3/6); weak platy structure; gleyed; clear smooth boundary.

60–180   Much less mottled dark red (10R 8/6) and much more gleyed white (5YR 8/1) silty clay; mottles more distinct and larger (to 5 cm diam.); weak platy structure; gradual smooth boundary

Table 4.8 continued

180–250 White (5Y 8/1) silty clay with frequent distinct dark red (10R 8/6) mottles to 5 cm diam.; same continues to undetermined depth, becoming progressively more gleyed with fewer red, but more clearly defined and firmer mottled areas

Physical and chemical analysis:

Depth	% sand	% silt	% clay	% C	CEC*	Ca	Mg	K	Na	pH	% free Fe oxides
					\multicolumn{5}{c}{Exchangeable cations (meq.100g⁻¹ soil)}			
0–18	66	25	12	0.51	2.0	0.5	0.2	0.03	0.08	5.0	0.40
18–25	60	22	20	0.22	2.7	0.5	0.6	0.02	0.00	5.0	0.40
25–35	42	22	40	0.11	4.2	0.8	0.3	0.02	0.00	4.8	5.26
35–60	29	20	53	—	6.9	0.5	0.5	0.04	0.06	4.6	6.43
60–180	32	18	52	—	10.2	0.4	0.7	0.07	0.01	4.8	4.47
180–250	23	21	57	—	10.6	0.4	0.8	0.12	0.02	4.8	9.75

Source: Ahmad and Jones (1969)

Notes: Soil colours determined by the use of Munsell colour charts
* Cation exchange capacity (CEC) determined by NH_4OAc extraction, making these values probable overestimates for a soil that probably has a large variable-charge component

Other tropical soil types

Although oxisols and ultisols are quantitatively dominant in most tropical regions, there exist also a wide variety of other soil types, some of which contain an agricultural potential far in excess of their small coverage, and therefore require special mention. Many of these soils are patchily distributed in tropical landscapes, occupying special geological or geomorphological features (Figure 4.12). Others may form units in topographically related soil sequences ('catenas'). A sensitivity to this small-scale heterogeneity is critical in land use planning in tropical regions (see Chapter 11).

Alfisols

Alfisols represent less intensely leached and more fertile soils that are characteristic of many drier areas in the tropics. They share many of the physical characteristics of ultisols, such as a clay-enriched B horizon, but are of much higher native fertility than these and consequently of greater agricultural potential. Alfisols are especially widespread in the Sahelian region of Africa, but are also present in the drier zones of the Neotropics and Asia, where they are frequently associated with dry forests. They correlate with luvisols and planosols in the FAO–UNESCO map legend, and are common in temperate areas as well as the tropics.

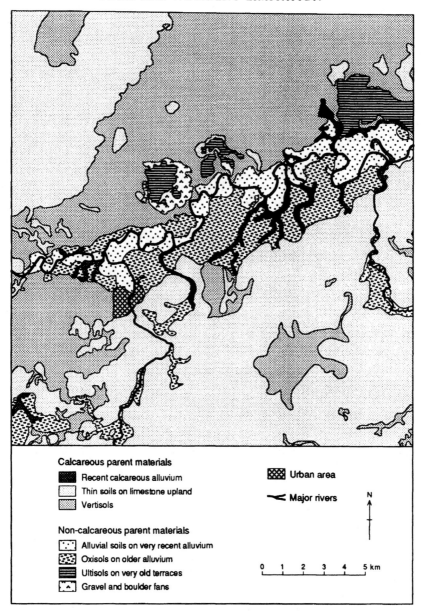

Calcareous parent materials

■ Recent calcareous alluvium

☐ Thin soils on limestone upland

▨ Vertisols

⊠ Urban area

⤙ Major rivers

Non-calcareous parent materials

⦂ Alluvial soils on very recent alluvium

▨ Oxisols on older alluvium

▤ Ultisols on very old terraces

⟁ Gravel and boulder fans

N

0 1 2 3 4 5 km

Figure 4.12 Generalized distribution of soil types in the upper Belize Valley, Belize. In this area, the Belize River runs through limestone uplands that are covered by thin limestone soils and vertisols. However, the upper reaches of the river lie in a region of non-calcareous bedrock, which has provided the material for alluvial terraces of varying age, on which more typical tropical soils have developed

Source: Jenkin *et al.* 1976: Map 3 (simplified)

85

Vertisols

Vertisols are recognized as a distinct soil order in both USDA and FAO–UNESCO soil systems, which is symptomatic of their specialized properties. These are soils whose clay mineral fraction is dominated by montmorillonite, which imparts special features to both their chemistry and their physical properties. Unlike most other tropical soils, vertisols contain a large permanent-charge CEC which is normally saturated with Ca^{2+} and Mg^{2+} ions, and are therefore of very high inherent fertility. However, their propensity to swell and shrink during wetting and drying cycles produces a physically unstable soil that often poses difficulties to agricultural operations. Swelling and shrinking also leads to slow self-churning in these soils, which prevents horizon development and causes the profile to be homogeneously grey-black. Vertisols tend to be found wherever montmorillonite formation is favoured. Consequently they are especially common in drier parts of the tropics, and occupy large surface areas in India, Australia and the Sudan. They are also common elsewhere, usually as small soil bodies in low-lying landscape positions. In limestone terrain they are frequently found on flat or low-lying sites, where silicate materials deriving from dust or ash falls accumulate, and a continuous supply of Ca^{2+} and Mg^{2+} is available during weathering (e.g. Figure 4.12).

Recent volcanic soils

Many soils developed on recent volcanic deposits are dominated by the clay mineral allophane, whose special physical and chemical properties have been described earlier. Although these soils occupy only about 1 per cent of the tropics, they support highly productive agriculture and dense human populations in parts of Central America and South-East Asia, as well as volcanic islands, such as Hawaii. In the FAO–UNESCO map legend they are described as andosols, and in the USDA system as andepts.

Limestone soils

Limestone represents a distinctive parent material for soil formation, as it lacks silicate minerals, except as impurities, and therefore does not weather to form secondary clay minerals. Instead, $CaCO_3$ dissolves completely when exposed to carbonic acid, leaving little mineral material for soil formation. As a result, limestone soils tend to be skeletal and composed of organic matter plus any silicate material that was present as impurities in limestone, or has arrived extraneously as dust and volcanic ash deposits (Figure 4.13). While thin and droughty, these soils are generally fertile, as any clay mineral present tends to be montmorillonite. Depressional areas within limestone terrain often contain vertisols (Figure 4.12). Although generally unsuited to

Figure 4.13 Shallow soils developed on coraline limestone, Barbados. The limestone is relatively pure $CaCO_3$, and most soil mineral material probably derives from volcanic ash falls coming from the Soufrière volcano on the nearby island of St Vincent

mechanized agriculture, these soils have been an important resource for traditional shifting cultivation systems, most notably that of the Maya of the Yucatán peninsula. In the USDA classification system limestone soils are usually placed within the much larger group of mollisols, which includes most temperate-area prairie soils. In the FAO–UNESCO system these soils are usually classed as rendzinas.

Recent alluvial soils

Young soils on recently exposed parent materials are estimated to cover over 22 per cent of the tropics (Table 4.6). Some of these are on sand deposits created during Pleistocene drought (e.g. Figure 2.1), but a significant proportion occur on recent river alluvium or coastal deltaic deposits. Soils developed on this alluvium represent a highly heterogeneous set, whose properties reflect both the geology of the watershed producing the alluvium and the relative age of the deposit. Watersheds composed primarily of quartzitic materials, such as the Guiana Shield in South America, often produce alluvium containing few easily weatherable silicate minerals and infertile soils that are gradually transformed into giant podzols. On the other hand, actively eroding mountain systems such as the Andes and Himalayas provide alluvium to their adjacent tropical river systems that are high in weatherable minerals and extremely fertile. River systems such as the Ganges and Mekong support some of the densest rural populations in the tropics, and intensive agriculture on the soils of flooded forests in Amazonia has sometimes been recommended as an effective land use (e.g. Sioli 1973). However, the recent recognition of the importance of these forests to the aquatic ecosystem (Forsberg *et al.* 1993) requires that these suggestions be reconsidered. Within floodplains, alluvium generally occurs as systems of terraces of varying elevations, each subject to differing frequencies of flooding. As the fertility of these soils depends upon continual deposition of new sediments, higher terraces experiencing less frequent flooding generally have poorer and more weathered soils than those closer to the river. Soils on very high terraces that are no longer subject to sediment deposition are normally transformed gradually into ultisols or oxisols (Figure 4.12). Coastal deltaic deposits pose special chemical problems of desalinization and often become extremely acid when drained. Their use has been primarily restricted to wetland rice cultivation, which preserves anaerobic conditions (Sanchez 1976).

PLANT NUTRITION ON SOILS OF VERY LOW FERTILITY

Weathering processes in the tropics lead to most better-drained soils being gradually transformed to oxisols or ultisols. These soils have few weatherable minerals remaining and are dominated by clay minerals of low exchange activity and variable charge, and plants growing on them are separated from fresh geological materials by many metres of weathered regolith. Under these conditions, atmospheric inputs become the main extraneous source of nutrients. While the total quantity of these nutrients is usually small (e.g. Kellman *et al.* 1982; Kellman and Carty 1986), sufficient retention mechanisms exist in soil to ensure their effective capture (see above). These atmospheric inputs are probably sufficient to offset small leakages of nutrients in percolating water, and indeed they appear to have been the nutrient source that has allowed re-establishment of forest on infertile savanna soils during Pleistocene interglacials (Kellman 1989).

Progressive ageing of soils poses a general nutrient supply problem for plants, but the problem is especially acute in relation to phosphorus. Iron and aluminium oxides have a strong affinity for this nutrient, and their increased abundance in older soils leads to increasingly effective phosphorus 'fixation' in a form not readily available to plants. Plinthite nodules have an especially large capacity to sorb phosphorus (Tiessen *et al.* 1991), and Crews *et al.* (1995) have shown that readily available phosphorus peaks at intermediate soil age (*c.* 150,000 yr) on Hawaiian volcanic flows before declining on the oldest soils. Because phosphorus inputs from atmospheric sources are very small, this progressive 'occlusion' of the element should lead to its becoming the primary limiting nutrient on very old soils. An analysis of litter fall and its nutrient content in a large number of tropical forests has indicated that phosphorus was the element being cycled most effectively, implying that this was limiting on plant growth (Vitousek 1984). In contrast, nitrogen appears to cycle rapidly in tropical forests and not to be limiting on plant growth (Vitousek and Sanford 1986), although it may become so in drier savannas (Hogberg 1989).

Because soil ageing is a gradual and widespread process in tropical environments, we would expect natural selection in tropical floras for genotypes best able to function under these conditions. Earlier, we described some of the plant root properties that contribute to more effective nutrient retention; here we will treat other characteristics that appear to have been selected by low nutrient availability.

Decreasing the demand for a nutrient by possessing a slow growth rate is one of the most effective means of coping with low soil fertility (Chapin 1980). For example, fast-growing tree seedlings of mahogany (*Swietenia macrophyllum*) failed within two years of being planted in low-fertility savanna soils in Belize (Kellman and Miyanishi 1982), while those of the very slow-growing *Xylopia frutescens* survived. Effective reabsorption of nutrients prior to tissue death is a further nutrient conservation tactic, but one that is available only to certain elements. Calcium resides primarily in cell wall structures and so cannot be readily translocated prior to tissue death. However, more mobile elements can be partially removed from senescing tissues, and reabsorption of phosphorus from leaves appears to be especially common on more infertile soils (Vitousek and Sanford 1986). Because leaf tissues are nutritionally costly to construct, leaves of greater longevity provide more effective photosynthetic returns per unit of nutrient investment (Chapin 1980), and we would therefore expect greater leaf longevity to be strongly selected for on infertile tropical soils. In the tropics, this tactic is precluded in dry forest by obligate deciduousness, but is available in moister environments where nutrient supplies are likely to be less available. The disadvantages of such a tactic in the tropics include a greater exposure to herbivory, and thus the need to possess chemical or physical protectants (see Chapter 5), as well as an accumulation of epiphylls in humid environments, which may lead to decreased photosynthetic efficiency.

Large biomass is not, in itself, a nutrient conservation tactic, although large quantities of nutrients may be sequestered in high-biomass tropical forests. Rather, large plant biomass allows effective competition for light. However, the existence of large-biomass forests on low-fertility soils indicates that large biomass is not precluded by low soil fertility, although its re-establishment after dissipation of the nutrients that it contains may pose special problems. In savannas, the development of a large biomass is constrained by frequent fires, but in their absence a forest's nutrient biomass can be gradually accumulated from atmospheric sources (Kellman 1989).

However, once established, a forest whose biomass contains a large sink of nutrients can serve to drive an enriched cycle of nutrients even on infertile soils, provided that recirculation of these nutrients is rapid.

Highly weathered tropical soils present one further common nutritional problem for plants: aluminium toxicity. Under acid conditions (pH <5.5), aluminium compounds become increasingly soluble, and Al^{3+} ions enter the soil solution and exchange complex. Aluminium is toxic to the roots of many plant species, and the roots of sensitive species may be inhibited in very acid soils. Plants native to these soils are tolerant of these conditions, but many crop plants are less so. Breeding for Al-tolerant crop plants that also have low phosphorus demands represents an important stategy in tropical agriculture (Sanchez and Buol 1975).

NUTRIENTS IN TROPICAL AQUATIC SYSTEMS

Nutrients commonly limit primary productivity in tropical aquatic systems for the same reasons that they do in terrestrial systems: there is an abundance of solar energy and a year-round growing season, allowing productivity to continue until limited by some other factor, which commonly is a nutrient element. As in terrestrial systems, most nutrients used have been recycled from the decomposition of dead organic tissues rather than being new inputs. Photosynthesis in aquatic systems takes place at, or close to, the water surface, where there is an abundance of solar energy. However, in non-flowing water bodies, such as lakes, most mineralization of nutrients takes place in bottom sediments, where organic detritus from the surface falls and is decomposed. Bottom water is therefore nutrient-rich but light-deficient and also frequently anoxic as a result of oxygen consumption during decomposition.

In temperate lakes, seasonal recirculation of nutrients to the surface takes place as a result of turnover in the water column induced by winter cooling of the surface water. This usually results in a burst of primary productivity in the spring as temperatures rise. In contrast, in tropical lakes, continual heating of surface water by high solar radiation loads preserves thermal stratification and suppresses water column turnover, leaving surface waters low in nutrients. This situation can be alleviated by wind stirring of the water body, which is especially effective if this is shallow. Wind stirring may be a seasonal phenomenon induced by intensified trade winds in the dry season, or it may be more sporadic as a result of storms. In the latter case, violent stirring can also bring oxygen-deficient water to the surface, which may induce sporadic fish kills. Other sporadic nutrient enrichment can be induced by wet-season floods. As a consequence of these processes, primary productivity in many tropical water bodies is low but subject to sporadic increases. Beadle (1981) provides an excellent review of these processes in tropical lakes, to which the reader is referred.

The nutrient status of river water is largely a reflection of the geology of the watershed that it drains, and can be very variable between sub-basins of a single large watershed, such as that of the Amazon (Furch 1984). Watersheds such as that of the Rio Negro are dominated by quartzitic materials and nutrient concentrations are very low. These rivers are frequently 'blackwater' rivers with a high content of humic substances which are not effectively removed during percolation through the sandy soils (St John and Anderson 1982). In such rivers there may be almost no photosynthetic activity, and food chains are based on organic detritus deriving from adjacent terrestrial communities, such as flooded forests. In less oligotrophic rivers productivity is higher. However, in these, productivity is usually greatest during the dry season, rather than the wet season when more nutrients are available, because turbidity at that time limits light availability (Rai and Hill 1984). Much of the productivity in large river floodplains is concentrated in floodplain lakes, which receive large influxes of nutrients (Rai and Hill 1984). Coastal zones where these rivers discharge are often also zones of high marine productivity (Longhurst and Pauley 1987).

5

TROPICAL BIOTAS

Tropical biotas have evolved in an environment that possesses several unique properties (Chapters 3 and 4) and that has had a long and complex geological history (Chapter 2). As was mentioned in Chapter 1, a diversity of life forms and a complexity of biological interactions are two of the unique features of tropical ecosystems. In this chapter, we will discuss the functioning of biological populations in the relatively aseasonal environment of the tropics and examine how natural selection and organic evolution appear to operate in such places. We will then review the major patterns of biological diversity that exist globally and within the tropics, and some of the hypotheses that have been proposed to account for the high levels of species diversity that are characteristic of many tropical communities. Finally, we will discuss the conservation and management problems that are presented by tropical biotas, including those involved in human diseases.

Temperatures permit biological activity throughout the year in most parts of the tropics, but seasonal drought may limit activity in some areas. Furthermore, the nutrient substrate available to organisms, especially those occupying moister climatic zones, is generally poor relative to that available in temperate areas, and may limit the levels of organic productivity. However, superimposed upon this matrix of generally low-fertility soils are smaller soil bodies of higher fertility, especially in certain floodplains and in areas of recent volcanic activity. We have also seen that tropical climates were much more widespread in the past, and that most tropical environments have probably been subject to less severe disruptive changes during the Quaternary than were high-latitude sites.

THE REGULATION OF BIOLOGICAL POPULATIONS

All organisms have an inherent capacity to increase their numbers exponentially, although at rates that differ from species to species. This intrinsic rate of natural increase, usually referred to as r_{max} in population growth equations, is influenced by several intrinsic properties of the organism, among which are birth rate, longevity, and the speed with which it achieves reproductive maturity. Of these, the last is especially important, and organisms with short generation times have an enormous capacity to increase their numbers rapidly, especially when reproductive activity can be continued throughout the year. However, despite their inherent capacity for increase, most biological populations remain constrained at numbers that may be relatively steady or, more commonly, fluctuate even to the point of local extinction. How populations are regulated represents one of the fundamental questions addressed by theoretical population biologists, but is one that has received no widely accepted single answer (Krebs 1994). The question is also of fundamental concern to applied scientists who must recommend means of constraining pest outbreaks and disease epidemics.

The most obvious constraint on population size is scarcity of resources, which may take several forms depending on the organism involved. For plants, the availability of light, moisture and mineral nutrients is finite in any habitat, and defines an upper threshold for the quantity of biomass that can be sustained at that place. Paradoxically, individual plants in resource-rich habitats often grow to very large sizes, *reducing* the numbers of individuals that can be sustained there (Harper 1977). For organisms of higher trophic levels (herbivores, carnivores, parasites, etc.), the resource base consists of populations of other organisms, which are themselves subject to limitation by their predators. This sets the stage for complex and potentially extreme predator–prey oscillations in time and space. One factor helping to stabilize these oscillations is that predator–prey systems rarely exist in isolation: each organism is influenced also by many independent interactions with other 'third parties', such as a parasite preying upon the predator, as well as abiotic factors. Many of these interactions operate in a 'density-dependent' way, in which negative feedbacks upon the populations intensify as their densities increase. Moreover, behavioural traits, such as migration and predator territoriality or switching to alternative secondary prey prior to extinction of the primary prey, can contribute to stabilizing extreme oscillations in multi-species communities (Strong 1992).

Many parasitic micro-organisms, including human diseases, have very short generation times and a potential for explosive population increase. Where such organisms are not lethal to their hosts they may coexist stably with a host population that remains in a debilitated, but viable, state. However, more lethal micro-organisms have the potential rapidly to eliminate their hosts and subsequently crash owing to resource disappearance. These sorts of organisms may be expected to coexist with hosts in cycles of periodic epidemics and host crashes, followed by later host recovery. The conditions initiating an epidemic in a host population are of considerable practical concern. Epidemics result when infection levels achieve a threshold needed for self-sustaining population growth (Bailey 1975). A critical factor influencing this threshold is usually density of host organisms, with higher densities increasing the probability of rapid inter-host transmission. Conditions external to the host–parasite relationship also influence diffusion of parasites among their hosts. For example, human disease outbreaks in the tropics are characteristically more common in wet seasons than in dry seasons, as a consequence of the greater ease of spread of micro-organisms either directly (e.g. many intestinal parasites) or with the aid of arthropod vectors that are more abundant in the wet season (e.g. malaria).

Competition between populations of different species that all make use of the same resource will also limit population size. A widely cited theoretical prediction, based on early competition experiments by Gause (1934), is that species utilizing the same resource cannot coexist indefinitely, as the more effective resource user will eventually monopolize this, driving the less effective user to extinction. In view of this, the successful coexistence of large numbers of species that draw upon the same resources, such as higher plants, presents a fundamental biological enigma, especially in tropical communities where the numbers of coexisting species may be very large. We will return to this issue below.

In contrast to population regulation mediated by the availability of resources and competition for these, populations may also be regulated by periodic catastrophic events, which eliminate the population or reduce it to a small size or dormant phase. This has commonly been called 'density-independent' population regulation, as it is achieved by a factor un-influenced by the density of the population concerned. This form of population regulation is usually non-selective and may affect groups of coexisting populations. At a global scale, by far the most pervasive form of density-independent regulation is achieved by seasonal winter conditions in higher latitudes. There, long-lived organisms capable of entering a dormant state can survive the low temperatures and resume activity in the subsequent summer, but the populations of many short-lived organisms, such as insects and annual plants, are reduced to zero and must redevelop from dormant over-wintering diaspores. As a consequence, these organisms spend much of each summer season in a population re-expansion that rarely achieves densities sufficient to regulate their prey or competitors before the subsequent autumn crash. Similar processes may operate in tropical areas possessing severe dry seasons, but elsewhere climatic conditions allow a potentially indefinite population expansion, until limited by the availability of resources and competition for these. This condition represents one of the fundamental ways in which tropical environments differ from those of higher latitudes, as it ensures that pressure from short-lived predators, pathogens or competitors is an omnipresent force helping to regulate populations and shape the course of evolution in the tropics. It also ensures that pest and parasite pressure on domesticated plants and animals is likely to be much more severe in tropical than in temperate environments, and that human diseases are more likely to be persistent problems in the tropics.

Coping with persistent predation in the tropics

The term 'predation' is used here to refer to any interaction involving one organism preying upon another, whether the result is lethal, or merely debilitating, to the prey. Defined in this way, the term includes all forms of herbivory and parasitism, as well as seed and seedling consumption. In communities that contain natural enemies of predators, predation may be constrained to levels that do not cause host extinction, allowing coexistence of the two populations in some sort of dynamic equilibrium. However, in the absence of these natural enemies, and in an environment in which predatory organisms are capable of continual population expansion, only two means exist by which prey populations can avoid eventual extinction: defences that repel potential predators or limit the extent of their effects; and avoidance of the predatory organisms.

Physical or chemical defences are likely to be especially important for long-lived sessile organisms, such as perennial plants, that lack any behavioural mechanisms permitting avoidance. A comparison of results from temperate and tropical forests (Coley and Aide 1991) has shown that both leaf toughness and content of alkaloids (a group of secondary compounds) are significantly higher in tropical latitudes. Other work has shown that tropical alkaloids are significantly more toxic than those in temperate plants (Levin 1976). Despite this protection, levels of leaf herbivory were still found to be significantly greater in tropical than in temperate forests (10.9 per cent as against 7.5 per cent), with young leaves being especially vulnerable (Coley and Aide 1991). Interestingly, Coley and Aide (1991) did not find defences in tropical dry forest leaves to be significantly greater than those in temperate forest leaves. As leaf longevity in both forest types is relatively short (9.6 and 6.0 months in dry tropical and temperate forests, respectively), while that of shade-tolerant trees in wet forests is much longer (32.2 months), it suggests that exposure time has been crucial in selecting for chemical defences.

Other, more intricate forms of defence against predators have also been described in tropical communities. Among the best-known are various myrmecophytes, plants which host ant colonies in a mutualistic relationship. The swollen thorns of some species of small Neotropical acacias support ant colonies that survive on extrafloral nectaries and protein-rich leaf structures provided by the host plant. In turn these both provide protection from herbivores and attack vines that attach themselves to the host plant (Janzen 1966). A similar ant–plant symbiosis has been demonstrated in the tropical pioneer tree *Cecropia* by Schupp (1986). Here, *Azteca* ants living in stem internodes attacked both herbivores and vines, and experimental removal of these ants resulted in slower plant growth rates. The African savanna tree *Acacia drepanolobium* hosts aggressive *Crematogaster* ants which appear to offer some defence against large mammalian browsers such as giraffes (Madden and Young 1992; Figure 5.1).

Figure 5.1 Ants on an *Acacia drepanolobium* branch, Kenya. Aggressive ant colonies are housed in the swollen thorns produced by these small savanna trees, which are often subject to browsing by ungulates. These ants responded quickly to a slight shaking of the branch by the photographer

Defences, whether physical, chemical or mutualistic, all represent a cost to the plant, and would be expected to be selected for only if the benefits provided, in terms of less leaf loss and greater growth, compensated for the costs. This aspect of the interaction has recently been examined experimentally by Sagers and Coley (1995) using clones of an understorey rain forest shrub that varied in leaf toughness and chemical defence. When protected from herbivory, plants with more chemical defences grew more slowly than others with fewer defences, indicating a metabolic cost of the defences. However, when exposed to herbivores, plants of different clones showed no significant differences in growth rates, suggesting that defensive investment and more rapid growth rate could be traded off as alternative means of coping with herbivory. A survey of defensive characteristics of 46 tree species in tropical forest has shown that pioneer tree species have fewer defences than non-pioneers (Coley 1983). The author hypothesized that the greater tolerance of herbivory by pioneers was due to faster growth rates and lower cost of producing the short-lived leaves that are characteristic of pioneer trees (Coley 1988). Leaf longevity is characteristically greater in plants growing upon soils of low fertility (Chapin 1980), and nutritional costs of replacing leaf tissue at these sites are proportionately higher, so investments in anti-herbivore defences should be higher at these sites (Coley *et al.* 1985). McKey *et al.* (1978) have found that the leaves of tropical forest trees growing on giant podzols contain approximately twice the concentration of phenolics as compared with leaves from trees on more fertile soils, but a more complete data set is required for the testing of this prediction. However, we can tentatively add well-developed anti-herbivore defences as a further feature characteristic of plants growing upon low-fertility tropical soils.

Avoidance represents the second major way of coping with persistent predation in the tropics; this may take the form of avoidance in time as well as in space. Young leaves produced during tropical dry seasons, when herbivore populations are low, suffer less herbivory than those produced during the wet season (Aide 1992), and leaves produced synchronously suffer less herbivory than asynchronously produced leaves, presumably because of herbivore satiation in the former case (Aide 1993). Predator avoidance in space, by means of low prey densities that diminish the probability of prey location, has received considerable attention in tropical forests. In 1970, Janzen (1970) and Connell (1971) independently suggested that predation upon seeds and seedlings of tropical forest trees may limit successful recruitment near to conspecific adults and thus maintain low-density populations in these communities. While these authors stressed the potential of this mechanism to prevent competitive exclusion and sustain species-rich communities (see below), the mechanism clearly has more general implications for the functioning of tropical ecosystems, especially those managed systems that seek to sustain high-density agricultural populations.

Evaluations of the Janzen–Connell hypothesis over the past quarter-century have produced equivocal results (see Burkey 1994 for a recent review); while the process appears to be responsible for maintaining low densities in some species, it is clearly not as universal a process as was once thought. Hubbell (1980) has pointed out that even where distance-dependent predation has been demonstrated, the very large seed fall that normally occurs beneath parent trees may be sufficient to override a high proportionate seed loss there and allow successful recruitment near conspecifics. The most complete test of the hypothesis has been carried out by Condit *et al.* (1992a) using data on recruitment patterns of 80 tree and shrub species in a 50-ha plot in Panama. Thirty-eight of these species showed no consistent pattern of recruitment near to conspecific adults, and a further 27 showed preferential recruitment near to these; only 15 species showed some tendency to be repelled by adults. Among these, the distance over which the repulsion effect operated was usually less than 5 m, and only the most common tree species showed a large recruitment reduction within 10 m of adults. The authors concluded that the Janzen–Connell effect was present but not common in this forest. Despite this, the fact that a localized effect was apparent among some of the most densely distributed species holds important implications for coping with predators in managed ecosystems. In these, crop plants usually exist at far higher densities than those found in natural populations and so are potentially susceptible to severe predation, but the very localized effect found by Condit *et al.* (1992a) suggests that even limited isolation may impart some predator protection in these systems.

EVOLUTION IN THE TROPICAL MILIEU

Organic evolution represents the consistent change through time in the genetic structure of a population of organisms. These changes can lead ultimately to changes in the appearance and performance of the organisms making up the population, as these characteristics are usually under genetic control. Changes that are sufficiently profound can leave organisms of the changed population reproductively isolated from those in the original parent population, and effectively a new species. These evolutionary changes can result in the wholesale transformation of a species population into one with a changed genetic structure or, more commonly, the splitting off of new species from existing ones, in a process usually termed 'speciation'.

Organic evolution involves natural selection operating on the inter-organism genetic variability that exists in all populations that have not derived by cloning (asexual reproduction) from another organism. In this, organisms possessing genetic traits that give them some advantage relative to other organisms in the same environment will have a higher survivorship and leave more offspring, leading ultimately to the dominance of their genetic traits in that population. Because the genetic variability on which natural selection operates is produced randomly, it leads also to a high degree of randomness in the whole evolutionary process. Some aspects of natural selection may be sufficiently pervasive (e.g. competition among plants for light), and the availability of a favourable genetic response sufficiently common (e.g. large plant size), to lead to convergence on a common 'adaptation' in the same sort of environment (e.g. trees in moister climate zones). However, each species population has a unique ancestry and may exhibit a great deal of idiosyncratic detail in form or function that often confounds simple patterns of evolutionary convergence.

The production of organic diversity by means of natural selection leading to speciation is usually assumed to involve a cycle of population expansion and range fragmentation leading to isolated subpopulations; natural selection acting on these isolated subpopulations over long time periods may lead to their transformation into new species no longer able to interbreed with the original population when reunited with it. This has been termed 'allopatric' or 'geographic' speciation. While many species appear to have originated in this way, a variety of mechanisms exist that may allow non-isolated populations to undergo rapid genetic change and become new species; this has been termed 'sympatric' speciation. There is no clear consensus among evolutionary biologists about the degree and persistence of isolation necessary for the formation of new species, other than that greater isolation which persists for longer time periods is more likely to result in the formation of new species. Indeed, it is probable that different species may have rather different evolutionary histories. For example, some extant plant species, such as *Ginkgo biloba*, are morphologically indistinguishable from mid-Tertiary fossils, while other lineages, such as the genus *Eucalyptus*, underwent explosive speciation in Australia as that continent desiccated during the late Tertiary.

Despite differences in rate, evolution represents a process that has the potential to diversify the biota of any region. Through time, we would expect the continual appearance of new species in any region, and the total number of species in the biota to increase until balanced by the disappearance, due to extinction, of an equal number of species. Individual lineages appear to have undergone periods of rapid expansion associated with the acquisition of new properties imparting particular competitive advantages (e.g. the angiosperms in the early Cretaceous; Crane *et al.* 1995), or the appearance of new environmental opportunities, such as the emergence of subtropical arid zones in the Tertiary (see Chapter 2).

Several special characteristics may be manifested in evolution in the tropics. In the absence of seasonally low temperatures, little natural selection for tolerance of low temperatures may be expected, and the sensitivity of many tropical plant species to occasional frost at their range limits (e.g. Silberbauer-Gottsberger *et al.* 1977) suggests that this has been a particularly difficult evolutionary challenge to overcome. The existence of many region-specific tropical and temperate plant families and genera, and the paucity of the former in high-altitude mountain areas in the tropics (see Chapter 6), further suggests the difficulties of evolutionarily traversing the 'frost barrier'.

In contrast to the absence of seasonally stressful temperatures in the tropics, both persistent biological interactions and chronically low soil nutrient resources may be expected to be important agents of natural selection in these habitats. Moreover, a prolonged growing period should permit short-lived organisms to achieve many more generations per year, providing greater potential for rapid evolutionary change among these groups. The replacement of generalized climatic stresses as agents of natural selection in temperate areas, with potentially very specific biological stresses in the tropics, led Dobzhansky (1950), in a widely cited paper, to suggest that tropical organisms are more narrowly specialized than their temperate counterparts. However, in a recent comparative review, Price (1991) found no evidence for more specialization among tropical herbivores and parasites, nor any evidence for more competition and fewer vacant niches in tropical communities than in temperate systems. He concludes that while tropical biotic interactions may be more diverse than in temperate areas, they may not cause any stronger natural selection. This suggests that the diversity of many tropical biotas may not be the product of a unique degree of self-augmenting inter-species specialization in these places, but must be sought in other mechanisms. However,

99

this does not preclude the possibility that certain lifestyles may be possible in aseasonal tropical environments that would be precluded in temperate areas. For example, the existence of certain intricate plant–insect mutualisms, such as that exhibited by ants and acacias, may be more feasible in an environment where the insect does not need to re-establish new populations annually. Similarly, the perennial vine habit, which is a prominent and diverse component of many tropical woody communities, is almost absent from high latitudes, and it has been suggested that the large-diameter xylem cells required in these elongate plants make them especially vulnerable to embolism due to freezing in low-temperature habitats (Teramura *et al.* 1991).

BIOLOGICAL DIVERSITY IN THE TROPICS

One of the commonest reactions among temperate-area biologists on first visiting the tropics is to comment on the great diversity of life forms present there. In part, this reflects the existence of life forms not commonly seen in temperate areas, such as woody vines, but more generally it reflects the great variety of species of any life form type (such as trees or insects) that may be encountered in even a small area. While we shall see that these casual observations about diversity can often be corroborated quantitatively, it is also necessary to recognize that they come primarily from tropical forested environments, and that other lower-diversity regions and communities can also be found within the tropics. This requires us to begin any discussion of tropical diversity, and its potential explanations, by examining the diversity patterns that exist, including both temperate–tropical contrasts and diversity patterns within the tropics.

Patterns in biological diversity

Assessments of diversity patterns are usually made for broad taxonomic or functional groupings of organisms (e.g. higher plants, woody plants, corals, birds) rather than for all organisms. This is because data do not exist on the diversity of many groups, and because many of these differ so much in the absolute numbers of species contained that pooling of data would result in a masking of patterns by the most speciose groups. To be comparable, statements about diversity must also be made relative to some common area, and most data come from two very different spatial scales. At one extreme, museum collections provide information about the total numbers of species in some group that have been encountered in large spatial units, such as forested areas or savanna areas within a country. We shall refer to diversity measured at this scale as 'regional diversity'. These sorts of data can be combined to provide data on species numbers in spatial units of larger size, such as continents, but cannot provide much information about diversity patterns *within* the original collecting regions, because of incomplete censuses.

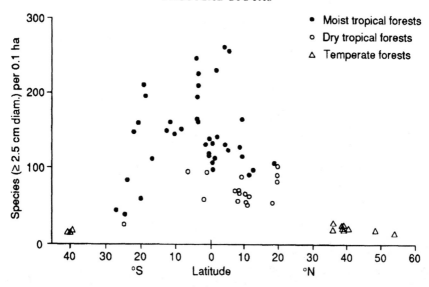

Figure 5.2 Variation with latitude in the numbers of species of plants ≥ 2.5 cm diameter in standardized 0.1-ha forest samples

Source: Gentry 1988: Figure 2

At the other extreme of scale, complete field enumerations have provided relatively accurate information on numbers of species in very small areas (e.g. 1 ha or less), especially for sessile organisms such as plants. We shall refer to diversity measured at this scale as 'local diversity'.

Diversity at small and at large spatial scales need not necessarily be correlated, although in the tropics they often are. A large regional diversity could, in theory, be produced by a diverse assemblage of locally species-poor communities. However, while some low-diversity communities do exist within the tropics, high regional diversities are usually matched by high localized diversity. The latter has sometimes been referred to as 'within-community' or 'alpha' diversity and is especially well developed in tropical forests. A wide variety of different measures of biological diversity have been used in the literature, each employing different scaling procedures. However, most of these measures are highly collinear, and in the discussion that follows we will use the simplest and most straightforward measure: numbers of species per unit area, also known as 'species richness' or 'species density'. When this is applied to terrestrial biotas, the areas referred to can usually be specified, but when applied to oceanic areas (e.g. Figure 5.3), the area is often more indefinite.

The numbers of species of a wide variety of different taxonomic lineages show an increase at low latitudes (Fischer 1960). Groups as different as woody plants (Figure 5.2), ants (Table 5.1) and planktonic formanifera (Figure 5.3)

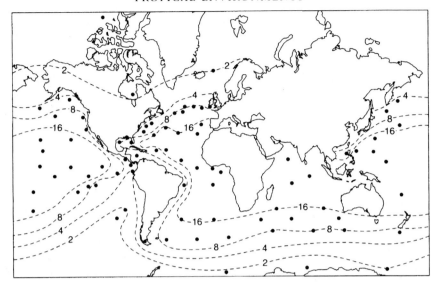

Figure 5.3 Global diversity pattern for recent species of planktonic formanifera
Source: MacArthur 1972: Figure 8-10

all show this pattern. Moreover, palaeontological data indicate that this is a very ancient pattern (Crane and Lidgard 1989; Jablonski 1993; Stehli *et al.* 1969). While not all lineages show this pattern (Price 1991), it is of sufficient generality to suggest that tropical environments have been especially favourable places for the generation of new species or the persistence of these once formed (or both); we will return to this topic below.

Superimposed upon this general pattern of increasing diversity with decreasing latitude are some major intra-tropical diversity patterns. The highest levels of local diversity of woody plants in tropical forests occur in wet climate zones,

Table 5.1 Change in the number of ant species with increasing latitude in South America

Region	Approximate latitude (S)	Number of ant species
São Paulo, Brazil	20°–25°	222
Misiones, Argentina	26°–28°	191
Tucumán, Argentina	26°–28°	139
Buenos Aires, Argentina	33°–39°	103
Patagonia, in total	39°–52°	59
Patagonia, humid west	40°–52°	19
Tierra del Fuego	43°–55°	2

Source: MacArthur 1972: Table 8–1

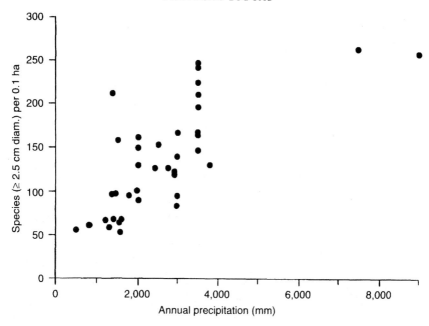

Figure 5.4 Variation with precipation in the numbers of species of plants ≥ 2.5 cm diameter in standardized 0.1-ha lowland Neotropical forest samples

Source: Gentry 1988: Figure 3

and diversity decreases with both rainfall and altitude (Figures 5.4 and 5.5). Both the absolute numbers of vascular plant species and the densities of these differ between continents (Table 5.2). Embedded within high-diversity tropical forest communities there may also be other low-diversity communities. While some of these, such as mangroves and seasonally flooded forests, have an obvious association with very specialized habitat conditions, others do not, and considerable debate continues on the explanations for these islands of simplicity (Connell and Lowman 1989; Hart *et al.* 1989). Tropical forest

Table 5.2 Estimated vascular plant diversity in the world's three major tropical forest regions

Region	Total area (ha × 10⁶)	Estimated number of vascular plant species
Neotropical	400	90,000
Indo-Malayan	250	35,000
Malaya	13	7,900
African	180	30,000

Source: Huston 1994

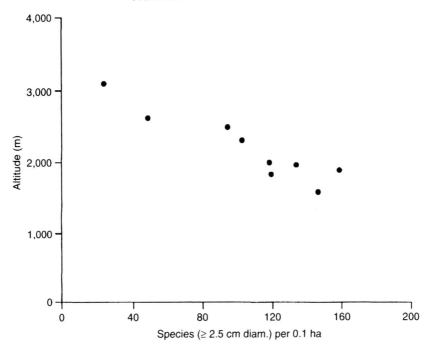

Figure 5.5 Variation with altitude in the numbers of species of plants ≥ 2.5 cm diameter in standardized 0.1-ha forest samples in the Andes

Source: Gentry 1988: Figure 10

diversity also decreases on isolated oceanic islands, forming part of a larger pattern of biotic impoverishment in these places. The rain forests of Hawaii, which are dominated by a single species (*Metrosideros polymorpha*), represent an extreme example of this process (Figure 5.6). Other intra-tropical diversity patterns are more idosyncratic. Although as a community type mangroves are of low diversity, the mangrove communities of the South-East Asian tropics are considerably more diverse than those of other areas. Reef corals are also much more diverse in the Indo-Malayan region than elsewhere in tropical oceans (Figure 5.7), and diversity of ungulate animals is notably higher in Africa than in other tropical regions (Table 5.3). Mares (1992) has analysed the patterns of mammal diversity in South America, and concluded that this is related primarily to areas of 'macrohabitat'; Amazonian forests possess mammal diversities that are no greater than those of other regions when the area covered is allowed for. The existence of these latter patterns, which lack any obvious habitat correlates, suggests that explanations for diversity may not reside exclusively in contemporary environmental conditions, but may have an historical and biogeographical component.

Figure 5.6 Metrosideros (ohi'a)–*Cibotium* (tree fern) forest, Hawaii

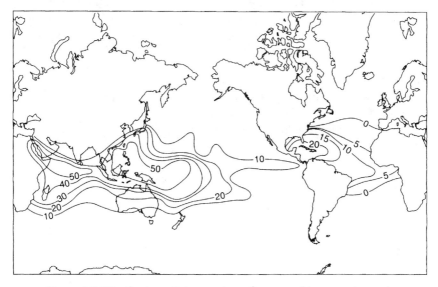

Figure 5.7 Distribution of the number of genera of hermatypic corals
Source: Huston 1994: Figure 12.22

Table 5.3 Numbers of species of ungulates recorded in some national parks and protected areas in the tropics

Location	Habitat type	Number of ungulate species
Africa		
Tarangire Game Reserve, Tanzania	Open *Acacia* savanna	14
Kafue National Park, Zambia	Treed savanna	19
East Tsavo National Park, Kenya	Open woodland	13
Serengeti National Park, Tanzania	Open and treed savanna	20
Ruwenzori National Park, Uganda	Open savanna and thickets	11
South Asia		
Gir Forest, Gujarat, India	Deciduous woodland and treed savanna	6
Wilpattu National Park, Sri Lanka	Open forest and scrub	7
Kanha National Park, India	Open forest	10
South America		
Estación Biológica de los Llanos, Venezuela	Savanna	2

Source: Huston 1994

The generally high diversity of tropical forest communities is now well documented, but data on the diversity of savannas are much less complete, and the very different enumeration procedures usually employed in the two community types makes comparisons difficult. Data on tropical forest diversity have usually been gathered in areas up to 1 ha, but with counts limited to trees over a certain size (e.g. diameters ≥ 5 cm or ≥ 10 cm). In contrast, most savanna data come from much smaller plots and comprise total species counts. As a result, realistic comparisons between the two community types can be made only in terms of regional diversity. Data on regional plant diversity of several Neotropical savanna areas are provided in Table 5.4. These show moderate diversity at the smallest spatial scales, but regional diversities that are low relative to those of forests. For example, in the cerrado region of Brazil, covering approximately 2 million km^2, fewer than 1,000 plant species are recorded, while regional diversity of Neotropical forests (covering approximately 4 million km^2) is estimated at 90,000 species (Table 5.2). Menault (1983) argues that African savannas are much more diverse than those of the Neotropics, and estimates that plant species richnesses as in standardized areas of 10,000 km^2 are 1,750 and 2,020 for savanna and forest respectively (Figure 5.8). However, the degree to which the savanna totals are inflated by the inclusion of forest outliers, such as gallery forests, is unknown. We conclude from these scattered data that tropical savannas are of lower plant diversity than tropical forests by a degree that probably varies from moderate to extreme, depending on the continental region.

Table 5.4 Floristic diversity of various Neotropical savanna regions

Region	Area (km²)	Numbers of species			
		Trees and shrubs	Subshrubs, half-shrubs, herbs, vines, etc.	Grasses	Total
Cerrado	2,000,000	429	181	108	718
Rio Branco savannas	40,000	40	87	9	136
Rupununi savannas	12,000	~50	291	90	431
N. Surinam savannas	~3,000	15	213	44	272
Venezuelan llanos	250,000	43	312	200	555
Colombian llanos	150,000	44	174	88	306

Source: Sarmiento 1983

The high local diversity that characterizes many tropical communities necessitates a number of other associated conditions within these communities, and some enigmatic consequences. Most obviously, mean population size and density per species must be lower. Distribution of individuals among species is normally highly unequal within diverse communities, with no common species, a few species of intermediate abundance and very large numbers of rare species. For example, more than one-third of the 303 woody species ≥ 1 cm diameter that were measured on a 50-ha plot in Panama occurred at a density of less than 1 stem per hectare, while the most abundant species (an understorey shrub) made up less than 17 per cent of all stems (Hubbell and Foster 1986a). Another 50-ha plot in Malaysia, enumerated in an identical way, has yielded 825 species, with the most abundant of these accounting for only 2.5 per cent of all stems (He *et al.* 1996). How the many rare species that exist in these communities are able to avoid extinction and maintain their populations presents a fundamental enigma. Since much of the diversity that is characteristic of the tropics is contributed by these rare species, this enigma is at the heart of the larger issue of how such diverse biotas have evolved and been maintained in tropical environments.

Explanations for tropical diversity

Biological diversity is the product of an evolutionary process that is gradual and cannot be observed over human lifespans, except in short-generation organisms. Moreover, it cannot be replicated experimentally, because the process has a random component. As a consequence, 'explanations' for the products of evolution are inevitably speculative, and not amenable to the normal scientific evaluation process. This constraint has not inhibited widespread speculation about potential explanations for the diversity of low-latitude

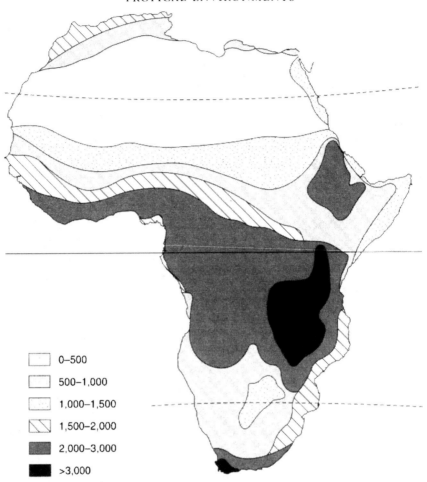

	0–500
	500–1,000
	1,000–1,500
	1,500–2,000
	2,000–3,000
	>3,000

Figure 5.8 Regional plant diversity of the African continent, expressed as the number of species per 10,000 km²

Source: Menault 1983: Figure 6.2

biotas, which requires us to devote some attention to the topic. Rohde (1992) provides a recent critical review of many of these hypotheses. Many putative explanations for tropical diversity have been cast as unique solutions but, given the complexity of the biological world, such simplicity seems quite unrealistic. Moreover, processes that facilitate coexistence of diverse species assemblages in the tropics (e.g. tree-fall gaps in tropical forests, river dynamics in floodplains) have sometimes been cited as putative explanations for this diversity, while failing to explain why the same process in extra-tropical areas has not promoted equivalent diversity. With this in mind, we concentrate our discussion in this section on processes that are likely to be *unique* to the tropics, or especially intense there, rather than attempting to review the very large number of processes that have been cited as contributing to diversity, but are not unique to low latitudes. We also emphasize those hypotheses for which some supporting circumstantial evidence exists.

As we have seen, tropical forest communities, especially those in moister environments, probably represent the most diverse ecosystems in the tropics and it is therefore not surprising that much of the theorizing about tropical diversity has been focused upon these. However, there is a further reason to justify an emphasis on tropical forest plants in a search for explanations of tropical diversity. A diversity of habitat opportunities is normally assumed to be necessary for the existence of a diverse assemblage of organisms. As plants provide the resource substrate for higher trophic levels, a diversity of plants will provide a diverse set of opportunities for herbivores feeding upon them and hence a plausible explanation for diversity among herbivores and, in turn, the organisms feeding upon these. This places plants in a key role in a chain of explanation about tropical diversity. Because plants are themselves trophically unspecialized, their diversity is also especially enigmatic.

Two alternative, but not necessarily contradictory, assumptions about the explanation for organic diversity may be identified, each of which carries with it certain implications. The first assumption holds that diversity is limited by the ability of a community to contain coexisting populations of different species. When applied to the problem of latitudinal gradients in species diversity, this assumes that, for some reason, tropical communities have greater 'coexistence potential' than those of temperate areas. The most important implication of this assumption is that evolution will inevitably provide the species to fill the vacant niches available in the community, which exists in a fine-tuned 'saturated' state. This assumption has dominated much of the theorizing about diversity among ecologists, and has led to explanations being sought for why tropical communities should have greater 'coexistence potential' than those of temperate areas. The alternative assumption about biological diversity is that this is little affected by community coexistence potentials, but rather reflects a balance between regional processes promoting speciation or extinction. The implications of this assumption are that communities are probably not saturated, that evolution is gradual and may not necessarily result in the filling of all vacant niches, and that communities are probably loosely organized and flexible. This approach to the explanation of community diversity has received increasing attention in recent years (e.g. Ricklefs and Schluter 1993).

Hypotheses about coexistence

Following Pianka (1966) there was a plethora of attempts in the ecological literature to identify mechanisms that would allow coexistence of large numbers of species in tropical forests, and prevent exclusion by a few effective competitors. We will concentrate upon those that invoke mechanisms likely to be unique to the tropics and which also have accumulated some empirical support. Although reference is often made in the ecological literature to the possibility that physical heterogeneity in tropical environments may contribute to species coexistence (e.g. Salo *et al.* 1986; Huston 1994), we do not believe that there is any convincing evidence for greater local heterogeneity in tropical than in temperate environments, and thus exclude this topic from the discussion that follows. The major feature of tropical diversity is its high within-community manifestation, often in habitats of extreme physiographic homogeneity. Moreover, even when physiographic heterogeneity exists, it appears to contribute little to the overall pattern of diversity (He *et al.* 1996).

The Janzen–Connell hypothesis was based on the assumption that seed and seedling predation were likely to be especially intense in aseasonal environments, limiting the densities of individual tree species and thus preventing competitive exclusion. As a consequence, it was assumed that opportunities would exist for other species to coexist with effective competitors, promoting the existence of diverse communities of trees. As we have seen, the evidence in support of this form of density-dependent population regulation is equivocal; it appears to limit the densities of only a small number of common species, but does not influence the distribution of most rare species (Condit *et al.* 1992a). At most, it would seem to provide a plausible explanation for the rarity of communities of very low diversity in the tropics, but be unable to account for the existence of large numbers of rare species in most communities.

In a later hypothesis, Connell (1978) suggested that periodic disturbances could prevent competitive exclusion, and maintain diverse assemblages of effective and ineffective competitors in a non-equilibrial condition. Although proposed as an explanation for diversity in tropical rain forests and coral reefs, this 'intermediate-disturbance' hypothesis did not propose any uniquely tropical mechanism, and while it has received widespread acceptance as a general community-diversifying process it does not in itself address the issue of the latitudinal diversity gradient. An important extension to the intermediate-disturbance hypothesis was made by Huston (1979), who pointed out that rates of competitive displacement would depend also on site productivity, making sites of low productivity the most favourable for coexistence of diverse assemblages of species throughout most of the range of disturbance frequencies. Huston went on to demonstrate a negative correlation between tropical forest diversity and soil fertility in Costa Rica (Huston 1980), and has since suggested that the generally low fertility of tropical soils may slow rates of competitive exclusion in tropical forest communities and preserve diverse communities there (Huston 1994). While having much intellectual appeal, and considerable empirical support at local scales, Huston's hypothesis has been questioned as an explanation for the latitudinal gradient in species diversity. Wright (1992) has argued that the Costa Rican soil fertility gradient was a spurious correlation deriving from a more general correlation between rainfall and high species richness, and that species-rich tropical forest communities can be found on soils of a wide fertility range.

While the above hypotheses have been proposed to account for circumvention of a process of competitive exclusion that is assumed to be relatively rapid, Hubbell and Foster (1986b) have suggested that the rate of competitive exclusion among species that are very similar functionally may be so slow as to permit the almost indefinite coexistence of different species, once these have formed. Using a simple stochastic model and some realistic assumptions, these authors show that populations of species comprising several thousand individuals are effectively immune from extinction except over geological time spans, over which time speciation would be expected to compensate for any losses. The important implication of this process, if correct, is that plant communities may have very high levels of saturation, and many may be effectively unsaturated. The existence in tropical forest of regional gradients of species diversity that are unrelated to conditions in the physical environment adds weight to this argument. For example, the tree floras of Central America are notably less diverse than those in similar environments in South America (Gentry 1990), suggesting that the former are unsaturated. There is also a growing body of evidence suggesting that biological communities in general are unsaturated (Cornell and Lawton 1992). The important implication of this conclusion is that diversity levels are more likely to be set by regional and historical processes of speciation and extinction than by processes that limit coexistence among species in a community.

Regional and historical hypotheses

The possibility that the tropics have been little affected by climate change, and thus that populations there have been less subject to extinctions, has frequently been cited as a process potentially contributing to diversity in low latitudes (e.g. Dobzhansky 1950; Fischer 1960; Pianka 1966). More recent demonstration of Quaternary climate change in the tropics led to the reverse hypothesis, that forest fragmentation during this epoch promoted allopatric speciation (Simpson and Haffer 1978). However, the latitudinal gradient of species diversity is a very ancient one that certainly pre-dates the Quaternary (Crane and Lidgard 1989; Jablonski 1993), and we have seen in Chapter 2 that the extent of tropical forest fragmentation in the Pleistocene was probably much less than once assumed. The original hypothesis, that the tropics may have acted as a refuge experiencing low extinction rates, thus remains tenable as a potential contributing process to the high biological diversity of low latitudes.

More speciation in the tropics could also contribute to high diversity there, and recent palaeontological evidence (Jablonski 1993) indicates that the tropics have been a major source of evolutionary novelty throughout much of the geological record. Potential reasons why this should be so include a tropical environment that is more ancient, or has been more extensive, than other sorts of environment, or some unique properties of the environment that are unrelated to its age or area. Gentry (1986) has argued that speciation among Neotropical plants is ongoing and rapid in response to soil variability. However, there exist no convincing arguments in support of why this process should be confined to the tropics, other than the theoretical expectation that a year-long growing season should permit more rapid evolution among short-lived organisms. Nor is there strong evidence to suggest that the tropics are uniquely ancient as a type of habitat. However, there is convincing evidence that the tropics have been a very common type of habitat during geological time.

Terborgh (1973) has suggested that speciation may be an area-dependent process, with habitat types that cover large areas tending to accumulate more species as a consequence of species' range expansions, followed by isolation of subpopulations and allopatric speciation among these. Rosenzweig (1995) has recently elaborated upon this idea. Area-dependent speciation would also be expected to be most successful if the process was unconstrained by community saturation, as Cornell and Lawton's (1992) recent analysis suggests that it may be. While Terborgh invoked this idea only in relation to contemporary habitat conditions, arguing that the tropics are a globally common habitat type, the argument can be extended considerably if one recognizes that speciation is also time-dependent, and considers the geological history of tropical environments (Chapter 2). We now know that humid tropical climate conditions dominated a large proportion of the earth's surface in the Cretaceous and early Tertiary, when the angiosperms were diversifying and coming to dominate world vegetation. If speciation is both an area-dependent and a time-dependent process, maximum number of lineages in humid tropical environments would be the expected outcome of this geological history.

This interpretation of the diversification process would also be compatible with the lower diversity of geologically more recent regional habitat types, such as those of savannas and dry forests, and the low diversity of community types occupying small, specialized and geologically ephemeral habitats such as floodplains, montane areas and specialized soil types. However, the literature also contains many interpretations invoking the difficulties of overcoming the physical and chemical stresses of these habitat types as reasons for their low diversity (e.g. Wright 1992), and the evidence in support of either interpretation is equivocal. On the one hand, specialized, but reasonably extensive and probably very ancient, habitat types, such as the coastal zones occupied by mangroves, appear never to have supported large numbers of species, suggesting difficulties in adapting to this environment. On the other hand, some very diverse biotas, such as the Cape flora of the southern tip of Africa (Figure 5.8), have evolved and persisted in an environment that is certainly 'stressful' for plant life. We are left, then, with the suggestion that successful diversification of lineages in new habitats may be a very stochastic process, with the probability lowered when these depart significantly from that of the source habitat, but not necessarily precluded by this.

Area dependence has also entered into recent interpretations of how diverse biotas are maintained in tropical environments. Periodic recensus data from a 50-ha plot on Barro Colorado Island in Panama have shown that many of the rarer species are declining, with 10 having gone extinct (replaced by nine new species) on the plot between 1982 and 1990 (Hubbell and Foster 1992; Condit et al. 1992b). Reasons for the decline appeared to include disappearance of a seed source for pioneer tree species in nearby previously disturbed areas, and mortality caused by a severe El Niño drought. These results indicate a remarkable degree of dynamism in forest composition, and dependence of plot diversity on migratory interaction with surrounding forest. As such, they confirm the importance of regional biotic diversity in setting levels of local diversity (Ricklefs 1987; Ricklefs and Schluter 1993; Eriksson 1993). While superficially these results appear to contradict the conclusion that communities are unsaturated (Cornell and Lawton 1992), they are probably more usefully interpreted as indicating that tropical forest communities are loosely organized and capable of temporarily sustaining many different species, each of which may have relatively large 'home range' requirements. Provided that such home range areas are available regionally, a rich regional biota can be sustained, producing diverse local communities of considerable compositional dynamism. The implications for conservation of biodiversity in such systems will be treated in the next section.

Conclusions

As emphasized at the beginning of this discussion, any conclusions about explanations for the diversity of tropical biotas must inevitably remain tentative, given the nature of the evolutionary process. Despite that, available evidence, especially that coming from the palaeontological record, suggests that certain putative explanations are probably more correct than others. Explanations positing some unique properties for the tropical environment that allow more species to coexist there than elsewhere have been notably lacking in strong supporting evidence (e.g. Latham and Ricklefs 1993). Of these, the Janzen–Connell hypothesis appears to have some empirical support as a mechanism limiting the development of very depauperate tropical forest communities, but fails to account for the very large numbers of rare species that contribute most to the diversity (Condit et al. 1992a, b). In contrast, increasing evidence supports the idea that most communities may have very high ceilings of saturation, with many being effectively unsaturated. Under these conditions, diversity is more likely to be set by the larger-scale and longer-term processes that promote the evolution and persistence of a rich regional biota.

Speciation in such an unconstrained biotic system is more likely to reflect opportunities than limitations, and we have argued that the great extent of humid tropical conditions at a time when a major new biotic lineage (the angiosperms) was diversifying may have been a critical historical event. When combined with less climatic disruption and potentially fewer extinctions during the Quaternary, this process appears to offer the most plausible contemporary explanation for the very high biotic diversity of most humid tropical environments.

CONSERVATION OF TROPICAL BIOTIC DIVERSITY

A large proportion of the world's biotic diversity exists in low latitudes, and conservation of as much of this as is possible in the face of human population growth represents a major challenge to the science of conservation biology. Much of the conservation effort has been focused on tropical forest communities because they contain the highest diversities in terrestrial habitats. The demographic structure of the present human population means an inevitable continued population increase for at least another generation (see Chapter 9). However, optimistically, we can look beyond this to an eventual stabilization of the human population, followed by population reductions to levels that allow reasonable standards of living within the available resource base. In this scenario, the short-term conservation problem could be described as a 'holding operation': how to preserve as many species as possible over several centuries in a shrunken area of protected or lightly used habitat. The longer-term conservation problem is likely to focus on how to restore communities, using the biotas that have survived the earlier period of devastation. Short-term preservation will involve species survival in patches of natural or semi-natural habitat, many of which are likely to be small and scattered, and some used for activities such as forestry and animal husbandry. We will address problems of species preservation in the latter areas in the two chapters that follow, and concentrate here on species survival in less intensely used natural habitat remnants.

In most tropical regions, these remnants may include a few relatively large blocks, but most are likely to be made up of systems of small fragments of the original habitat (Figure 5.9). Some species are unlikely to survive except in very large fragments, because of large home range requirements; included among these are many large vertebrates, especially carnivores, whose preservation may depend upon captive breeding and maintenance in zoos. Many other rare species with patchy distributions are likely to be lost owing to 'sampling effects' during rapid habitat shrinkage. However, for the large numbers of other species making up tropical biotas, we can ask whether survival for several centuries in systems of isolated small habitat fragments is feasible.

There are some reasons for cautious optimism about the prospects for preserving species under these circumstances. Communities of reduced

114

Figure 5.9 An intensively used mountain landscape in eastern Honduras. Despite the steeply sloping terrain, much of this landscape is covered by fields and pastures, and the original forest is now restricted to remnants on the most inaccessible or unusable terrain, such as ridge crests and gullies

diversity occur naturally within the tropics, indicating that some loss of diversity does not necessarily initiate a process of community collapse, as is sometimes implied in the popular literature on conservation. Moreover, we know that tropical biotas survived considerable disruption due to Quaternary climate change, indicating that they are less 'fragile' than is sometimes supposed (Chapter 2). In particular, some forest biotas appear to have survived temporary range shrinkage during glacial periods, and thereafter successfully re-established extensive ranges. While we lack good historical data on other types of communities during these events, it seems reasonable to infer that they also survived these disruptions successfully. Even where range re-expansion was of limited extent, such as that by the rain forests of eastern Australia, a rich forest biota survived these events (Adam 1992). However, it must also be recognized that the speed with which change occurred during these Quaternary events was much slower than for the changes precipitated by human activities, and this may have allowed processes such as migration to shrinking refuge areas to occur. A further cause for optimism is the possibility that many tropical communities may be unsaturated, relatively open to invasion and compositionally flexible. This raises the prospect of more species being able to be 'packed' into small habitat remnants than exist there naturally (Kellman 1996). Furthermore, rates of extinction, at least among perennial plants, may be sufficiently slow (Hubbell and Foster 1986b) that we can envisage their temporary survival through several centuries in situations that would not permit their indefinite existence.

In contrast to these reasons for optimism, there exist many causes for concern about our ability to conserve biotic diversity, even for short time periods, under conditions of habitat fragmentation. Given the rarity of many tropical forest species, a significant proportion of these are likely to be lost without their existence even having been documented (Gentry 1986). Packing of species into communities of increased diversity inevitably means smaller average population sizes in these places, and an increased probability of extinction during random population fluctuations. Many species, including both animals and plants, may require large 'home ranges' and an ability to migrate easily within these. The isolation imposed by fragmentation may limit the ability of these species to maintain viable populations, especially in fragmented systems where species packing leads to high rates of local extinction. Terborgh (1992) has warned that the loss of certain seed predators, or the carnivores that prey upon these, could lead to unpredictable floristic changes in fragmented forest systems, and a potential further loss of biodiversity. Moreover, species densities are low in the very ancient monsoon forest fragments of northern Australia, which have probably been isolated for the past 20 million years (Bowman and Woinarski 1994; Russell-Smith 1991). While the persistence of these patches is a remarkable testament to the survival of some form of tropical forest community under conditions of fragmentation, their depauperateness indicates that the size of this biota may be gradually reduced.

Predictions about whether fragmented systems can sustain most of a region's biota for several centuries face the fundamental problem of there being few naturally occurring systems of this sort whose composition and functioning would provide clues; most communities were until recently part of more extensive community types that covered large landscape areas. In the absence of many naturally occurring fragments, most projections have resorted to the use of oceanic islands as analogues of fragmented terrestrial systems, and drawn upon a widely discussed theory about the dynamics of island biotas (MacArthur and Wilson 1967). This theory proposes that diversity in local areas is maintained by a balance between local extinctions and the arrival of new immigrants. When applied to oceanic islands, it predicts that small islands will sustain lower equilibrium numbers of species than large islands, owing to a higher extinction rate among smaller populations, and that remote islands will sustain fewer species than islands close to a source of colonists, because of reduced immigration rates. Applied to the fragmentation problem, the theory proposes that isolation of patches of continuous habitat should be followed by reductions in species numbers as a consequence of a drop in immigration rates from adjacent communities, while local extinction rates continue. The theory has been the subject of much debate, especially about its applicability to oceanic islands (see Gilbert 1980), but recent data suggest that it may capture some essential elements of the dynamics of at least a subset of a region's biota. The high species turnover rates recorded on the 50-ha plot in Panama (Condit *et al.* 1992b) suggest that at least a subset of the tree species act as ephemeral migrants. Species of this subset would presumably be most likely to suffer reductions in diversity following isolation, as a result of reductions in immigration rates.

In view of this, it is surprising that some of the few naturally occurring fragments of tropical forest that exist and have been examined show local species richness that is essentially identical to that of continuous forests in the same region (Kellman *et al.* 1994; Meave and Kellman 1994; Tackaberry and Kellman 1996). The most likely explanation for this is that these fragments remain functionally connected to continuous forests in the region, by reason of proximity and gallery forest corridors that facilitate movement of seed dispersers (Kellman *et al.* 1994). Despite their limited degree of isolation, these natural tropical forest fragments are useful as indicators of the community reorganization that may be necessary in artificially fragmented systems. While their tree flora is composed of typical tropical forest species, bird-dispersed species are somewhat better represented than in continuous forests, suggesting some selection for increased dispersability (Meave and Kellman 1994). In contrast, pioneer trees are under-represented, probably because of the small numbers of large canopy gaps that occur in these forest patches. The communities are organized 'centrifugally' with most species showing greatest concentrations either in the patch interiors or at the edges (Meave 1991), suggesting that a major habitat gradient exists between interior and edge, along which species are specializing. Both changing light conditions and disturbance frequencies have been identified as contributing to this diversity-enhancing gradient (MacDougall and Kellman 1992; Kellman and Tackaberry 1993). The formation of a protective edge zone of fire-tolerant trees appears to be especially important in preventing highly intrusive edge effects (Kellman and Meave 1997). In sum, these observations suggest that these fragmented communities have undergone considerable reorganization in structure and function, but are nevertheless capable of sustaining large species numbers.

Some research has also sought experimental evidence as a basis for predictions about the consequences of habitat fragmentation (Bierregaard *et al.* 1992). While this work has provided important information about the immediate dangers of fragmentation, such as intrusion of external atmospheric conditions into forest fragment interiors and short-term adjustment to the fragmented condition, it is unlikely to indicate much about the long-term consequences of fragmentation until many decades have elapsed.

Management implications

Although our understanding of how best to conserve biotic diversity in shrunken and fragmented systems remains far from complete, sufficient information exists to provide some first approximations about appropriate management strategies. A widely dispersed fragment system is likely to maximize habitat heterogeneity (Bell *et al.* 1993), and dispersal of subpopulations among functionally independent subunits is likely to reduce the probability of global extinction, provided that inter-patch dispersal is possible (Goodman 1987). Consequently, a fragment system that encompasses regional habitat heterogeneity but also preserves the potential for significant inter-patch migration represents an ideal (and idealistic) solution. Migration of propagules between fragments is especially difficult to monitor, making prescriptions on how best to achieve this the subject of much speculation and some controversy (e.g. Simberloff *et al.* 1992). In theory, migration is likely to be facilitated by patch proximity, the presence of an intervening habitat matrix that is not completely inhospitable, and the existence of movement corridors (such as strips of gallery forest) between patches. In cases of extreme isolation, periodic reintroductions of propagules may be necessary. Rapid movement of fragments to a reorganized but stable condition is likely to be important in preventing a period of prolonged readjustment during which species extinctions occur (Kellman 1996). The available information on forest fragmentation suggests that intrusive edge effects are especially problematic, and that the development of a specialized edge community that can protect forest interiors may be of special importance (Kellman *et al.* 1996).

While the planning of biological reserves prior to the fragmentation process is an ideal procedure, it is one that is increasingly unlikely to happen except in the case of well-planned land colonization schemes (see Chapter 11). Elsewhere, community fragments are likely to occupy terrain unsuitable for agricultural activities, such as steeply sloping stream banks and mountain tops (Figure 5.9). Huston (1993) has argued that these less productive sites are likely to be especially rich biologically and so to have conservation benefits disproportionate to their area. However, problems of species persistence in these small and highly isolated fragment systems remain a major challenge in tropical conservation biology (Schelhas and Greenberg 1996). Many are likely to be subject to some human utilization, and preservation of regional biotas within them may require major interventions by way of species transplantation, edge formation, etc.

Identification of species most at risk of extinction, and hence potential candidates for more active management intervention, is also necessary; these may often not be simply the rarest species. For example, Karr (1982) has found that the tropical forest bird species most susceptible to extinction following isolation were those whose populations were subject to wide oscillations, not the chronically rare species. This conclusion parallels the rarity of pioneer tree species found in forest fragments (Meave and Kellman 1994), as these too are subject to large population fluctuations. However, in deciding upon strategies for management intervention, a form of 'triage' may be necessary. While it may be repugnant to consider the loss of *any* species, attention may need to be directed to those with the greatest probability of survival, or we run the risk of wasting scarce resources on futile rescue efforts.

At the other extreme, identification is needed of those species which have little difficulty in surviving in fragmented systems and whose preservation is most likely to ensure the survival of other species. Huston (1994) refers to these two groups as 'structural' and 'interstitial' species, respectively. The facilitative role played by structural species in permitting the entry of other species into successional communities is well known, but their role in stable communities has received less attention (see Wilson and Agnew 1992). In fragmented tropical forest systems, tree species that are pre-adapted to withstand occasional fires by virtue of possessing such characteristics as thick bark (e.g. Uhl and Kauffman 1990) are likely to be particularly important structural components of fragmented systems. They are also promising species for use in any future forest restoration programme.

THE RESOURCE POTENTIAL OF DIVERSE BIOTAS

Diverse biotic communities, such as tropical rain forests, provide limited potential for the support of large populations of people or other animal species (Kikkawa and Dwyer 1992). This is primarily because only a small proportion of the plant species present have value to particular animal species, and these normally occur at low densities and are widely scattered. For example, while extractive forestry in temperate forests can often make use of a high proportion of standing timber, only a small proportion of the tree species present in tropical forests are normally considered valuable and worth extracting. Their extraction must be achieved at proportionally high cost, because of their low densities. Similar arguments hold for other non-timber forest products, such as natural rubber or brazil nuts. Thus, diverse communities do not, in themselves, normally represent a major resource for extractive use, and it could be argued that their diversity reduces their economic worth.

The diverse biotas of the tropics are also unlikely to be potentially important sources of many major new plant and animal domesticates, although they may provide potential sources of many lesser domesticates, such as pasture legumes. In the past century, the only major domestication of a tropical forest plant has been that of the rubber tree (*Hevea brasiliensis*), a native of the Amazonian rain forests. The history of domestication has usually witnessed the early domestication of a limited number of plant and animal species, involving strong selection on these for useful characteristics, rather than the continual introduction of new domesticates from wild populations. This probably reflects a widespread screening for usefulness of wild populations by early human groups, followed by concentration on those species showing greatest promise for selection of desirable traits. Thus while many indigenous groups in the tropics make use of, and may even cultivate, a range of locally derived wild species (e.g. Salick 1989), their primary subsistence comes from a limited range of widespread tropical crops and animal species that have usually been domesticated elsewhere.

In contrast to their limited role as potential sources of major new domesticates, diverse tropical biotas probably have a much larger potential role to play in providing templates for pharmacological and pesticidal compounds and sources of genetic material imparting resistance to pests and diseases. The prevalence of predatory interactions in tropical biotas has led to selection for a diverse suite of defensive compounds. Indigenous human populations that have lived in these communities for many millennia have engaged in prolonged screening of the biota for pharmacological properties, and now usually possess a rich pharmacopoeia of natural drugs for coping with human diseases (e.g. Boom 1989). This knowledge is now much sought-after by drug developers (e.g. Balick 1990), and is increasingly protected as intellectual property by indigenous peoples, who possess few other economic resources (e.g. Dickson and Jayaraman 1995). Most perennial tropical plants possess genes imparting resistance to pests and diseases, and those that are close relatives of crop plants are a potentially important gene source for breeding resistance into these crops. Consequently, a conservation strategy that targets close taxonomic relatives of crop plants is likely to be especially valuable. However, since the closest relatives of many crop plants are 'weedy', this does not necessarily imply a conservation strategy focused upon the preservation of pristine plant communities.

A further economic rationale for conservation of diverse tropical biotas is their potential for providing a source of income through nature tourism, which forms a significant component of one of the largest and most rapidly growing industries in the world (Giannecchini 1993; UNEP 1991). Parks and reserves which draw visitors because of their biota and other natural features can provide a valuable source of foreign exchange. Large numbers of tourists are attracted to Kenya and Ecuador because of their diverse wildlife resources, and tourism provides the second largest source of foreign exchange in both countries (Barnes et al. 1992). A survey of tourists in Belize, Costa Rica, Dominica, Ecuador and Mexico found that between 50 and 79 per cent of those questioned cited parks and protected areas as important in influencing their choice of destination (Boo 1990).

Nature tourism theoretically has the potential to be environmentally and economically sustainable, but careful planning and management are essential to achieve this. Although tourism is often considered to be a non-consumptive activity, disturbance by tourists can cause detrimental environmental effects and endanger the resource base on which it depends. Serious negative impacts are especially likely to occur when the focus of visitor interest is a sensitive site or species, or one which is of most interest during critical periods of its life cycle, such as nesting green turtles on Costa Rican beaches (Jacobson and Figueroa Lopez 1994). Ecuador's Galápagos islands (Figure 5.10) have experienced a phenomenal growth in tourism in recent years, with visitor registration increasing from 7,788 in 1977 to 50,000 in 1994 (Boo 1990; Honey 1994). Ninety per cent of the islands have been designated as a national park and many controls have been put in place. For example, visitors are restricted to trails at designated visitor sites throughout the islands and must be accompanied by a trained and licensed guide. However, without limits on the numbers of visitors, it is inevitable that environmental stresses will be placed on the islands, owing to the sheer volume of tourist traffic. Although most Galápagos animal species appear to be accustomed to human presence, disturbance can interfere with reproductive activities (Burger and Gochfeld 1993). Human population increases due to an influx of immigrants from mainland Ecuador seeking a share of the economic benefits also add to the environmental stresses placed on the islands. Research to determine the types and volume of tourist activities that are environmentally sound, as well as planning and management to ensure that these limits are adhered to, are extremely important to ensure that biodiversity-based tourism is carried out in a sustainable manner in any area.

Although tourism usually provides economic benefits, these are not always received by people living at or near tourist destinations. Local populations are those most subject to any negative effects of tourism such as social disruption, increased risks to safety and livelihood as a result of the protection of dangerous or crop-destroying animals, and environmental degradation. Ultimately, for a tourism industry to be an effective means of conserving the diverse biological resources of the tropics, it must benefit local people and receive their support. It is important that socio-economic as well as environmental considerations should be incorporated into tourism planning and management (Kenchington 1989; Nelson 1994).

TROPICAL PESTS AND DISEASES

We include here organisms that infect domesticated plants, animals or people, feed upon crops (mainly insect herbivores) or compete with them as weeds. The diversity of tropical biotas, and the propensity for year-long biological activity, leads to pest and disease problems in tropical environments being especially problematic; not only is there usually a far richer suite of predatory organisms available than is present in temperate areas, but many of these organisms are potentially active throughout much of the year, which could lead to severe infestations. Thus coping with pests and diseases can be added to soil infertility as the second major problem faced by human populations in the tropics.

Figure 5.10 Nature-based tourism can provide an important source of income in tropical countries such as Ecuador. The Galápagos Islands are the major attraction for many tourists because of their spectacular and easily approached wild animals, such as this blue-footed booby (*Sula nebouxi*) (a), which has nested on a tourist trail. However, rapidly increasing numbers of visitors may pose challenges for the management of the islands. Development of alternative sites with natural features of interest, such as this biologically diverse Amazonian rain forest (b), can help to spread the economic benefits and reduce the negative impacts on the most popular sites

The variety of agricultural pests and diseases

Given the diversity of most tropical biotas, one would expect these to have the potential to provide a greater diversity of agricultural pest organisms than those of temperate areas. This has been confirmed by Wellman (1968) in a comparison of diseases of crops grown in both tropical and temperate parts of the New World (Table 5.5). This diversity may be expected to impose an added economic cost on tropical agriculture generally, as it requires that a greater variety of pest control strategies be kept available for deployment when necessary.

Several authors have examined the patterns of pest diversity on individual crops grown in different parts of the tropics (Strong 1974; Strong *et al.* 1977; Banerjee 1981; Kogan 1991). Most plant-feeding insect pests are of localized distribution; for example, Kogan (1991) finds that 86 of 106 herbivore pests on soybeans occur in only one of the six major world regions in which this crop is grown, while more than 85 per cent of the 1,905 herbivore pests

Figure 5.10(b)

reported for cacao are known from only one cacao-producing area (Strong 1974), and a similar pattern is found in sugarcane (Strong *et al.* 1977). For both cacao and sugarcane, pest numbers were found to correlate only with the local area of crop production, and the authors concluded that there had been a rapid accumulation of local pests to asymptotic levels after crop

Table 5.5 A comparison of the numbers of diseases recorded on crops that are grown in both temperate and tropical parts of the New World

Crop	Temperate zone	Tropics
Citrus spp. (oranges, etc.)	50*	248
Cucurbita pepo (pumpkin-squash)	19	111
Ipomoea batatas (sweet potato)	15	187
Musa acuminata (banana)	8*	180+
Oryza sativa (rice)	54	550–600
Phaseolus vulgaris (beans)	52	253–280
Saccharum officinarum (sugarcane)	35–56*	450+
Solanum spp. (potatoes)	91	175
Zea mays (maize)	85	125

Source: Wellman 1968

Note: Includes diseases caused by viruses, bacteria, fungi, phanerogams, algae, lichens and nematodes
* Results from southernmost states in the continental USA

introduction to the region. However, Banerjee (1981) has found that over a 150-year history of tea cultivation, age of the cultivated area is more closely correlated with pest numbers than area, although area plays a secondary role in accounting for pest diversity. However, this author also found that at local scales, herbivorous pests accumulated rapidly on individual tea plants, reaching asymptotic levels of over 200 pest species after 35 years. These data do not address the relationship between local and regional patterns of pest diversity, but work by Lawton *et al.* (1993) on the herbivores of bracken (a weedy fern of worldwide distribution) suggests that local herbivore complexes represent a proportional sample of the region's complex, with richer regional complexes promoting greater local diversity on this plant.

While the reasons for the positive correlation between regional numbers of pests and the area under cultivation remain obscure, the data do indicate that, at least for herbivorous insects, most regions including islands have a sufficiently rich biota of potential pests to impose severe pest problems on introduced crops within a relatively short period of time. Hence, being an exotic may bring only temporary relief from a variety of pest organisms, although it may provide protection from particularly lethal pests of the indigenous range. For example, plantations of rubber trees in the Neotropics have been repeatedly infested with the native South American leaf blight fungus *Microcyclus ulei*, which is specific to members of the rubber genus *Hevea*; but the same tree has been grown free of this disease in Africa and Asia, where it was introduced disease-free and has since been rigorously quarantined (Wastie 1975). In these areas the tree is subject to infection by a number of local pathogenic fungi, none of which is as lethal as the South American leaf blight (Wastie 1975).

Rather different patterns appear to exist among tropical weed complexes. At local scales, these usually comprise mixtures of locally derived and exotic species, with the latter being especially abundant in the most intensely farmed areas (Kellman and Adams 1970; Kellman 1973). Exotic weeds are often the most noxious, reflecting both their prolonged selection in an agricultural milieu and their freedom from the natural enemies that control their populations within their native range. For example, the shrub *Lantana camara* occurs as a minor component of second-growth vegetation in the Neotropics, where it is native, but has become a major pasture weed elsewhere in the tropics, after having been introduced as an ornamental. As a gradual infiltration of exotic weeds throughout most agricultural areas in the tropics may be expected, we can probably anticipate a continuing problem of noxious new weed introductions.

The failure of large-scale geographic isolation to provide permanent protection from pests means that pest control will ultimately have to be achieved in most tropical areas by the implementation of measures at the local scale.

Local pest infestations

Most noxious tropical weeds are short-lived herbaceous plants, capable of rapid population expansion, but incapable of competing indefinitely with larger plants. Most of these weedy species probably originated as short-lived ephemerals in patchily distributed disturbed areas, from which they were displaced eventually by larger plants. Survival of these species required short life cycles, effective means of seed dispersal and, often, seed capable of prolonged viability in soil. In tropical agricultural situations, year-long growth is possible, disturbance is continuous, and protection is provided from displacement by larger plants, allowing weed populations to develop and persist at levels that cause severe competition with coexisting crops. Moreover, weed seed can accumulate to levels of several thousand seeds per square metre in tropical agricultural soil (Kellman 1974a), ensuring a rapid recovery of the population after a weeding operation. It is therefore not surprising that the difficulty of weed control represents one of the most frequently cited reasons for field abandonment in shifting-field agriculture (see Chapter 8).

Wild tropical plant populations cope with predation by herbivores or pathogens by one or more of the following means:

1 They exist in communities that contain sufficient natural enemies of the predators to constrain their numbers.
2 They avoid infestations by maintaining densities at low levels.
3 They resist predation by means of chemical or physical defences.

The high diversity of most natural plant communities means that predator control by the first two mechanisms is likely to be naturally facilitated. However, most agricultural systems are of relatively low crop diversity, which removes such potential natural protection and leaves the crops vulnerable to local eruptions of predators. Moreover, selection for anti-predator defences is unlikely to have been strong during prehistoric domestication, when avoidance of infestation by field relocation was probably the norm. Indeed, chemical and physical defences in plant tissues used for food are likely to have been selected against, if the defences made the plants hard to digest, poisonous or difficult to harvest. Consequently, there is no reason to suppose that most domesticated plants will be better defended than wild plants, and they may well be less so.

The most rudimentary form of dealing with local pest infestations, whether by weeds, herbivores or diseases, is to avoid these by abandoning the agricultural activity and moving elsewhere. As we shall see in Chapter 8, this forms an important component of the rationale for shifting-field cultivation. However, where human population densities preclude passive avoidance as a viable strategy for coping with pest infestations, other means must be invoked. Increased labour inputs (e.g. weeding) have been a traditional response to this problem, followed more recently by technological inputs (e.g. mechanical weed control, chemical insecticides). However, the limited long-term success of these techniques, and their increasingly marginal economic returns, has led to a search for lower-input means of controlling pest infestations.

Our current understanding of what controls populations naturally, suggests a number of strategies as potentially most promising. The imposition of a physical stress that mimics the density-independent control of short-lived populations in temperate areas represents the first possibility. In tropical areas with a severe dry season this occurs naturally, but its effect can be eliminated as a means of pest control by the introduction of dry-season irrigation. At the other moisture extreme, pest infestations in wetland agriculture (notably of rice) should be much reduced by an annual period of dry-field conditions that provides a radically different habitat for pests. Secondly, the importance of natural enemies in controlling wild populations implies considerable promise for strategies that seek to incorporate biological controls as permanent components of agricultural systems. Avoidance of pest buildup by means of isolating susceptible hosts provides a further potential strategy for coping with local infestations. For host-specific pests, crop rotation can provide isolation in time and space, but is likely to be applicable only to groups of short-lived crops of differing pest susceptibility. Alternatively, host isolation could, in theory, be provided by maintaining low crop densities in mixed-crop systems. Mono-specific crop systems are likely to be highly vulnerable to predatory pest infestations, but in natural plant communities only the most abundant populations appear to be subject to density-dependent limitation (Condit *et al.* 1992a). This suggests that crop populations of moderate density, grown in agricultural systems of a diversity much lower than that of natural tropical plant communities, may be sustainable without a high risk of infestation. Avoidance of a buildup of weed populations represents a special problem, as the interactions between weeds and crops are competitive, rather than predatory. In theory, a crop, or combination of crops, that pre-empts plant growth resources represents the most viable control strategy. The incorporation of a vine 'cover-crop' that is capable of rapidly occluding light reaching the soil surface is a particularly promising alternative. Finally, the selective breeding of domesticated varieties possessing greater resistance to pests and diseases offers a further strategy. However, the capacity of pests and diseases to continually evolve resistance to these defences means that such breeding programmes must be viewed as ongoing activities and not finite projects, and as such may be prohibitively costly in many tropical countries. In Chapter 9 we will return to a more detailed discussion of these agricultural strategies and explore the prospects for their elaboration in the future.

Human diseases in the tropics

Tropical environments also support a diverse group of organisms which cause or contribute to the transmission of human diseases. The social and economic costs of these diseases are high, and have significant consequences for human habitation and management of tropical environments. Diseases caused by pathogenic organisms kill millions of people in tropical countries each year, and are responsible for more than half of human deaths in the most severely affected areas (Walsh 1990). Additional hundreds of millions are affected by sublethal infections which result in much human suffering and lost productivity, and may lead to chronic illness or increased susceptibility to secondary infections.

The widespread belief of earlier decades that infectious diseases would soon be conquered by advances in medical science was fuelled by successes with the use of new antibiotics and vaccines, and, later, the global elimination of smallpox. Unfortunately, the eradication of smallpox has proved to be the exception rather than the rule. The previous optimism has been replaced by increasing concern about the resurgence of familiar diseases such as malaria and tuberculosis, and the emergence of others including AIDS and haemorrhagic fevers such as Lassa fever, dengue fever and Ebola (Garrett 1994; Morse 1993, 1995; WHO 1994). Although some emerging diseases may be caused by new organisms which have evolved from mutations or genetic recombinations, it is thought that most are existing diseases which have taken on a new significance to human populations by crossing species boundaries and infecting a different host, by moving geographically from a small isolated population into a larger one, or by range expansion (Le Guenno 1995; Miller 1989; Morse 1993).

The causative agents of most communicable diseases which infect humans are micro-organisms including viruses, bacteria, protozoa and fungi, and macro-parasites such as helminths and arthropods. The life cycles of these pathogenic organisms and the methods by which they infect humans are often very complex. Some can be transmitted directly from person to person (e.g. influenza and measles), or are passed through some medium such as contaminated food or water (e.g. cholera and infectious hepatitis). Others involve interactions with one or more other species which act as vectors, intermediate hosts or reservoirs.

Many tropical diseases are transmitted by arthropod vectors. Various species of mosquitoes transmit serious diseases such as malaria, yellow fever, filariasis, dengue fever and some types of encephalitis. Trypanosomes that produce different forms of sleeping sickness in humans are carried by tsetse flies in tropical Africa, and triatomine bugs in the Neotropics. Some diseases undergo a period in an intermediate host which allows development of the agent during an earlier part of its life cycle. Schistosomiasis, a disease caused by several species of parasitic trematode worms, is a major health concern in many regions of the tropics, particularly in Africa. Schistosome eggs hatch in fresh water, and their life cycle subsequently passes through their intermediate host, which is one of several species of snails, then back into water, and later into humans, where the adult stage develops. Eggs are passed into water with faeces or urine for the cycle to begin again (Malek 1985). Reservoir species are those in which zoonotic infections – those diseases which normally occur in animal species but which are transmissible to humans – are maintained. These are primarily wild mammals, birds, arthropods and domestic stock.

The abundance of human diseases in the tropics is a reflection not only of the diversity of disease organisms, their hosts and vectors, but also of characteristics of human populations and their environment. In the following sections we will discuss these factors as they relate to the incidence and management of tropical diseases.

Human population demography and disease

The course of any disease depends on the duration of the infection, the ease with which it can be spread, and its virulence. Threshold density, the critical value of the population size which is necessary for an infection to be maintained in a group, varies among diseases and may change with host density and behaviour (May 1993). Two human population trends in many areas of the tropics which are certain to increase the severity of disease problems are rapid population growth (Chapter 9) and urbanization. Both of these trends increase the probability of threshold densities being reached, allowing for the maintenance of disease organisms endemically in a population. Migration of people from rural areas into urban centres exposes large populations to diseases which were previously confined to isolated groups, and facilitates the spread of infection among different regions. These processes are accelerated by overcrowding and lack of medical services and sanitation in many rapidly expanding cities.

The development of new patterns of human movement on a larger scale also facilitates the spread of disease. The great diversity of infectious diseases originating in the Old World, particularly in Africa, is thought to be a result of their having evolved along with human populations for long periods of time in those areas (Culliton 1990; McNeill 1993; Miller 1989). Smallpox, measles and other diseases were introduced into the Americas with the arrival of European explorers. It has been estimated that New World human populations were reduced to about 10 per cent of their original pre-1492 size primarily as the result of introduced diseases. Some groups may have been reduced to as little as 2 or 3 per cent of their estimated original number, and others disappeared altogether. This process continued in more recent times as contact with isolated populations in the Amazon basin has occurred (Black 1992, 1994). Access to increasingly remote regions is becoming much easier and faster, breaking down geographic barriers between formerly isolated groups. The development of modern air transportation networks now provides the potential for infectious organisms to be spread anywhere in the world within a few days. Agents of disease can also be introduced into new areas by the transportation of goods or domestic livestock.

Environmental changes and human disease

Environmental conditions affect the relationships between populations of pathogenic organisms, their hosts and vectors at various points in their life cycle, and changes in these conditions frequently alter the incidence and severity of diseases for human populations. Numerous vector species have an aquatic stage in their life cycle, and diseases are often more common during and after periods of unusually heavy rainfall. The Rift Valley fever virus is carried by various mosquito species of the genus *Aedes*. This disease affects both humans and livestock, and sudden epidemics occur throughout many parts of the African continent. In Kenya these mosquitoes are often found in shallow depressions called dambos which fill with water during the wet season. The eggs produced by infected mosquitoes can maintain the virus for years, and outbreaks of the disease occur when abundant rains provide favourable conditions (Walsh 1988). Periods of heavy rainfall related to the El Niño Southern Oscillation (Chapter 3) have been associated with epidemics of West Nile fever in southern Africa, encephalitis in Brazil, and Japanese encephalitis and malaria in Ecuador, Peru and Bolivia (Nicholls 1993).

Land management practices can influence the prevalence of human diseases. Water impoundment projects in particular are notorious for their association with increased incidence of many serious illnesses. Shifts in the prevalence of various schistosome species and a corresponding increase in human disease have been recorded after dam construction in several regions of Africa. *Schistosoma haematobium* infections increased from 2–11 per cent to 44–75 per cent in four areas investigated before and after construction of the Low Dam at Aswan, Egypt. In Ghana, rates of urinary schistosomiasis in children living in the region of the Akosombo dam rose from 5–10 per cent before impoundment to over 90 per cent in some lakeside communities afterwards (WHO 1985a). The emergence of *Schistosoma mansoni* infections in eastern Uganda appears to be associated with the introduction of irrigated rice-growing (Bukenya *et al.* 1994). These types of mega-projects can also indirectly affect the incidence of disease by encouraging migration of human populations to these higher-risk areas, mixing of people and their diseases from different regions, and easier transmission owing to increased population density in the area.

Water impoundments are also frequently associated with an increase in the vector populations responsible for the transmission of arthropod-borne diseases. Areas of the tropics which are flooded for wet rice cultivation provide optimal conditions for vectors such as malaria-transmitting anopheline mosquitoes (Figure 5.11). The increased warmth of the sunlit water encourages mosquito breeding, shortening their generation time, and increased local humidity lengthens their lifespan (Desowitz 1980). A study of species composition and numbers of mosquitoes biting humans in Kenya showed a major difference between non-irrigated areas, in which anopheline mosquitoes were

Figure 5.11 Young rice plants in a flooded field in Bali, Indonesia. The amount of open water is maximized during this stage of wet rice cultivation, which provides an ideal habitat for disease-carrying mosquitoes

poorly represented, and irrigated areas, where they formed much higher proportions of the collections (Chandler *et al.* 1976). The incidence of mosquito-transmitted Japanese encephalitis in Asia is higher around rice fields (Morse 1993). It has been suggested that several epidemics of Rift Valley fever might be attributable to construction of the Aswan Dam in Egypt and the damming of the Senegal River in Mauritania (Le Guenno 1995; Walsh 1988). Even very small-scale increases in the availability of standing water can be beneficial for insect populations whose larval stages need this habitat to breed. Small water storage containers are thought to contribute to the high prevalence of dengue haemorrhagic fever in Asia, and other mosquito-borne viruses in Africa (Monath 1994; Morse 1993). The provision of breeding sites may not only increase vector density, but also prolong the period of vector activity. Peaks in African *Anopheles gambiae* mosquito population densities have been reported in the middle of the dry season in areas with artificial water impoundments (Coluzzi 1994).

Other forms of habitat alteration such as deforestation also have the potential to create new opportunities for disease organisms to emerge into human populations by altering ecological relationships. Replacement of forests with agricultural land or urban settlements may provide new habitats for populations of disease carriers such as rodents and arthropods. As greater numbers of humans make inroads into wilderness areas, the possibility of

130

contact with a wider range of animal populations becomes higher, thus increasing the risk of exposure to new zoonotic diseases. It has been suggested that the reduction or elimination of reservoir fauna due to habitat destruction may result in parasites entering a purely human cycle (Desowitz 1978).

Emerging viral diseases are of particular concern because their effects are often very severe, they are difficult to treat, and they generally lack susceptibility to pharmaceutical drugs (Miller 1989). Their speed of replication and mutation allows for great evolutionary potential and rapid adaptation to environmental changes, and pathogenic variants can easily arise (Le Guenno 1995; Morse 1993). Zoonoses are the source of most emerging viral diseases, and it has been suggested that their emergence is frequently due to changes in human behaviour which increase the opportunities for transfer of viruses from their natural reservoir to humans (Morse and Schluederberg 1990). Oropouche fever, a zoonotic viral disease, is becoming an increasing health problem in South America. The disease is transmitted to humans by *Culicoides paraensis*, a biting midge. The maintenance cycle of this virus is uncertain, but it is thought that primates, sloths and possibly birds act as vertebrate hosts, and an arthropod as vector. Outbreaks of Oropouche occurred in the Brazilian Amazon shortly after construction of the Belém–Brasilia highway. It was surmised that colonization of the area and clearing of the forest for crops provided new human host populations for the virus and new habitat for the midge, which breeds in discarded cacao hulls and banana stumps (LeDuc and Pinheiro 1988; Tesh 1994). Rodents can carry numerous diseases, so changes in land use or human settlement patterns which provide more favourable habitats for these animals or opportunity for contact may also favour viral emergence in humans; this process has been suggested as a potential contributing factor in diseases such as Argentine, Bolivian and Venezuelan haemorrhagic fevers (Morse 1993; Tesh 1994).

Close contact with domestic animals can facilitate disease transmission to humans. Waterfowl, particularly domestic and wild ducks, are important reservoirs of new strains of influenza. The transmission of genetic material from avian influenza viruses to those of humans can occur in pigs, which can act as mixing vessels for reassortment. New viruses may then be passed to humans and become pandemic (Scholtissek and Naylor 1988). Cattle, sheep and goats can act as amplifiers for diseases such as Rift Valley fever and Crimean–Congo haemorrhagic fever (WHO 1985b). People migrating from rural areas into cities may be accompanied by their domestic livestock, and the more crowded conditions increase the potential for the transfer of zoonotic diseases.

Management of human diseases

Control of human diseases in the tropics is a difficult problem owing to their diversity, the complexity of interactions between pathogens, people and other organisms, and the lack of sufficiently detailed knowledge about some diseases. However, what is currently known indicates that it is possible to improve control strategies. Many common tropical diseases were formerly more widespread worldwide but have been largely controlled in other regions by improvements in immunization, medical treatment, hygiene, nutrition, housing and water supply (Walsh 1990). Improvements in these areas within the tropics can also be expected to reduce the prevalence of endemic diseases.

Practices which minimize human contact with disease vectors are key to disease prevention programmes. Control of some diseases can be achieved by environmental management to reduce or eliminate vector habitats such as by careful storage of domestic water supplies and disposal of wastes, or manipulation of reservoirs and irrigation systems (Ault 1994). Improvements to housing which reduce contact with rodents and arthropods, and situation of domestic livestock away from homes can also serve as control measures. Although expense is often a limiting factor in disease control programmes, implementation of even relatively simple preventive measures may prove cost-effective over the long term. For example, it has been suggested that provision of waterproof boots and gloves for rice field workers could reduce the incidence of schistosomiasis and avert the more serious costs of treating the disease and its complications (Bukenya *et al.* 1994). In some instances it may be feasible to integrate control programmes with economically beneficial community projects. A control programme in India for filariasis, a disease caused by nematode parasites, involved removal of the aquatic plants which provided breeding habitat for the mosquito vectors. These plants had previously been used as a source of fertilizer for coconut trees, so they were replaced with a nitrogen-fixing sunhemp which was grown under the trees. Several species of edible fish, including some weed-eating species, were reared in the ponds instead to provide a source of income (Panicker and Dhanda 1992).

The numerous examples of disease emergence associated with environmental disruption in the tropics suggest that potential health effects should be considered during the planning stages of large-scale projects. Microbial impact assessments, analogous to environmental impact assessments, could be used to predict and mitigate possible disease effects of development projects (Monath 1993; Morse 1993). Ongoing surveillance programmes in areas where there is a high potential for disease emergence could be valuable in providing early-warning systems to aid in their containment. These could concentrate in areas of high biological diversity such as tropical rain forests and regions which are undergoing major environmental or demographic changes (Morse 1995). For example, in areas where Rift Valley fever outbreaks are likely, remote sensing data could be used to monitor vegetative changes which would provide conditions favourable to the mosquito vectors. When these occurred, appropriate vector control or vaccination programmes could be implemented (Walsh 1988).

The evidence linking specific ecological changes to their potential epidemiological consequences is circumstantial in many instances, as the search for causal factors of most epidemics is necessarily done after the disease has been identified and efforts to control its spread are under way. More detailed knowledge of the relevant ecological conditions which lead to outbreaks of these diseases is essential for their management, but observations prior to outbreaks are often not available, so in these cases the exact causes remain speculative. Even basic information such as the natural host is unknown for some diseases, such as the viruses which cause Marburg and Ebola haemorrhagic fevers. A few studies have been initiated to attempt to provide a controlled test of the effect of environmental disturbance on human health. One of these is an investigation of the consequences of tropical rain forest logging in Papua New Guinea (Gibbons 1993). Baseline health data were collected from people in four villages: two in the area to be logged, and two in undisturbed areas. Monitoring of human health in the study sites, as well as investigation of potential disease-transmitting species, has been planned to continue for several years after logging. A similar study in Brazil is examining human health in four experimental areas, of which three would be deforested to various degrees for the planting of different combinations of agricultural cash crops and native fruit trees, while one was to remain a control area (Gibbons 1993). These types of studies are essential to provide a better understanding of ecological factors which contribute to disease problems, and therefore to an improved ability to mitigate them.

6

TROPICAL FORESTS AND FORESTRY

Tropical forest represents one of the two major plant formations of the tropics, the other being tropical savanna. Today the two formations cover roughly 40 per cent and 60 per cent of uncultivated tropical land surfaces, respectively; the propensity for forests to degrade to savannas under intense human activities (see Chapter 7) precludes any exact assignment of proportions under 'natural' conditions. Forests exist in each of the three geographical regions of the tropics: the Neotropics, Africa and the Asia-Pacific area (Figure 6.1). Among regions, forests often share many common structural traits and plant life forms, although few common species. Taxonomic similarities are found primarily at the level of genera and families, although even here differences between the regions exist. For example, tree species of the family Dipter-carpaceae (dipterocarps) dominate the forests of South and South-East Asia, but are absent from other tropical forests.

The tropical forest formation encompasses a variety of forest types. These may be thought of as comprising a 'central' type, evergreen lowland tropical forest (also called 'rain' forest), which becomes modified along gradients of

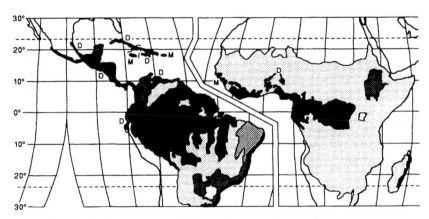

Figure 6.1 Distribution of the main types of tropical vegetation

134

altitude or moisture availability. Embedded within this matrix are smaller areas of specialized forest type, which usually are found in habitats that are extreme for plant growth. In general, tropical forests are of high diversity relative to those elsewhere in the world (Figure 5.2), and potential explanations for this diversity were discussed in Chapter 5. Here, we will focus on the structure and dynamics of tropical forests, and their potential to act as a renewable source of timber and other useful products. We will also evaluate the potential for tree plantations to be used as an alternative source of timber.

TROPICAL FOREST STRUCTURE

The floristic complexity of most tropical forests, and the absence of distinct 'associations' of species in these, led many early tropical forest ecologists to emphasize the physical structure of these forests, rather than their floristic composition. The preparation of a 'profile diagram' of a narrow strip of forest as a means of analysing this structure was pioneered by Davis and Richards (1933–4), and has since been used by many other workers to describe both tropical and extra-tropical forest (Figure 6.2). One consequence of this early preoccupation with structure, and the search for orderly patterns within it, was the development of the misconception that tropical forest canopies were stratified into distinct layers. Richards (1952), whose pioneering text had a profound influence on tropical forest ecology, cited data for tropical forest tree height distributions that showed no tendency for the existence of distinct modes, yet went on to suggest that stratification was readily apparent to an observer. Reasons for why such stratification should exist are lacking, and we now know that most tropical forests are far too dynamic for such rigid structuring to develop and persist.

Although any profile diagram of necessity represents an extremely small sample of the forest that it seeks to describe, most diagrams capture many

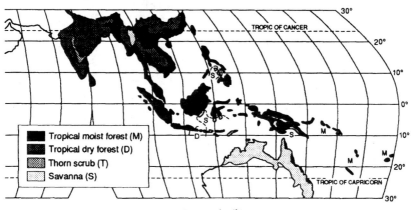

Figure 6.1(b)

135

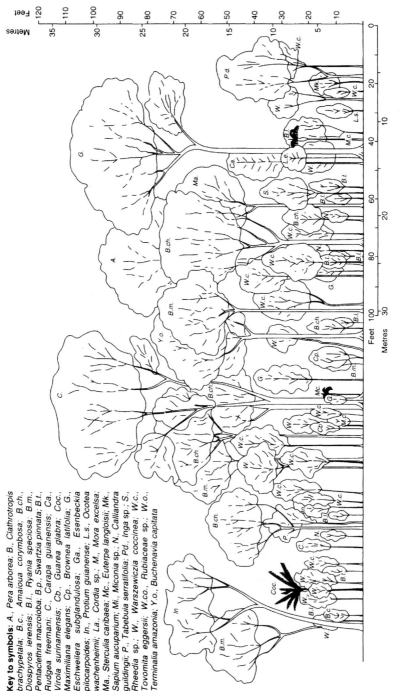

Key to symbols: A., *Pera arborea*; B., *Clathrotropis brachypetala*; B.c., *Amaioua corymbosa*; B.ch., *Diospyros ierensis*; B.l., *Ryania speciosa*; B.m., *Pentaclethra macroloba*; B.p., *Swartzia pinnata*; B.t., *Rudgea freemani*; C., *Carapa guianensis*; Ca., *Virola surinamensis*; Cb., *Guarea glabra*; Coc., *Maximiliana elegans*; Cp., *Brownea latifolia*; G., *Eschweilera subglandulosa*; Ga., *Esenbeckia pilocarpoides*; In., *Protium guianense*; L.s., *Ocotea wachenheimii*; La., *Cordia sp.*; M., *Mora excelsa*; Ma., *Sterculia caribaea*; Mc., *Euterpe langloisii*; Mk., *Sapium aucuparium*; Mi., *Miconia sp.*; N., *Calliandra guildingii*; P., *Tabebuia serratifolia*; Pd., *Inga sp.*; S., *Rheedia sp.*; W., *Warszewiczia coccinea*; W.c., *Tovomita eggersii*; W.co., *Rubiaceae sp.*; W.o., *Terminalia amazonia*; Y.o., *Buchenavia capitata*

Figure 6.2 Profile diagram of a mature tropical evergreen forest in Trinidad. The forest area shown is 61 m long and 7.6 m wide

Source: Beard 1946: Figure 2

Figure 6.3 Buttressed *Terminalia superba* trees, Lancetilla Botanical Gardens, Honduras

of the distinctive structural features of tropical forest communities. These features include a dominance by woody plants, and a tendency for trees to be tall and unbranched except in the canopy, which is often quite compact. Certain life forms, such as understorey palms and lianas, that are absent or rare in other forests, are prominent features in many tropical forests as are certain plant structures. The most obvious of these is the tendency of many trees to develop plank buttresses, representing vertical upgrowths of the main lateral roots (Figure 6.3). Buttresses tend to be especially common on upslope

137

sides of trees, or opposite crown asymmetries, where they appear to perform a tension support function (Henwood 1973; Warren *et al.* 1988; Young and Perkocha 1994). However, no consensus exists as to why they should be prominent in the tropics but absent from trees in non-tropical forests. In most lowland evergreen forests the main canopy occurs at heights that vary from 25 to 45 m, with occasional tall emergents protruding above this in some places, and a variety of shorter trees beneath the main canopy; the latter group usually includes both immature canopy trees and others of smaller stature that remain in the subcanopy even when mature.

Two aggregate measures are often used in comparisons of forest structure: the density of stems ≥10 cm diameter at breast height (DBH), and the summed basal area of these stems at the same height, assuming a circular stem cross-section. Collation of data on these measures in a large number of tropical lowland evergreen forests (Meave and Kellman 1994) indicates means (and standard deviations) of 497.4 (± 135.0) stems per hectare ($N = 99$), and 36.6 (± 11.5) $m^2.ha^{-1}$ ($N = 58$).

The micro-climate at the forest floor of tropical evergreen forests is much modified by the forest canopy. Except in tree-fall gaps, light levels at the forest floor are generally between 0.5 and 2.0 per cent of those experienced above the canopy (MacDougall and Kellman 1992). Temperature extremes are buffered by the forest canopy, there is little wind, and humidities are generally high. The forest floor is covered by a litter layer whose thickness varies with soil properties; on many soils it may comprise only 1–2 cm of recently fallen litter, while on infertile soils high in aluminium, a thick root mat up to 15 cm deep can exist (Chapter 4; Figure 4.1). Despite the presence of a continuous fuel layer provided by this litter, micro-climatic conditions beneath evergreen forests rarely allow this to dry to the point of flammability (Uhl *et al.* 1988c) and fire is rare, except where the canopy is destroyed by events such as hurricanes. The precipitation received as throughfall beneath a tropical evergreen forest canopy averages 80–85 per cent of external precipitation, with stemflow down trunks contributing 1–2 per cent, and the remainder retained by the canopy and re-evaporated (Bruijnzeel 1989). Although throughfall drop size in tropical forest is usually larger than that of direct rainfall, drops reaching the soil surface normally fall from heights much less than the 8 m needed to achieve terminal velocity. This, combined with the greater water storage provided by multiple canopy layers, reduces the erosion potential to about 40 per cent of that at uncovered sites (Brandt 1988). Erosion beneath forest is further suppressed by the protection provided by litter layers and root mats. The high organic matter contents of surface soils, and a well-developed system of macro-pores, normally result in high infiltration rates and little overland flow.

Small plants occupying the floor of tropical forests usually comprise low-density populations of shade-tolerant tree seedlings, whose numbers and species composition may vary continuously in response to irregular seed production by parent trees. Intermingled among these may be small numbers of specialized shade-tolerant forbs, vine seedlings and ferns. Greater understorey plant biomass is normally found only where canopy gaps have occurred (see below).

Animal populations

Both species diversity and total biomass of tropical forest animal communities tend to be higher than those of temperate forests (Huston 1994). The latter condition is most directly attributable to the year-round availability of food items in the tropics, which contrasts with the severe bottleneck that the absence of these during the winter imposes on most temperate animal populations. Only birds, and a limited number of flying insects, are able to avoid this temperate limitation by engaging in long-distance seasonal migration to tropical environments. Within tropical forests, lower biomass and diversity of mammals tend to occur in forests occupying poorer soils (Emmons 1983).

Accompanying the increased biomass and diversity of animals in tropical forests are more subtle changes in taxonomic composition and functional grouping. Among mammals, a greater proportion of species in tropical forests are of small body size, with increased numbers of bat species being especially important (Flemming 1973). The abundance of bat species leads to aerial species becoming a prominent functional group, while arboreal species, notably primates, comprise a group largely absent from temperate forests (Flemming 1973). Most importantly, an entirely new group of specialist frugivorous (fruit-eating) mammals appears in tropical forests, whose existence is precluded in temperate forests by the absence of year-round fruit availability. This group plays an especially important role in the dispersal of seeds of tropical forest plants, many of which are large (Figure 6.4). Large seed size is likely to be strongly selected for in the shady environment of tropical forests, and the coevolution of this trait with that of seed dispersal by large frugivores during the early Tertiary is seen by Upchurch and Wolf (1987) as a key event in angiosperm diversification at that time (see Chapter 2). Large mammals are most likely to be adversely affected by forest reduction and fragmentation, so considerable concern exists about the fate of these species and the large-seeded plant species whose dispersal may be dependent on them.

Altitudinal change in forest structure

Increased altitude in tropical mountain forest systems is accompanied by consistent changes in forest structure and floristic composition, although there are many regional variations on these changes, and a large number of different systems have been developed by forest ecologists for classifying them. Increased altitude is accompanied by decreases in tree stature and lowered

Figure 6.4 An unidentified seedling germinating from a very large seed in montane forests of the Sierra de Lema, southern Venezuela

canopy height, as well as reductions in tree species diversity (Figure 5.5). Accompanying these changes are usually subtle shifts in floristic composition away from species belonging to exclusively tropical families and genera towards those that are members of temperate taxa. For example, in the Neotropics, species of alders (*Alnus*) and oaks (*Quercus*) become increasingly prominent at higher altitudes, while in South-East Asia, species of *Lithocarpus*, an oak relative, become more abundant in montane forests. In montane

Figure 6.5 Cloud forest at *c.* 1,500 m elevation, Mt Apo massif, Philippines. Both the abundance of epiphytic mosses on trees and the presence of tree ferns are characteristic of tropical cloud forests

forests of Australia and New Guinea, species of the southern beech (*Nothofagus*) become increasingly prominent, while throughout tropical montane areas, species of the southern-hemisphere gymnosperm genus *Podocarpus* can be found.

In the zone of increased precipitation and cloud envelopment that occurs on all tropical mountains (cf. Figure 3.11), trees become loaded with mosses and other epiphytes, and moisture-dependent plant forms such as tree ferns become common (Figure 6.5). This forest type is often referred to as 'cloud forest'. Above the zone of maximum precipitation trees become increasingly stunted, and closed-canopy forest gives way to forms of shrubland at altitudes between 3,000 and 4,000 m. On the highest mountains, this shrubland eventually gives way to a form of alpine tundra. On lower peaks and exposed ridge crests, cloud forests are often made up of stunted and deformed trees, forming what has sometimes been termed 'elfin woodland'. Elfin stature has been attributed to wind stress in these locations (Lawton 1982). On isolated peaks, the changes in forest structure described above tend to be much compressed – a phenomenon usually referred to as the 'Massenerhebung effect'. This appears to correlate with a greater frequency of low-elevation cloud incidence on these isolated peaks.

Because tropical montane forests often occur in areas of steep terrain, and in a cloudy environment which people consider unattractive for agricultural usage, they have not been subject to intense human disturbance until the past half-century. Unfortunately, the pressures of increasing human populations are radically changing this condition, and these forests are now under intense pressure in many areas. Relatively fertile soils may exist in mountainous volcanic regions, whose forests are now often cleared for agriculture, despite high risk of severe erosion owing to the steep slopes. Even where not used for agriculture, fuelwood collecting in these forests has put increasing pressures on what are often relatively small bodies of forest (Smiet 1992; Figure 5.9).

Moisture-related changes in forest structure

A decrease in tree stature and lowered tree diversity are also found along gradients of decreasing annual rainfall and an increasingly prolonged dry season. However, the flora throughout these gradients tends still to be dominated by tropical families and genera, suggesting that adaptation by tropical plants to increasing drought has been more readily achieved than the adaptation to the lower temperatures in tropical mountains. The most visible manifestation of this drought adaptation is deciduousness among trees, which becomes increasingly frequent as one moves into drier climate zones. In areas with a short dry season both evergreen and deciduous trees coexist, with the latter being more shallowly rooted (Jackson *et al.* 1995). In areas with a more prolonged dry season, all species may become deciduous, and such specialized plants as cacti (in the New World) may appear as a forest component (Figure 6.6). These dry deciduous forests often occur intermingled with savanna to form landscape mosaics, with forest restricted to more fire-protected sites. Such mosaics are especially widespread in Africa (Menault *et al.* 1995). This pattern suggests a delicate fire-mediated balance between the two formations, which will be more fully discussed in Chapter 7. Finally, in the driest tropical climates that have not been subject to intense grazing or severe human disturbance, dry forest stature is reduced to an open woodland of thorny shrubs and succulents (Figure 6.7). This often comprises small vegetation pockets in rainshadow areas and on especially droughty soils, but in the arid north-east of Brazil it forms a larger regional vegetation type locally called 'Caatinga' (Figure 6.1).

The seasonal drought that imposes deciduousness on plants in dry forests also affects animal populations. Insect populations are much reduced in areas with a severe dry season, and insects move into moist local refugia (Janzen 1973). There may also be some intra-regional movement of populations between dry and moist forests. The dry season also represents a period of food scarcity for vertebrates, although some plant species fruit during this time. Frugivores cope with this scarcity by local migration or switching to alternative food items (Leighton and Leighton 1983).

Despite the limitation to agriculture posed by seasonal drought, dry forests have suffered disproportionately from clearance for this usage (Murphy and

Figure 6.6 Tropical deciduous forest under dry-season conditions, Guri, south-eastern Venezuela

Lugo 1986). This probably relates primarily to the more fertile alfisols with which dry forests are often associated, but a reduced incidence of pest and disease problems may also have augmented this selection. A major consequence of this clearing history is that there is now proportionately much less of this forest type surviving than of moister forests, and it has been much less fully studied than moister tropical forest types.

Specialized forest types

Most tropical regions possess smaller areas of more specialized forest types; space permits mention of only the more common of these.

Mangrove forests

Along tropical shorelines that are protected from severe wave action, and especially where large quantities of river-borne sediments are being deposited, a unique tropical forest community exists in a habitat that elsewhere supports only herbaceous salt-marsh vegetation. The habitat is uniquely stressful for plant life both because of waterlogging of the soils, which may be flooded continually or on a diurnal tidal cycle, and because of the salinity of sea water. Salinity imposes a severe osmotic gradient that inhibits water uptake by plants, and salts at the concentrations found in sea water are often toxic

Figure 6.7 Thorn woodland with cacti, Tehuacán Valley, southern Mexico, under wet-season conditions. This very dry zone represents a rainshadow area in the lee of the Sierra Madre Oriental mountains

to plants. Although this habitat must be a very ancient one, only a very limited number of plant species in a small number of genera have evolved tolerance of these harsh conditions, and these form low-diversity forest communities. The trees comprising these communities often contain specialized morphological structures that facilitate root aeration, such as stilt roots or pneumatophores (Figure 6.8). These communities are often made up of a few species that are widespread within the ocean basin, and locally form mono-specific zones parallel to the coast. At a global scale, the most diverse mangrove floras exist in the South-East Asian region (Chapman 1977).

Mangrove forests have been used as sources of firewood and charcoal but their very specialized soils have resulted in little clearance for agriculture except in some parts of South-East Asia where the sites are used for flooded rice-farming under specialized management techniques (Sanchez 1976). In this region, much more mangrove clearance has taken place for fish and shrimp ponds. Undisturbed mangrove communities are important habitats for some birds and terrestrial vertebrates as well as for the juvenile stages of many coastal fish species, especially shrimp, which are provided with shelter and feed on the detritus of the forest community. A number of major tropical coastal fisheries, such as those of the Gulf of Mexico, are essentially based on the adjacent mangrove forests (Qasim and Wafar 1990), indicating the importance of preserving these systems.

Figure 6.8 The soil surface of a mangrove community in Belize, showing both the stilt roots of *Rhizophora mangle* and the pneumatophores of *Avicennia nitida*

Freshwater swamp forests

Tropical forests occupying habitats that are subject to inundation by fresh water have received much less study than have mangrove forests. In part this reflects the destruction of these forests in many parts of the tropics, especially in the major river valleys of South-East Asia, where they have been cleared for wet rice-farming. The habitat is a heterogeneous one, making it difficult to generalize about this forest type; hydrological regimes may vary from persistent inundation to the extreme differences experienced by flooded forests on the Amazon and Orinoco river systems (Figure 3.14), while soil properties reflect the inherent heterogeneities of tropical alluvium. In comparison to non-flooded forests, floras are usually of lower diversity and more specialized (Junk 1989). Although the alluvial soils that these forests occupy make them attractive for agricultural activities, their potential importance to local fisheries (Forsberg *et al.* 1993) indicates that this clearance may not be free of hidden costs.

Heath forests

Throughout the tropics, sites of extreme infertility that are not covered by savanna support a form of stunted forest that is given differing local names,

145

such as 'keranga' in Borneo and 'bana' in Venezuela. A limited flora showing extreme specialization owing to the paucity of nutrients is usually involved (e.g. Sobrado and Medina 1980). These forests are found most often on giant podzols, although a variant of them occurs also in peat swamps (Whitmore 1984). The poverty of the soils has usually resulted in these sites not being cleared for agriculture, although the peat forests of Borneo suffered severe destruction from fires during the extreme El Niño drought of 1982–3 (Malingreau et al. 1985).

TROPICAL FOREST DYNAMICS

The absence of annual growth rings in the woody tissue of tropical trees makes it impossible to determine the ages of these with accuracy, or to examine the growth responses of these trees to climatic variability or community disturbances. This has posed a fundamental difficulty when studying the growth dynamics of individual trees in tropical forests, the dynamics of the tree populations, and that of the forest community as a whole. The large size of many tropical forest trees led to an early assumption that these were of great age, and that the forests composed of these trees were relatively immutable. However, recent results that draw upon observations made in tropical forest plots over multiple years or decades suggest much more dynamism in these forests than was earlier believed.

Observations of both tree mortality/recruitment rates, and of the rate of creation of forest canopy gaps, have been used to derive estimates of stand dynamics. Annual mortality rates for trees in tropical forests (normally measured for trees ≥ 10 cm DBH) usually lie between 1 and 2 per cent per year (Swaine et al. 1987; Hartshorn 1990a). If the forest is assumed to be in a steady state, these data may be used to estimate tree turnover time, or the time necessary for all the trees originally measured to die. Estimates of this sort by Rankin-de-Merona et al. (1990) for a number of Neotropical forests indicate turnover times of between 50 and 120 years. Annual rates of canopy gap creation may also be used to estimate the time necessary for the whole canopy to undergo turnover. For several Neotropical forests, these rates tend to average about 1 per cent per year, producing canopy turnover rates of between 60 and 160 years (Hartshorn 1990a). Both sources of data suggest forests that are remarkably dynamic, and a recent study has suggested that turnover rates may have increased in recent decades, possibly as a consequence of elevated atmospheric CO_2 levels (Phillips and Gentry 1994). These high tree turnover rates provide the opportunities for the relatively rapid changes in species composition that have been documented in tropical forests (Hubbell and Foster 1992; Condit et al. 1992b).

The dynamism of tropical forests means that trees are dying at relatively frequent intervals, freeing resources for utilization by other trees. Light, rather than soil nutrients, has been shown repeatedly to be the critical resource

Figure 6.9 Tropical forest canopy gaps photographed from above, near Camarata, southern Venezuela. The small upper gap probably represents the canopy position of a fallen tree, while the larger lower gap represents the location where the canopy of this tree fell

made available at these times (Uhl *et al.* 1988b; Vitousek and Denslow 1986; Denslow *et al.* 1990), making the creation of canopy light-gaps the central event in forest dynamics (Figure 6.9). The canopy gap regime of tropical forests is dominated by small-sized canopy gaps, with decreasing frequencies of larger-sized gaps (Figure 6.10). One consequence of this is that while small canopy gaps may be well represented in small forest areas, relatively large forest areas are required to 'sample' the infrequent large-gap regime of tropical forests. This means that small forest fragments may experience a large canopy gap only rarely, and species dependent upon these forms of gap for successful regeneration may thus be at increased risk of extinction in small fragments.

Canopy gaps vary not only in size but also in shape, orientation, and the degree of disturbance caused during the death of the tree. The smallest canopy gaps are created by branch breakage or the death of individual small trees, while larger gaps are created by the death of larger trees, especially when these fall and damage surrounding smaller trees. The largest canopy gaps are created by multiple tree blowdowns (Nelson *et al.* 1994) or hurricane damage. Large gaps created by the synchronous death of many adjacent trees have usually been reported only for forests of low diversity or mono-specific stands (Mueller-Dombois 1986). The heterogeneity of micro-environments created

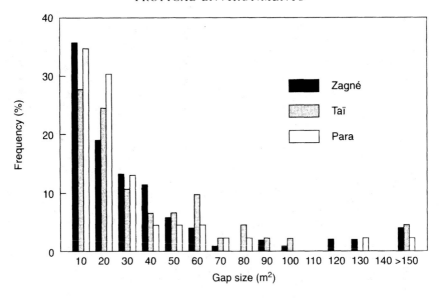

Figure 6.10 Frequency distribution of canopy gap sizes in three tropical forests in Côte d'Ivoire

Source: Jans *et al.* 1993: Figure 2

by canopy gaps varies with both gap size and type of tree death. Trees that die standing create only increased light levels and a gradual increase in fallen plant debris. Those that are snapped and fall normally create elongated gaps with damage to nearby trees, especially where the canopy falls (Figure 6.11), while those that fall with an uprooted root mat also expose bare mineral soil in a hummock and hollow couplet. There has been much speculation about the role that heterogeneity in gaps, whether within gaps, among these or in the timing of gap creation, plays in facilitating coexistence of large numbers of plant species in tropical forests (e.g. Hartshorn 1980; Denslow 1987; Brandani *et al.* 1988). While there is convincing evidence that this heterogeneity does contribute to slowing rates of competitive exclusion and providing a variety of opportunities for different plant species, it exists also in the gap regimes of low-diversity temperate forests, indicating that gap heterogeneity is not in itself sufficient to account for the diversity of tropical forests.

Developments taking place at the sites of canopy gap creation depend both on the type of gap, on the properties of plant species available to take advantage of the increase in the light resource, and on a large element of chance. The smallest gaps may be filled by in-branching of adjacent tree canopies, but in all other gaps one or more new trees are recruited to the canopy. Tropical forest tree species tend to form two relatively distinct groups in

Figure 6.11 Recent tree fall in a gallery forest, Apure, Venezuela. In tree-fall gaps, the largest canopy opening usually occurs where the canopy of the felled tree has fallen, damaging smaller trees

terms of seedling establishment: shade-tolerant species capable of germinating and surviving at least temporarily in the dense shade of forest understoreys, and light-demanding pioneer species that establish only in well-illuminated sites (Swaine and Whitmore 1988; Whitmore 1989). The first group is often associated with large seeds that contain abundant carbohydrate reserves but are of short longevity (Ng 1978; Figure 6.4); their seedlings are slow-growing and form populations in the forest understorey that are sometimes referred to as 'advance regeneration'. Lieberman *et al.* (1985) suggest that this group may be further subdivided into slow-growing understorey trees which may have short or long lifespans, and those capable of reaching the forest canopy by virtue of being able to respond opportunistically to canopy gaps. In the Costa Rican forest studied by these authors, half of the species fell into the shade-tolerant group that did not require gaps for initial establishment. In contrast, gap-requiring pioneer species tend to have small seeds (Ng 1978), which are often readily dispersed and capable of prolonged viability while buried in the forest soil (Holthuijzen and Boerboom 1982). These species also tend to grow rapidly in well-illuminated conditions, but have low-density wood and are of relatively short lifespan. The proportion of the forest tree flora that is made up of pioneers appears to vary widely. Ng (1978) concludes that only a small proportion of Malayan forest tree species are pioneers, while Lieberman *et al.* (1985) conclude that half of the Costa Rican forest trees studied require gaps for establishment.

Smaller canopy gaps tend to be filled by the release of shade-tolerant seedlings and saplings already present in the forest understorey, whose growth accelerates in response to increased light levels (Augspurger 1984; Brokaw 1985). Normally there is intense competition among these juveniles, and few successfully achieve canopy status. In contrast, with increased canopy gap size, increased numbers of pioneer tree seedlings can be found establishing, especially when tree-throw creates soil disturbance; many of these probably derive from seed already present in the soil (Holthuijzen and Boerboom 1982; Putz 1983; Figure 6.12). The rapid growth of these seedlings results in their overtopping and temporarily suppressing slower-growing advance regeneration, but their short lifespan means that they fill the canopy gap only temporarily and are eventually replaced by the more slowly-growing species. Very large gaps created by major disturbances may be covered temporarily by stands of short-lived pioneer species, which often form mono-specific stands (Figure 8.7). However, intermingled among these are normally large numbers of sprouts from snapped stems (Vandermeer *et al.* 1995), which ensure a rapid redevelopment of the forest to its original composition. In large gaps, herbaceous pioneer plants may also establish from soil seed banks, and temporarily compete with tree seedlings. However, they are normally overtopped and suppressed by tree seedlings within a few years. Less easily suppressed are populations of vines which may also appear in large gaps, deriving either from seed or from vegetative parts that have fallen with the tree canopy that created the gap. These can envelop tree seedlings and saplings and severely affect their growth; they pose a potentially severe problem in tropical forestry (Putz 1991) that will be discussed further below.

As a result of these gap-regeneration processes, tropical forests may be thought of as complex mosaics of former gaps in differing stages of development (Figure 6.13). Few gaps are likely to be spatially congruent with an

Figure 6.12 Young pioneer trees (*Cecropia* sp.) establishing in a large tree-fall gap in evergreen forest, Palenque, southern Mexico

earlier gap, so the mosaic pattern can be exceedingly complex (Figure 6.14). However, several authors have noted a tendency for certain areas to be more gap-prone than others, suggesting a more persistent mosaic pattern (Young and Hubbell 1991; Yavitt *et al.* 1995). The floristic composition of the young tree population that is available to fill a gap is likely to be highly unpredictable from place to place. It depends upon the composition of the advance regen-eration, which may be expected to vary spatially as a consequence of location of parent trees, and temporally in response to fruiting periodicities. It also depends on the population of buried viable seed present in the soil, and the availability of seed immediately after gap creation. While the unpredictability of the gap regeneration process serves to slow competitive exclusion among species, and promote their continued coexistence, it also poses difficulties in the management of tropical forests for production of desirable timber.

TRADITIONAL TIMBER EXTRACTION FROM TROPICAL FORESTS

The large volume of woody tissue that exists in tropical forests represents one of the major immediately exploitable resources of many tropical land-scapes. The possibility that this woody tissue can be renewed continually by tree growth makes this a potentially renewable resource. Tropical forests have

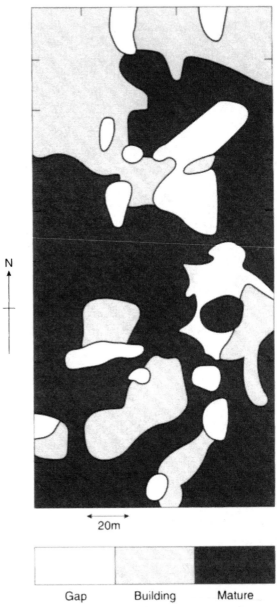

N

20m

| Gap | Building | Mature |

Figure 6.13 Map of the distribution of three phases of canopy development on a 2-ha plot in evergreen dipterocarp forest, Malaya. The extensive building phase at the north end of the plot is related to regrowth from earlier forest clearance

Source: Whitmore 1990: Figure 2.20

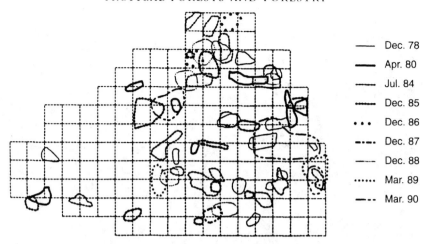

——	Dec. 78
——	Apr. 80
———	Jul. 84
·······	Dec. 85
• • •	Dec. 86
—··—	Dec. 87
——	Dec. 88
······	Mar. 89
—·· —	Mar. 90

Figure 6.14 Distribution of canopy gaps appearing in a 1.5-ha area of secondary forest in Panama over a 13-year period. Squares are 10 m × 10 m quadrats

Source: Yavitt *et al.* 1995: Figure 1

traditionally acted as a source of wood and other building materials, and of fuelwood, for indigenous human populations. More recently, tropical forests have been used as a source of timber to be traded in either national or international markets. When used in the latter way, tropical forests have a fundamental problem: most are composed of a large variety of tree species with differing wood qualities and economic values. The only major exception to this generalization is in South-East Asia, where forests are dominated by tree species of one family, the Dipterocarpaceae, whose woods are generally of high utility. This leads to high volumes of merchantable timber being available naturally in these forests, and has resulted in their being the most intensively exploited forests of the tropics over the past half-century.

The most rudimentary form of commercial exploitation of a tropical forest's timber resource can be termed 'high grading' or 'creaming': selective removal of limited numbers of stems of the commercially most valuable timber species from forested areas. These are often high-value cabinet woods sold in international markets, such as the mahoganies (*Swietenia mahogani* and *Swietenia macrophylla*) in the Neotropics. Initial tree extraction may be followed several decades later by further removal of other stems of the same species that have grown to merchantable size or, where few of these remain, of economically less desirable 'secondary' timber species that may be used for construction in local markets. Today, the two logging phases are increasingly being collapsed into one relatively intensive logging phase (e.g. Uhl and Viera 1989), although even in this a relatively small proportion of trees are still extracted. For example, in logging of this sort in the Brazilian Amazon, only 2 per cent of

Figure 6.15 Log loading deck in forest being logged, Belize. These are usually the places of most severe disturbance during selective logging operations in tropical forests

trees were harvested (8 per hectare) (Uhl and Viera 1989). Logging of this sort is often conducted with few controls on the extraction process, except for some minimum tree bole diameter limit. Today logging operations are usually heavily mechanized, with logs being dragged to truck loading decks by crawler tractors or 'skidders' (Figure 6.15) or, where merchantable tree densities are high, by winching the bole to these sites from a 'spar tree'. Because of the low densities of valuable trees in most tropical forests, large areas tend to be exploited diffusely, with forest disturbance by felling and log extraction that is severe locally, and widespread throughout the area exploited (Figure 6.16). Log extraction trails are often made many hundreds of metres, or even kilometres, into forest to remove single valuable stems, resulting in considerable damage per tree extracted. Because extraction operations are essentially ephemeral, log extraction trails and road networks are viewed as temporary only, and constructed as cheaply as possible, with few precautions taken to reduce runoff by careful grading or installation of culverts (Figure 6.17). As a result, a period of severe localized soil erosion and river siltation often accompanies these operations (Douglas *et al.* 1992).

While a single low-density high-grading operation in tropical forest results in some severe localized disturbance, damage to the forest as a whole need not be extreme unless the operation has been conducted in a particularly careless way. For example, removal of 2 trees per hectare from forest in Gabon

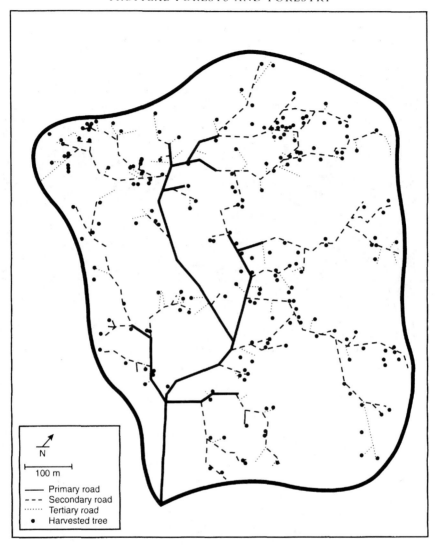

Figure 6.16 Distribution of logging trails and harvested trees in a selectively logged 52-ha forest in Pará, Brazil

Source: Uhl and Viera 1989: Figure 2

resulted in a 10 per cent decrease in canopy cover, and a loss of 10.8 per cent of trees and lianas due to incidental damage (White 1994). Isolated large gaps are created at the sites of tree-felling, and a system of disturbed soil along drag trails and logging roads is created. However, the former are usually filled by the normal processes of gap regeneration, while populations

Figure 6.17 Logs being extracted by crawler tractor during a selective logging operation, Belize. Immediate damage to soils can be reduced by logging only in the dry season, but soil erosion on sites such as this in the subsequent wet season can be severe

of fast-growing pioneer trees establish on the disturbed soil. Effects on animal populations vary, depending upon their dependence on the tree species extracted. For example, relatively intact vertebrate frugivore and folivore populations survived in heavily logged diptercarp forest in west Malaysia because members of that family are almost unused as a food source (Johns 1988), while lemur populations in dry forests in Madagascar actually increased after light selective logging, apparently in response to an increased fruit supply and higher leaf protein contents (Ganzhorn 1995). In the Budongo Forest Reserve of Uganda, which has been selectively logged since the 1920s, Plumptre and Reynolds (1994) examined densities of five primate species in relation to logging history and found no evidence of negative effects of this activity on two species, and increased densities in the other three. Johns (1992a) has stressed the importance of retaining small unlogged forest patches in logged areas as a means of providing temporary refugia for vertebrate species. A system of unlogged patches comprising less than 5 per cent of a logged area in Sabah was found to contain populations of most species intolerant of logging.

Although the effects of a single light selective cut of timber in tropical forest may be relatively benign, repetition of this operation at intervals of several decades can lead to a gradual degradation of forest structure. Yields

Table 6.1 Tree damage caused by the logging of dipterocarp
forest in Malaya

Type of damage	% of trees affected
Killed	
Timber trees	3.3
Destroyed by road-building	4.8
Destroyed by spar tree and loading-deck construction	3.6
Destroyed by felling and log dragging	39.2
Remaining	
Standing but damaged	6.0
Undamaged	43.1

Source: Johns 1988: Table 1

of the most valuable timber species often decline precipitously owing to initial over-cutting and the failure to leave sufficient seed trees (e.g. Gajaseni and Jordan 1990). Leaving trees of the poorest form may also degrade the gene pool of valuable species if the form has a genetic basis. Forest composition may change toward a forest composed of low-value trees, and a change in the composition of animal communities may be expected (Johns 1992b). Less directly, the access provided by logging roads facilitates hunting activities, and where land is scarce the areas may be invaded by settlers who complete the forest clearance.

More intense logging operations, such as those which are typical in the dipterocarp forests of South-East Asia (Johns 1988) and, increasingly, in tropical forests in other areas (Uhl and Viera 1989), are much more damaging. Johns (1988) found that half of the trees in a dipterocarp forest in Malaya were killed in a logging operation that removed only 3.3 per cent of tree stems (Table 6.1). Uhl and Viera (1989) found that a logging operation that removed 8 stems per hectare caused death or damage to 26 per cent of the trees in the forest; total canopy cover was almost halved. Such severe damage, especially if spread over large areas, may be expected severely to affect animal populations. In the Kibale Forest of Uganda, intensive logging operations tend to promote abundant herbaceous undergrowth; this attracts elephants, which damage tree regeneration and help perpetuate the open forest conditions (Struhsaker *et al.* 1996). The severe canopy opening associated with intensive logging operations also increases drying of forest fuels and greatly increases the risk of damaging forest fires (Uhl and Buschbacher 1985; Woods 1989).

MANAGEMENT ALTERNATIVES FOR SUSTAINED
FOREST UTILIZATION

Ad hoc exploitation of tropical forest timber reserves by high grading leads to an inevitable degeneration of a country's forest estate, often followed by its transformation to subsistence rural land use, with remnant forest tracts being further degraded by use as a fuelwood source. In the absence of a stabilization of the human population, such degradation is inevitable, but in its presence we may ask what prospects exist for sustained management of tropical forests. Because most forested land in tropical countries is government-owned (Gillis 1992), governments have enormous latitude in setting forest use policies, including the decision whether to preserve forested areas or release these for agricultural settlement.

The preservation of a significant proportion of a tropical landscape under forest cover can be justified, theoretically, on several grounds. There are obvious hydrological benefits to preserving a forest cover, especially in mountainous terrain. Added to this are the benefits for tourism, and of preserving potential genetic resources, that were mentioned in Chapter 5. However, these justifications are difficult to quantify economically, and are, in themselves, unlikely to be enough to ensure preservation of untouched tropical forests, except in relatively small areas. Preservation of larger areas of tropical forest is likely to require that these also provide tangible resources for the society preserving them. It has sometimes been argued that the development of tropical forestry plantations (see below) will relieve the pressure for timber products from natural tropical forest. However, this argument fails to recognize that the financially constrained governments of most tropical countries are unlikely to forgo any form of forestry revenue, and will therefore be likely to see tree plantations as an additional, not alternative, timber resource.

The density of persons supportable by extractive forestry activities will inevitably be lower than that supportable by some form of agriculture, so forest preservation for extractive use is unlikely to be able to compete with agriculture as a land use on the better soils and flat terrain best suited for agriculture. Rather, forestry is likely to be increasingly restricted to more marginal terrain that possesses poorer soils and steeper slopes. While this economic fact will assign forest cover to more marginal sites of lower biological productivity, these are likely to be sites where the hydrological benefits of a forest cover are greatest. Moreover, Huston (1993) has argued that such marginal sites are likely to be areas where most biological diversity exists, and that forest preservation in these areas will have benefits in biodiversity conservation that are disproportionate to the area preserved.

Resources that can be extracted in a sustainable way from tropical forest comprise timber plus a range of other minor forest products, both plant and animal. Given the biological diversity of tropical forests, multiple use of these resources is likely to be more complementary than competitive, with the preservation of a functioning forest community representing a unifying objective of the several uses. Timber is likely to be the primary resource extracted from most managed tropical forests, therefore management for sustained timber yield is likely to set the basic pattern in these forests. Consequently, we begin by discussing the prospects for managed and sustained timber extraction from tropical forests.

Sustained timber production from tropical forests

Management of tropical forests for sustained timber production poses a suite of theoretical and practical problems that has sometimes led to the abandonment of this as a realistic goal, and the resort to tree plantations (cf. Figure 6.22). However, the many problems that tree plantations face (see below) and the urgent need to conserve tropical forests in some form have led to renewed interest in this activity. The high diversity of tropical forests makes the management of these inherently complex, especially when autecological information about most tree species does not exist. Also, tree growth is a relatively slow process, requiring that extended time periods must elapse before the success of particular management strategies can be evaluated. Added to this is continual pressure to convert forested areas to agricultural use. As a consequence, few tropical forest management plans have been applied for time periods sufficient to evaluate fully their practicability. With this caveat in mind, we can review some of the principles guiding management schemes and examine, as far as is possible, their success.

Although the absolute ages of tropical trees cannot be determined with accuracy, sequential observation of tree growth rates on permanent plots shows that individual stems of canopy trees tend to follow a sigmoid growth pattern (Whitmore 1990). This involves an initial period of slow growth when young which, in shade-tolerant species, may be very protracted. Following this is a relatively brief period of rapid growth that is stimulated by release in a gap, after which slower growth returns as canopy status is achieved. As a consequence of this growth pattern many trees in tropical forests spend much of their lifespan in slow-growth phases. This raises the possibility of enhancing overall tree growth in forest stands by creating gaps at relatively frequent intervals, or by releasing mature trees from competition. The timber productivity of natural tropical forests is also limited by the low density of usable trees, so a further means of increasing it is represented by management techniques designed to change forest composition towards economically valuable species.

The primary management technique involves manipulation of timber extraction procedures to control the degree of canopy opening and hence the species of juvenile trees most likely to be favoured. In effect, this represents a form of controlled canopy gap creation. The severe damage caused by uncontrolled logging (Table 6.1) means also that considerable improvement can be achieved in protecting advance regeneration by more careful logging practices. For example, felling trees in directions that minimize damage, minimizing the numbers of drag trails used and planning their location prior to felling, and avoiding upslope log extraction can all minimize damage both to advance regeneration and to soils. Various other, more labour-intensive management procedures also exist, such as climber cutting and thinning and planting, but the high cost of these operations has generally led to their abandonment in recent years (Whitmore 1990).

Monocyclic systems

Two management systems for tropical forest, each of which envisaged single logging operations at long intervals, were formulated in the period after the Second World War: the Malayan Uniform System for the lowland dipterocarp forests of Malaya, and the Tropical Shelterwood System for the forests of West Africa. The Malayan Uniform System was designed to manage forests with a high volume of merchantable timber and abundant advance regeneration. It involved a single logging operation followed by poisoning of other unwanted trees to release abundant dipterocarp seedlings in the advance regeneration. Subsequent weeding and thinning treatments were envisaged at 10-year intervals until relogging of the site after 70 years. The system has now been abandoned for two reasons: the high costs of repeated treatments, and the displacement of forestry activities by agriculture from lowland areas to mountainous terrain, where adequate advance regeneration does not exist (Buschbacher 1990). In the Tropical Shelterwood System, advance regeneration was to be encouraged by canopy opening several years prior to the initial logging operation, followed by several weeding and thinning treatments after this. In practice, the system failed because of severe vine infestations caused by canopy opening, and the high costs of treatments (Kio 1979; Buschbacher 1990).

Polycyclic systems

In response to the problems of monocyclic systems, several systems that envisaged repeated light logging operations at intervals of several decades were developed. In practice, most have failed owing to excessive damage during logging operations, excessively short logging cycles, and the high costs of silvicultural treatments (Buschbacher 1990). However, one particularly innovative system is being formulated for forest management in Surinam, and could well serve as a model for comparable systems elsewhere. The CELOS system (de Graff 1986, 1991) is based on careful control of logging operations to minimize damage, and treatments that seek to induce accelerated growth among merchantable tree stems in the residual stand, rather than release of advance regeneration. Prior to logging, lianas are cut to minimize felling damage, and non-commercial trees above 20 cm DBH are poison-girdled and die slowly (Figure 6.18). Tree-felling directions are carefully planned, and logs extracted using a permanent system of logging trails. Accelerated tree growth has been recorded for 8–10 years after this treatment, and subsequent refinement treatments (poison-girdling) are envisaged, with relogging at 20-year intervals. There are several theoretical advantages of this system. The forest structure is changed gradually rather than drastically, reducing the risk of weed and climber infestations. Plant biodiversity is reduced only marginally, and the forest remains habitable to most animal populations, and usable for the extraction of other forest products. Nutrients are released gradually from poisoned trees, while management options remain flexible over relatively short time periods, because manipulations focus on trees that are approaching harvestable size, rather than on juveniles. Moreover, the preservation of a periodically used road infrastructure reduces the possibility of invasion by agriculturalists. It remains to be seen whether the system can be sustained economically and in the face of population pressures.

Clear-cut systems

'Clear-cutting' has become an emotive term among environmentalists, conjuring up images of devastated landscapes. However, the severe damage that can be caused by even light selective logging (Table 6.1) implies that in some circumstances complete forest removal becomes preferable to leaving a residual stand of severely damaged trees. The advantages to such a system increase where more stems in the logged stand are merchantable, and many tropical woods are now usable for pulp-making (Evans 1982), while virtually all are usable for charcoal. There have been few attempts to manage tropical forest by clear-cutting, and no published evaluations of its long-term sustainability. Saulei and Lamb (1991) have examined the recovery over 11 years of forested sites in New Guinea that were clear-cut for pulpwood. Logging areas were relatively large and soil disturbance by crawler tractors was severe. Not surprisingly, the regenerating forest was dominated by pioneer species, with non-pioneer trees coming mainly from stump-sprouting. Seedling densities were inversely related to the degree of soil disturbance. Saulei and Lamb (1991) make no attempt to estimate the long-term sustainability of the system, but suggest that smaller logging areas and less soil disturbance would improve seedling regeneration.

The ecological sustainability of clear-cut systems is likely to depend on careful control of scale of cutting in space and time, and upon ensuring successful regeneration of useful species on clear-cut areas. Hartshorn (1989, 1990b) has described a clear-cut management system being developed in the Peruvian Amazon that attempts to simulate some characteristics of large canopy gaps, and favours sustained production of gap-dependent tree species. Cutting takes place in forest strips 30–40 m wide, ensuring a seed source in

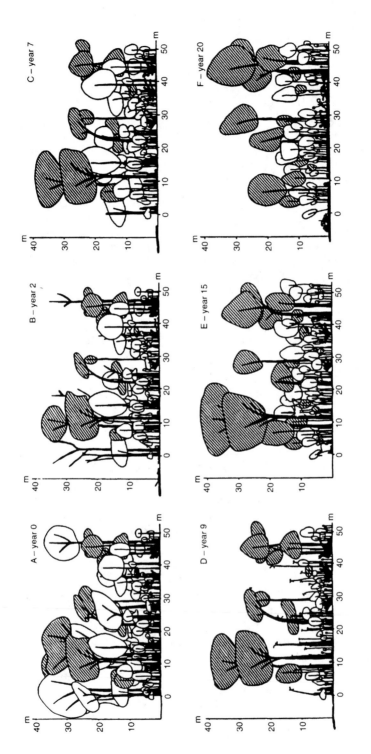

Figure 6.18 Anticipated changes in forest structure induced by the CELOS system. Hatched crowns are of commercially valuable species

Source: de Graff 1991: Figure 27.1

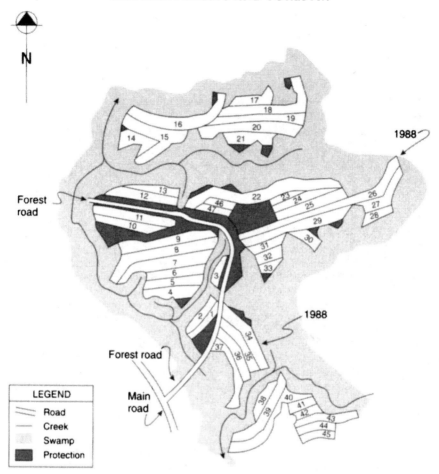

Figure 6.19 Logging plan for strip clear-cutting in a forest block being managed by the Yanesha Forestry Co-operative in Peruvian Amazonia

Source: Hartshorn 1990: Figure 8.2

adjacent forest (Figure 6.19). Tree boles and larger branches are extracted by animal power to minimize soil damage, and used for saw timber, posts and poles after treatment with preservatives, or transformed into charcoal. During the first cutting cycle, abundant regeneration from seeds and stump sprouts has been achieved, and a rotation time of 30–40 years is anticipated, with some intermediate cleaning and thinning treatments. The scheme is being managed as a co-operative effort by aboriginal peoples in the area, and net returns of US$3,500 per harvested hectare have been projected. Timber extraction from mangrove forests normally involves clear-cutting. Blanchard and Prado (1995) have recommended that, to ensure adequate regeneration of tree seedlings, this should be conducted in strip clear-cuts no wider than 20 m that are bordered by mature fruiting trees.

It is still too early to evaluate the ecological and economic sustainability of clear-cut systems, but experiments on these are likely to be increasingly necessary if an economic argument for preserving tropical forest is to be sustained. One special concern about clear-cut systems that extract most standing wood tissue is whether they can be sustained without artificial fertilization, given the relatively large quantities of nutrients that are likely to be removed from the site during each logging operation. This topic will be treated further below in relation to forestry plantations.

Non-timber forest products

Indigenous human populations have traditionally extracted a wide range of non-timber materials from tropical forests for subsistence use. Included among these are animal products (Chapter 10), various resins and gums, a variety of fruits and nuts, and a wide range of medicinal products. The intensity of extraction has usually been low, and these resources can usually be considered renewable. In some instances, these forest products have also become merchantable (e.g. natural rubber, brazil nuts, chicle, rattan), and there has been recent interest in the possibility that these can be harvested sustainably to supplement, or replace, timber production and provide an economic rationale for tropical forest preservation (Nepstad and Schwartzman 1992). Interest in this possibility has been greatest in the Brazilian Amazon, where an area totalling approximately 3,000,000 ha has been classified as 'extractive reserves' to be used for the production of non-timber products, mainly wild rubber and brazil nuts.

Despite the ecological attractiveness of these proposals, serious questions remain about their long-term commercial viability (Homma 1992; Browder 1992). Salafsky et al. (1993) have emphasized that extractive reserves may be more viable in some tropical forest areas than in others because of variabilities in both the forest resources and the sociopolitical context. In most species-rich tropical forests, useful products are diffusely distributed and large areas are required to supply significant quantities of the product. Supply of the product is generally inelastic, and over-exploitation may result when market demands increase (Homma 1992). For example, Nepstad et al. (1992) report potential over-exploitation of the brazil nut resource in one extractive reserve in Amazonia, where a complex set of interactions have reduced recruitment by the nut-producing tree (Bertholletia excelsa). Homma (1992) suggests that many natural forest products are eventually displaced by domestication of the product (e.g. rubber) or replacement by a synthetic product (e.g. rubber, chicle). Browder (1992) has emphasized the low standards of living experienced by most existing forest extractors, and the difficulties of providing social services to a scattered population. These difficulties have led Peters

(1992) to suggest that the greatest hope for economic viability of these activities is in a few low-diversity forest types that are dominated by species producing merchantable materials, such as palm hearts from the palm *Euterpe oleracea*, which dominates floodplain forests in eastern Amazonia.

In more typical species-rich tropical forests, it seems inevitable that this form of extractive activity will have to be carried out as a supplement to other activities such as timber extraction or tourism, rather than as an alternative to this. If well-planned (e.g. the CELOS system), there are no reasons why many of these extractive activities cannot be carried out concurrently with more typical forestry activities. For example, selectively logged tropical forests in Belize continue to be used for production of chicle, used in chewing gum, as the tree producing this material (*Manilkara zapota*; Figure 6.20) is not normally used for timber. Extractive activities of this sort can serve to reinforce the sustained occupancy and use of the forest by human populations, and thereby reduce the possibility of its being seen as un-occupied land that is available for agricultural settlement.

TROPICAL TREE PLANTATIONS

Tropical forests under natural management are estimated to be capable of producing a mean annual increment (MAI) of usable wood of between 0.5 and 2.0 $m^3.ha^{-1}$ (Leslie 1987). In contrast, plantations of fast-growing tropical trees grown in short rotations have achieved MAIs of 14–35 $m^3.ha^{-1}$ (Table 6.2). This very large difference, when combined with difficulties of developing adequate techniques for natural forest management, has led to increased interest in tropical tree plantations as a more efficient way of utilizing the wood-producing potential of tropical environments. In recent decades, there has been a steady increase in area of tree plantations in most tropical countries. Most plantations are mono-specific stands planned for short mono-cyclic rotations, and to date they have used a relatively limited number of tree species, especially pines and eucalypts (Evans 1982). These plantations provide not only potentially rapid growth, but also a simplified system that is more easily and efficiently managed, and a uniform wood type. One of the largest and most widely publicized plantation developments has been that at Jari in the eastern Amazon, where, between 1968 and 1986, 74,500 ha of tree plantations were established on land cleared of tropical forest (Fearnside 1988).

The primary purpose of most plantations has been the production of sawlogs or wood chips for pulp-making, with fuelwood production and protection being lesser objectives (Table 6.3). However, these objectives are likely to continue to change in response to new government initiatives. For example, the Brazilian government plans massive plantation development at Carajas in eastern Amazonia to provide charcoal for the smelting of iron mined in the area (Fearnside 1988). It has often been suggested that the

Figure 6.20 A tree of *Manilkara zapota* in Belize, showing the characteristic slashes used to promote latex flow in the wet season. The latex is the source of chicle, used in chewing gum, and represents an important non-timber forest product of many surviving Central American forests

creation of plantations will remove pressures on natural tropical forests (Baur 1968; Evans 1982), but, as mentioned earlier, governments are more likely to see these as additional rather than alternative sources of timber revenue. Indeed, when forested areas are converted to plantations, as at Jari, their creation results in disappearance of equivalent areas of forest.

Table 6.2 Average growth rates achieved in some tropical plantations

Plantation	Species	Mean annual increment $(m^3.ha^{-1})$	Rotation period (yr)
Fiji Pine Commission	*Pinus caribaea*	21	12–15
Jari Florestal, Brazil	*Pinus caribaea*	27	16
Scott Paper, Costa Rica	*Pinus caribaea*	40	8
Aracruz Florestal, Brazil	*Eucalyptus grandis*	35	7
PICOP, Philippines	*Albizzia falcataria*	28	10
Jari Florestal, Brazil	*Gmelina arborea*	35	10
CNGT, Papua New Guinea	*Araucaria* spp.	20	40
Seaqaqa plantations, Fiji	*Swietenia macrophylla*	14	30

Source: Evans 1982: Table 2.1

Table 6.3 Objectives of plantation establishment (%, by region)

Objective	Africa	Neotropics	Asia-Pacific	World
Sawlogs or pulpwood	65	50	52	54
Fuelwood and charcoal	9	29	4	18
Protection forest	6	10	36	17
Agroforestry	1	—	5	2
Minor products	1	1	—	1
Poles, posts	16	1	2	4
Other	2	9	—	4

Source: Evans 1982: Table 3.4

Most tropical forest plantations are recent, so neither their ecological sustainability nor their economic viability can be reliably evaluated yet. However, there appear to be some reasons for less enthusiasm about their prospects than has prevailed for several decades. The spectacular yields achieved by many tropical tree plantations are usually for those in their first growth cycle, and there are two reasons to suppose that these growth rates may not be sustainable indefinitely without large external inputs: soil nutrient reserves may be depleted, and pests and diseases may accumulate in plantations with time.

Ecological sustainability

Any harvesting of woody tissues from a forest results in some export of nutrients from the site, but under natural forest management these losses are modest. Nutrient concentrations are greatest in foliage and small branches, which are left on site to decompose. While the removal of bark from boles prior to extraction would reduce losses further, the normal 'bark plus bole' removal of low stem densities from tropical forests probably results in a nutrient export rate well below the rate of external inputs to the system from atmospheric sources or weathering of primary minerals. However, where almost all standing timber is extracted from sites at frequent intervals, as it is in short-rotation plantation forestry, losses are likely to exceed net inputs.

Table 6.4 provides estimates of the nutrients likely to be removed from sites when stands of *Pinus caribaea* are logged by alternative techniques. *Pinus caribaea* represents one of the most widely planted tropical plantation trees, and estimates are made for natural stands developing after fire suppression in its native savanna habitat and in two exotic plantations. Removals are expressed as a percentage of mean atmospheric inputs for nine tropical sites (Stewart and Kellman 1982). These data show both the large increase in nutrient losses as more types of tree tissue are removed, and the very large losses relative to inputs that may be expected in the short-rotation Jari plantation. Merchantable bole plus bark represents the most likely harvesting method, and use of this technique would result in a sustainable nutrient balance in both long-rotation systems if effective capture of atmospheric inputs was achieved. However, removals by this technique in the short-rotation Jari plantation would equal or exceed inputs for all elements, with an especially large deficit of phosphorus. Presumably first-rotation growth has been achieved in this plantation as a result of trees drawing upon soil reserves and the large quantities of nutrients that were mineralized when the original forest was cleared and burned. The data in Table 6.4 clearly indicate that comparable growth rates are unlikely to be sustained in later rotations without fertilization. The data also assume effective nutrient capture by plantation trees, which may not be realized.

An accumulation of pests and diseases in mono-specific plantation stands represents a further potential obstacle to sustaining high timber yields in plantations. The spatial isolation of individual trees of a species appears to offer at least partial protection from pest and disease epidemics in species-rich tropical forest (e.g. Wong *et al.* 1990), while some tree species have suffered severe infestations when grown in plantations. The case of the rubber tree (*Hevea brasiliensis*) has already been mentioned, and plantations of various tropical timber species in the family Meliaceae (mahoganies, tropical cedars) have suffered severe infestations by shoot borers of the genus *Hypsipyla* throughout the tropics, which has often forced plantation abandonment (Evans 1982). While some mono-specific tropical forests do occur naturally, severe herbivore infestations have been reported on several of these (Anderson and Lee 1995; Nasciamento and Proctor 1994). The natural populations appear to be able to sustain themselves through these sporadic outbreaks, and so too may plantations, but reduced growth rates may be expected.

Many plantations are of species exotic to the area and presumably free of the pests and diseases of their native habitat. However, given the tendency

Table 6.4 Estimated removals of nutrients in four forms of timber harvesting of natural or plantation stands of *Pinus caribaea*, expressed as percentages of estimated atmospheric inputs during the rotation period

Stand and type of harvesting	N	P	K	Ca	Mg
Savanna, 50-yr growth simulation					
Merchantable bolewood only	8	56	33	35	16
Merchantable bolewood and bark	19	78	40	43	20
Total tree less foliage	32	134	65	71	34
Total tree	56	256	107	85	48
Jamaica, 20-yr plantation					
Merchantable bolewood only	11	73	43	45	21
Merchantable bolewood and bark	24	102	51	57	26
Total tree less foliage	42	178	85	94	45
Total tree	73	334	139	112	63
Jari, Brazil, 12-yr plantation					
Merchantable bolewood only	42	282	165	177	80
Merchantable bolewood and bark	94	395	197	219	101
Total tree less foliage	162	685	329	513	174
Total tree	283	1,292	539	434	244

Source: Stewart and Kellman 1982: Table 9

for indigenous pests to adapt rapidly to exotic crops (Chapter 5), one can anticipate a continued accumulation of new pests on the plantation trees. At Jari, leaf-cutting ants have presented a problem in plantation establishment, requiring the use of insecticidal bait, while fungal infections on *Gmelina arborea*, the first tree species planted, have forced a shortening of the coppice rotation of this species from a planned 6 years to 3–4 years (Fearnside 1988).

If pest problems increase in tropical forest plantations in second and subsequent rotations, one can anticipate slower growth rates and occasional tree crop failures, increased costs of chemical pest control, or both. The options available for non-chemical pest control are largely precluded by the nature of the plantation system. Increased diversity of tree crop stands would eliminate one of the principal advantages of a plantation: an easily managed system that produces a uniform product. Moreover, the multiple-year rotation periods required in plantations reduce management flexibility in avoiding pest problems, by adopting such techniques as temporary site abandonment, or the switching of crops between years. The principal means of non-chemical control of pest problems are likely to be ongoing selection of pest-resistant varieties, or the introduction of new pest-resistant tree species, and the development of various forms of biological controls on pests (see Chapter 9). Both procedures indicate a need for ongoing research programmes where plantation systems have been established, which will add a further maintenance cost to these developments.

Table 6.5a Benefit:cost ratios for tropical forest under natural forest management and for tropical forest plantations over a range of discount rates

Regime	Discount rate (%)					
	0	2	4	6	8	10
Natural forest management	12.3	4.8	1.8	0.6	0.2	0.1
Plantation	1.3	0.9	0.7	0.5	0.3	0.2

Table 6.5b Assumptions made in calculating the above ratios

	Natural forest	Plantation
Mean annual increment ($m^3.ha^{-1}.yr^{-1}$)	1.8	18
Rotation period (yr)	60	20
Stumpage price (US\$.ha^{-1})	15	5
Establishment costs (US\$.ha^{-1}):		
Initial	50	1,000
2nd rotation	—	50
3rd rotation	—	100
Annual costs (US\$.ha^{-1}.yr^{-1})	1	1

Source: Leslie 1987: Table 3

Economic viability

While tropical forestry plantations are potentially highly productive, they also involve very substantial inputs for their establishment and maintenance (cf. Table 6.5). Considerable experimentation is required when selecting appropriate species to plant and the seed sources for these, as well as in continual genetic improvement. Tree seedling nurseries must normally be established. Site preparation and planting is costly, as is the establishment of an adequate infrastructure of roads; at Jari, over 3,500 km of roads have been established (Fearnside 1988). Developing stands must be weeded, thinned and protected from fire and pest outbreaks. These large development costs have resulted in most plantation establishment being the work of governments or large corporations, which have the necessary financial resources and can afford a delay in return on investment. Many of the operations are labour-intensive, and high labour costs have forced the abandonment of plantations in several areas (Figure 6.21).

Fearnside (1988) has attempted a preliminary evaluation of the economic sustainability of the Jari project over two decades, and concludes that these plantations have proven far more costly and less productive than originally envisaged. While an operating profit was recorded by the late 1980s, this was achieved only through subsidies by a profitable kaolin mine on the site, and the enormous establishment costs remain unrecovered. Fearnside estimates that the supply of charcoal for iron smelting at Carajas will require

Figure 6.21 A plantation of *Pinus caribaea* in Belize, whose maintenance was abandoned several decades previously owing to high labour costs. The result of this abandonment has been the development of woody vegetation that potentially competes with the pine, and the infestation of many pine canopies with woody vines

the establishment of plantations approximately 10 times the size of Jari's, and warns that the huge establishment costs that would be required for this mean that charcoal demand is likely to be met by continuing destruction of natural forests in eastern Amazonia.

Leslie (1987) has compared benefit : cost ratios for the natural management of tropical forest with those of forest plantations, and concludes that only where high discount rates are assumed are plantations economically superior to natural forest management (Table 6.5). When the other, less easily quantified benefits of maintaining a natural tropical forest cover are added, this makes a compelling case for the preservation of tropical forest covers. Thus where human population growth rates do not force deforestation for agriculture, there would appear to be little justification for transforming sites covered by managed tropical forest to forest plantations (Figure 6.22). The same arguments do not apply where plantations are established on degraded sites, such as secondary grassland, and indeed a significant proportion of tropical plantations have been established at such sites (Table 6.6). Here, establishment of a plantation cover brings benefits of watershed protection and possible re-establishment of animal populations, together with wood production. Tree growth rates are likely to be slower, and rotation times

Figure 6.22 Tropical forest being cleared for the establishment of conifer plantations, Queensland, Australia, 1981. Fortunately, tropical forest destruction for this purpose is becoming less frequent, and most plantation establishment is being concentrated on non-forested sites

Table 6.6 Previous use of land used for forestry plantations in the tropics (percentages, by region)

Previous land use	Africa	Neotropics	Asia-Pacific	World
Tropical forest	18	4	34	15
Scrub woodland	6	25	11	16
Abandoned land	3	18	8	12
Savanna and grassland	67	52	40	52
Pre-existing plantation	5	1	6	3
Mining spoil	1	—	—	1
Other land	—	—	1	1

Source: Evans 1982: Table 5.1

longer, than on sites cleared of forests, but the establishment of a plantation involves the creation of a new resource where none existed, rather than the replacement of an extant resource. It is on sites such as these that plantation establishment can be justified on ecological as well as economic grounds.

7

TROPICAL SAVANNAS

Savannas are normally defined as plant communities that are dominated by a stratum of grasses, or grass-like plants ('graminoids'), but which also contain some trees, usually as scattered populations that do not form a closed canopy. When applied in the tropics, the term is usually taken to include the full spectrum of graminoid–tree mixtures, from treeless grasslands (Figure 7.1) to open forests with a graminoid understorey (Figure 7.2). This spectrum can occur within short distances.

Figure 7.1 Treeless savanna, Gran Sabana, Venezuela. The line of trees in the middle distance is a narrow band of gallery forest along a river. This savanna receives 2,000–3,000 mm of rainfall per year, and pollen records indicate that it has existed for at least five thousand years (Rull, 1992)

Figure 7.2 Treed savanna, north Queensland, Australia. The tree layer is dominated by species of *Eucalyptus* and *Casuarina*

Savannas are widespread in the tropics (Figure 6.1). In the Neotropics, they consist of two large blocks to the south and north of the Amazonian forests, plus many other, more scattered areas. The largest savannna area occurs to the south of the Amazon on the ancient crystalline rocks of the Brazilian Shield, and is commonly referred to as the Cerrado. In the west this grades into the seasonally flooded savannas of the Pantanal, while in the north-east it grades into the dry thorn woodland of the Caatinga (Figure 6.1). North of the Amazonian forests, in Venezuela and Colombia, a second, smaller savanna region occurs on the plains between the northern Andean foothills and the Orinoco River; this is referred to as the Llanos. Smaller savanna areas are scattered elsewhere throughout tropical South and Central America, often on specialized geological substrates. For example, the Gran Sabana of south-eastern Venezuela occurs on sandstones of the Guiana Shield (Figure 7.1), while in Central America the most extensive savannas occur on coastal deposits of Pleistocene age that contain well-developed ultisols (Figure 7.3). Savannas are especially widespread in Africa, forming a broad, crescent-shaped zone around the West and Central African forest core, crossing the continent south of the Sahara Desert and swinging south throughout most of East Africa and western Madagascar to include large parts of southern Africa (Figure 6.1). Dry forests and woodlands are widespread in the southern savannas of Africa, and this zone has sometimes been mapped as dry forest rather than savanna. Dry forests and woodlands in this zone are often domi-

Figure 7.3 Coastal savanna, Belize. These savannas occur on alluvial and beach deposits along the east coast of Central America between Belize and Nicaragua. Most sites contain intensely developed ultisols, often containing iron concretions (cf. Figure 4.10), and may be seasonally waterlogged. The large trees in the middle distance are *Pinus caribaea*, which is native to these savannas

nated by species of *Brachystegia* and referred to locally as 'miombo' (Menault *et al.* 1995). In the Asia-Pacific area, the largest savanna zone comprises the *Eucalyptus* savannas of northern Australia and adjacent New Guinea. Elsewhere in this region, savannas are patchily distributed, with indications that many have been derived relatively recently from forest (see below).

The dominance of savannas by graminoids, and their lower floristic diversity as compared with forests, serves to create a superficial appearance of homogeneity that may obscure considerable heterogeneity in the composition and functioning of these communities. African savannas support large populations of native herbivorous mammals, and the carnivores that feed upon these, while those of the Neotropics supported almost none at the time of European contact. Within continental regions, there may be considerable compositional change in biotas along moisture gradients, while more locally there may be much change along physiographic gradients. For example, vegetation change along a topographic gradient in the Athi Kapti Plains of Kenya may consist of treeless grassland on the level uplands, *Balanites* tree–short grass cover on convex upper slopes, *Acacia drepanolobium* shrub–tall grass on concave lower slopes, and *Acacia xanthophloea* woodland in bottomlands and riparian zones.

At a global scale, savannas represent a major anomaly, as all tropical climate types, from wet equatorial zones to the semi-arid margins of the subtropical deserts, are capable of supporting some sort of forest cover (Chapter 6). Most large blocks of savanna tend to occur in areas with a pronounced dry season, and savannas only appeared as a distinguishable vegetation type in the fossil record in the Miocene (Wolfe 1985), when low-latitude climatic aridity was becoming widespread. Indeed, some early climate classifications even described seasonally dry tropical climates as 'savanna climates'. However, we now know that the assumption of such a simple deterministic relationship between climate and vegetation is inappropriate, and that many non-conformities exist. For example, while the Llanos savannas generally occur in a dry area that receives 800–1,500 mm of annual rainfall, some of these savannas occur in areas receiving over 2,100 mm (Medina and Silva 1990). Moreover, the Gran Sabana farther south receives up to 3,000 mm of rainfall with a very attenuated dry season (Dezzeo 1994). In contrast, dry tropical forest can be found in many areas of 'savanna climate' (Figure 6.6).

Savannas and forests, as the two major tropical vegetation formations, have often been conceived of as relatively immutable entities, with transformations from one to the other being limited to degradation of forest to savanna under the influence of human activities. However, there is increasing evidence that relations between the two formations have been very dynamic throughout the recent geological past, and especially during Pleistocene climatic change (Chapter 2). While degradation of forest to savanna as a result of human activities has been a widespread phenomenon recently (see below), we also know that forests have periodically replaced savannas in some areas.

SAVANNA STRUCTURE AND FLORISTIC COMPOSITION

Savannas are composed of two basic structural elements that differ radically in both morphology and function: graminoids and trees/large shrubs. The existence of two such different elements in savannas hints at a potential compositional instability in these communities that has been the focus of considerable theoretical interest. To these two basic life forms may be added non-graminoid herbs ('forbs') and small woody plants, both of which make up a relatively small proportion of the phytomass of tropical savannas.

The graminoid life form

The distinctive morphological feature of most grasses and their relatives (mainly sedges) is the absence of stem tissue, with above-ground phytomass made up of clusters of elongate leaves. This life form precludes most graminoids from achieving large stature and monopolizing light, which places them at a potential disadvantage relative to trees in competition for this

Figure 7.4 Grass resprouting a few days after a savanna fire. The singed upper leaf
blades are still visible at the ends of the freshly elongated leaves

resource. The major functional advantage of this growth form is the protected
location of meristems. Meristems comprise embryonic tissue whose cells are
capable of division to form new specialized tissues; in its absence vegetative
growth, or regrowth, is impossible. In the above-ground portions of most
plants, this tissue exists in vulnerable locations at shoot tips (the apical meristems), or beneath the bark of woody plants (the cambium layer). In these
plants, destruction of above-ground tissues can eliminate meristems, leaving
the plant incapable of regrowth, or severely limited in regrowth capacity. In
contrast, in grasses and sedges, meristematic tissue exists only at the bases of
leaves, at or below ground level, and is usually protected by a sheath of dead
leaf bases. In this location, these tissues are rarely affected when above-gound
tissues are destroyed, allowing the plant to regrow rapidly after damage,
provided that sufficient carbohydrate reserves have been stored in below-
ground organs (Figure 7.4). These plants are therefore optimally structured
to withstand repeated destruction of above-ground tissues by such agents as
grazing and fire.

The structural features that allow graminoids to withstand periodic defoliation also allow these plants to withstand periodic loss of above-ground tissues
due to desiccation. Most savannas are subject to an annual dry-season period
of reduced soil moisture availability, and graminoids respond to this either
by death, in the case of annuals, or by a period of reduced growth, dormancy

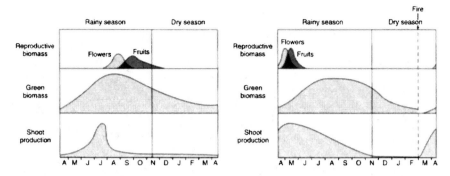

Figure 7.5 Phenology of two common grasses of the Venezuelan Llanos:
(a) *Trachygogon plumosus* and (b) *Leptocoryphium lanatum*
Source: Sarmiento 1984: Figures 19 and 20

or die-back of above-ground tissues. Within this annual growth periodicity, there may be further phenological specialization in terms of the period of maximum vegetative growth or flowering (Figure 7.5). The proportion of annual grasses in savannas shows no consistent pattern in relation to climate. In West Africa, annuals become increasingly prominent with lower rainfall (Breman and de Wit 1983), but the reverse has been found in the Venezuelan Llanos (Medina and Silva 1990).

The graminoid growth form also imparts one further important functional attribute: high potential flammability. Above-ground foliage is distributed in a way that maximizes rapid drying, ready combustion and easy fire spread. We will return to this topic in a later section.

Savanna trees and shrubs

Low-density populations of small trees or large shrubs are a normal component of most savanna landscapes. These woody plants are often short with a gnarled branching pattern, which contrasts with the tall, straight-boled trees characteristic of most tropical forests (Figure 7.2). The woody savanna flora is often of low diversity relative to that of forests, and made up of a limited number of species that typically belong to tropical families and genera. Species of *Acacia* are especially common in African savannas, while in northern Australia, species of the genus *Eucalyptus* are prominent in savannas, as they are elsewhere in the vegetation of that continent. While most of these woody species are savanna specialists, they can be joined by some species from local forests that are able to tolerate the savanna conditions. For example, the tropical oak *Quercus oleoides*, which occurs in dry forests in Mexico and Central America, can also be found in savannas there. Some regional patterns also exist in savanna tree floras. In Neotropical savannas, tree species richness

Figure 7.6 The mountainous *Pinus oocarpa* savannas of central Honduras

decreases northwards from relatively high levels in the Brazilian Cerrado, although numbers of other savanna plant species do not follow this pattern (Sarmiento 1983; see Table 5.4). The woody component of the southern African savanna zone is also richer than that of the northern zone (Menault *et al.* 1995). Scattered mountain savannas in both South-East Asia and Central America, which occur in steep terrain generally atypical of savannas, support single pine species: *Pinus merkusii* or *P. keysia* in South-East Asia and *P. oocarpa* in Central America (Figure 7.6). In addition, the lowland savannas of Central

America support *P. caribaea* var. *hondurensis* (Figure 7.3), a species that has become a major forest plantation tree throughout the tropics.

Like grasses and sedges, the woody plants in most savannas must tolerate a dry season that may be severe. However, more critical to their survival is an ability to tolerate both the burning and the grazing that is promoted by their existence in a matrix of graminoid vegetation. The way in which this is achieved will be discussed further below.

Savanna trees have the potential to create distinctive micro-environments around themselves because of their large size relative to plants in the surrounding herb layer. Their canopies provide shade which lowers solar radiation loading and evaporative stresses beneath them, as well as lowering soil temperatures (Belsky *et al.* 1993). Their larger size enables them gradually to establish an enriched nutrient cycle, in which greater quantities of nutrients, derived from biological nitrogen fixation, atmospheric accessions, inputs due to excreta from nesting or resting animals, or exploitation of greater soil volumes than grasses, are retained and recycled through the topsoil. This creates enriched soil bodies whose fertility levels can approach those found beneath tropical forest (Kellman 1979; Figure 7.7). Responses of the herb layer to these micro-environmental changes may be either positive or negative, depending on the balance between growth enhancement provided by greater soil fertility and less evaporative stress, and negative effects of competition with the trees. Belsky *et al.* (1993) have found increases in herb layer productivity beneath trees in arid East African savanna of 52 per cent and 95 per cent in areas receiving 750 mm and 450 mm of annual rainfall, respectively. Fertilization experiments showed nitrogen to be a limiting nutrient, and the authors attributed the increased growth beneath trees to a combination of more nitrogen and less evapotranspiration there. The reduced benefits to the herb layer at the wetter site suggest that growth enhancement may become negative in moist savannas.

Graminoid–tree balances

The apparently stable coexistence of two potentially incompatible growth forms in savanna communities, without competitive exclusion by one or the other, has evoked considerable theoretical interest. Walter (1973) assumed soil moisture to be the critical resource competed for and postulated a coexistence process based on differential rooting patterns between the two growth forms. Grasses were postulated to have an intensive root system allowing them to monopolize surface soil moisture resources, but leaving moisture that percolated to greater soil depth inaccessible to them. Trees, in contrast, were hypothesized to have deeper root systems, enabling them to gain access to these deeper soil moisture reserves.

Knoop and Walker (1985) have found some evidence in support of this hypothetical model in a dry southern African savanna. Although root systems

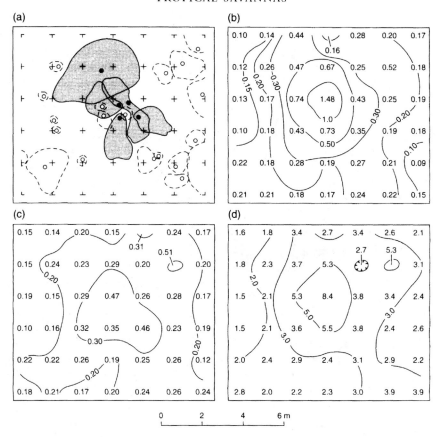

Figure 7.7 Enrichment of surface soil around an isolated small tree of *Byrsonima crassifolia* in the Mountain Pine Ridge Savanna, Belize. (a) tree canopy cover; (b) exchangeable calcium (meq. 100 g⁻¹; (c) exchangeable magnesium (meq. 100 g⁻¹); (d) percentage base saturation. Shaded area, *Byrsonima* canopy; dashed lines, pine sapling canopies. Excavation showed that the *Byrsonima* was very shallowly rooted, precluding accumulation of nutrients from deep soil sources

of trees and grasses were much less clearly differentiated than assumed in Walter's model, experimental removal of grasses led to increased soil moisture drainage to deeper soil layers at one site, although not at another. In contrast, Le Roux *et al.* (1995) have found no evidence to support the hypothesis in a humid West African savanna, where the roots of both trees and grasses were concentrated in upper soil layers, and most soil water for both life forms came from this zone. While the Walter model could, in theory, account for coexistence of mature trees and grasses under some conditions, it fails to address the issue of how tree seedlings could become established initially in the face of grass competition. Medina and Silva (1990) have postulated that savanna tree seedlings become established only episodically, during sequences of wetter years. While the absence of many tree seedlings in most savannas is one of their distinctive features, and contrasts with the abundance of these in tropical forests, there has been almost no detailed study of the population dynamics of savanna woody plants, and this hypothesis remains plausible, but untested.

Even if confirmed, this hypothesis about tree seedling recruitment raises the further enigma of why savanna tree populations do not increase gradually to displace the grass layer. While short tree longevity and a low rate of recruitment could account for this, most savanna trees appear to be of considerable longevity, often persisting clonally (Kellman 1979). A more plausible explanation is that tree recruitment is limited by either or both grazing and fire in savanna, and, as we shall see, there is considerable evidence that both can act to limit woodland development in these communities.

SAVANNA ENVIRONMENTS

The global pattern

The global pattern of distribution of savannas (Figure 6.1) shows no absolute correlation with any single environmental condition, but rather a tendency for savannas to be partially associated with certain broad-scale habitat conditions. The most obvious pattern is for savannas to occur in subhumid climates that are intermediate between the high-rainfall zones that are characteristically covered by evergreen tropical forest, and the very arid climates on the margins of subtropical deserts, which usually support a form of thorn woodland. This type of savanna environment is especially widespread in Africa. These subhumid areas normally receive between 500 and 1,500 mm of annual rainfall, with a pronounced dry season, and, in addition to savanna, may also support dry forest, sometimes in intricate landscape mosaics (Figure 7.12). The major exception to this pattern is the existence of some savannas in areas of much higher rainfall. Many of these wet-zone savannas, especially those in South-East Asia, appear to be the recent product of human activities (see below), but others appear to be of much greater antiquity (Figure 7.1).

A second major pattern is a tendency for savannas to occur upon soils that are in some way suboptimal for plant growth, by virtue of infertility or toxicity, or by physical constraints on root development, such as the existence of impenetrable soil horizons, or seasonal waterlogging. This association is most obvious in the Neotropics, where most savannas occur upon soils that are in some way limiting to plant growth. The Brazilian Cerrados occur on ancient shield rocks that have weathered intensely to form deep oxisols. While physically favourable for plant growth, these soils suffer from both extremely low fertility and aluminium toxicity (Lopes and Cox 1977a, b). In the Llanos, these nutritional and toxicity problems are further exacerbated by widespread plinthite (Figure 4.11). Other smaller Neotropical savannas also tend to be associated with specialized soil conditions. The Gran Sabana savannas occur primarily on sandstones of the Guiana Shield that have weathered to form quartzitic sands of extreme infertility, while the most extensive savannas in Central America (the Miskito, in eastern Honduras and Nicaragua) occur on deeply weathered quartzitic sands and gravels in a zone that experiences 2,600–3,500 mm of rainfall annually (Parsons 1955).

A final large-scale partial correlate of savannas is their tendency to be most common on landscapes that comprise extensive flat-terrain surfaces, with few topographic interruptions. Such flat terrain is often associated with especially intense soil development, and the existence of such phenomena as well-developed clay pans in ultisols, or plinthite accumulations, suggests some correlation between flat terrain and problematic soil conditions. However, we will argue below that this sort of terrain also provides optimum conditions for extensive spread of fire.

Correlations between each of the three habitat conditions that have been mentioned above (subhumid climates, problematic soils, flat terrain) and large-scale savanna distribution are asymmetrical, in that habitats possessing these characteristics can often be found supporting forest, rather than savanna, but it is rare to find savanna occurring naturally on sites that *do not* contain at least one of these conditions. This asymmetry suggests that certain conditions *predispose* an area to becoming savanna-covered, but do not necessarily require that this will happen. We will return to this issue in a later section, where we suggest a dynamic interpretation of the 'savanna anomaly'.

Habitat heterogeneities within savannas

Because of the extensive areas covered by savannas, it is not surprising that considerable habitat heterogeneity can exist within savanna-covered regions. In Africa, where savannas occur across a wide spectrum of rainfall conditions, a major distinction has been made between moist and arid savannas (Huntley 1982). The former occur upon poorer, more intensively leached soils, while the latter occur on more eutrophic soils which often contain calcium carbonate accumulations. Each tend to have a distinctive flora and assemblage of herbivores, the latter reflecting a major contrast in forage quality. Despite the better nutritional status of arid savanna soils, Breman and de Wit (1983) conclude that the low availability of nitrogen and phosphorus is a more serious limitation to plant productivity in them than is low rainfall, down to annual rainfalls of 300 mm in the West African Sahel. Scholes (1990) suggests that this pattern reflects the tendency for savanna production to comprise relatively short pulses of rapid growth following rainfall; while the *duration* and *frequency* of these growth spurts are controlled by rainfall, growth *rates* during the wet periods are still controlled by soil nutrient levels.

Figure 7.8 Complex local environmental variations in the Apure savannas, Venezuela. The low terrace in the middle distance is a sand-covered deposit of massive plinthite that supports scattered savanna trees where the plinthite outcrops. This landform gives way downslope to colluvial deposits supporting wet savanna, which in turn change to better-drained alluvium supporting gallery forest along the river

At more local scales, savanna community composition often changes along topographic gradients, which may be gentle, but can be sufficient to induce changes in soil properties, moisture supply and degree of flooding. In the dry savannas of Southern Africa, upland soils are often of low fertility, but lowlands contain more fertile soils (Scholes 1990). In the Apure savannas of the western Llanos, upland surfaces are covered by a sand sheet overlying massive plinthite which is covered by treeless grassland; this gives way on slopes to plinthite exposures where woody plants are more common, and then to a treeless wet seepage zone immediately upslope of narrow strips of gallery forest on river alluvium (Figure 7.8). Comparable topographic sequences have been described in the Cerrado by Eiten (1982), and variation in the density of the tree cover in this region has been shown to correlate with variations in soil chemistry (Lopes and Cox 1977b).

Finally, at very local scales, the activities of earth-burrowing invertebrates, notably termites and leaf-cutting ants, can create localized habitat differences to which savanna plants respond. Both termites and leaf-cutting ants excavate underground chambers and deposit large quantities of excavated soil on the surface, creating mounds that may reach 2 m in height and up to 20 m in diameter (Figure 7.9). These may reach densities of several hundreds per

Figure 7.9 A termite mound being colonized by woody plants, Gran Sabana savannas, Venezuela

hectare in some savannas (Spain and McIvor 1988; Figure 7.10). Mounds may persist even after abandonment by the colony, and form peristent small-scale habitat heterogeneities in savannas (Ponce and Da Cunha 1993). These colonial insects also transport large quantities of organic material to their nests, which can improve the nutrient status of the mound soil (Brener and Silva 1995). The response of savanna vegetation to mounds is most obvious in terrain subject to seasonal flooding, where mounds are preferentially colo-nized by woody species. However, even in terrain not subject to flooding, mounds may affect the distribution of herbaceous species (Spain and McIvor 1988).

HERBIVORES AND HERBIVORY IN SAVANNAS

Herbivory is a significant process in tropical savannas, as in all other tropical plant communities. Moreover, because a large proportion of savanna phytomass is leafy rather than woody, a much larger proportion of this is vulnerable to folivory than is forest phytomass. This serves to magnify the potential role of herbivores in the dynamics of savanna ecosystems, as well as to provide a major resource for human utilization by way of livestock husbandry. A major dichotomy exists in the form that herbivory takes in the savannas of the Neotropics and those of Africa. In the latter, herbivory is

185

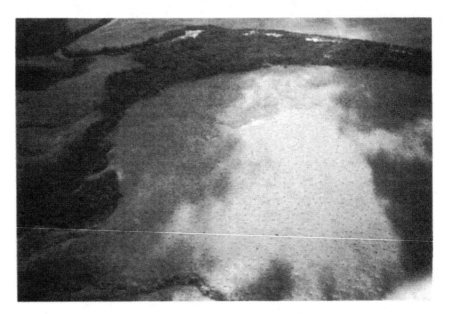

Figure 7.10 A termite mound field, Gran Sabana savannas, Venezuela

dominated by populations of large mammals, while in the former area, virtually no large native savanna herbivores exist. Redford and da Fonseca (1986) record only 11 endemic non-volant mammal species in the Cerrado savannas of Brazil, out of a total fauna of 100 species. Most of these mammals were found to be associated with gallery forests, rather than the savanna proper. A rich herbivorous biota of large mammals evolved in the Tertiary in South America, but these were concentrated primarily in the subhumid grasslands of temperate South America, not in the tropical savannas (Webb 1978). The North and South American temperate faunas intermingled after the Central American isthmian connection developed in the late Tertiary, and both faunas underwent mass extinctions in the late Pleistocene. However, there is no convincing evidence for the Neotropical savannas ever having supported large mammal populations comparable to those of the African savannas (Webb 1978). In Australia's northern savannas, kangaroos and wallabies form a large-vertebrate herbivore stratum, but do not achieve a regional biomass comparable to that of the African savanna herbivores.

While reasons for the paucity of large herbivores in the savannas of the Neotropics and Australia remain obscure, this absence has provided a major potentially unoccupied niche in these communities. McNaughton *et al.* (1993) have suggested that this niche may be being filled by leaf-cutting ants in the Neotropics, but insufficient data are as yet available with which to test this hypothesis. Anderson and Lonsdale (1990) have suggested that insects are probably major herbivores in Australian savannas, but here also there is a paucity of data. By analogy with African savannas, these authors consider grasshoppers to be the probable principal herbivores in Australia, with termites, despite their prominence, acting primarily as detritivores.

Most information on herbivory in tropical savannas has come from that by the large mammals of Africa. These tend to be classified as grazers and browsers, feeding respectively on the herbaceous and woody components of the savannas, with a smaller group of species playing both roles (Owen-Smith 1982). Browsers have less consumable phytomass available to them, but this is generally of persistently higher quality (especially in protein content) than is available in the herb layer. In contrast, grazers have much more phytomass available to them, but the quality of this is generally lower (Owen-Smith 1982) and varies seasonally. Herb-layer forage supports most wild and domesticated mammals in the tropics, and has been the focus of most research on herbivory in savannas. At a regional scale, above-ground productivity of this layer increases with annual rainfall (McNaughton *et al.* 1993), but accompanying this increased productivity is a general decrease in forage quality, especially its protein content (Breman and de Wit 1983; Figure 7.11). Moister savannas experience a brief period of high forage protein content early in the wet season, but this declines dramatically towards the end of the wet season and into the following dry season. In contrast, protein content of very arid grassland remains high even in the dry season. These seasonal patterns have important consequences for livestock production, as will be discussed further in Chapter 10.

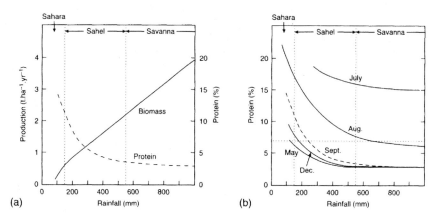

Figure 7.11 Changes in plant biomass and protein content along a moisture gradient in West Africa. (a) Mean rangeland production (t.ha⁻¹.yr⁻¹) and protein content (%) of this at the end of the wet season in September. (b) Mean protein content of rangeland biomass during different months. A minimum protein content of 7% is considered necessary to keep livestock in good condition

Source: Breman and de Wit 1983: Figures 3 and 4

The large herbivores of the African savannas are not simply passive users of the plant resource, but interact with this and can have major effects on the plant community. Most dramatically, elephant browsing can destroy tree covers and transform forest, or wooded savanna, to open grassland (Laws 1970), a process that is especially common when herd numbers increase and migration is constrained owing to encroachment by human populations. Other browsers can also affect woody phytomass, albeit less dramatically than elephants, and Bell (1982) suggests that this effect is most intense where forage quality is highest, leading to reduced tree densities and more open grassland on the most fertile soil areas in African savannas (i.e. the reverse of the Neotropical savanna pattern). Grazing also affects the morphology of the herb layer, with intensively grazed sites tending to develop short grass swards of high productivity and nutritional quality (McNaughton 1985; McNaughton and Sabuni 1988). Grazers have also been shown to be sensitive to the mineral nutrient content of forage, and to graze selectively in response to this both in space and in time (McNaughton 1988, 1990).

SAVANNA FIRE

Signs of previous fire, in the form of charred tree stems and grass tussocks or charcoal fragments, are virtually universal in tropical savannas, indicating the prevalence of this phenomenon in these communities. Functionally, fire may be thought of as an unselective 'herbivore/detritivore' because it consumes live and dead leaf tissues. During this process, much of the above-ground phytomass of the savanna community is consumed, and the nutrients contained in this are mineralized and deposited as ash (most nutrient elements), or volatilized (primarily nitrogen and sulphur). As in the case of conventional herbivory, this process can temporarily stimulate savanna productivity and provide a pulse of higher-quality foliage in the post-fire period.

The mechanics of fire

Burning involves the rapid oxidation of the carbon compounds that make up most plant tissues, involving consumption of oxygen and the release of carbon dioxide and the release as thermal energy of the chemical energy stored in fuels. In detail, the burning process involves the ignition of volatile gases released from biomass fuels due to pre-heating, which in turn serve to pre-heat adjacent fuels, by radiative and convective heat transfer, to the ignition point (Whelan 1995). As such, the burning process is a spatially step-wise reaction whose continuation is highly dependent on the characteristics of the available fuels. Not only must there be an adequate quantity of these to ensure the necessary pre-heating, but they must be in a sufficiently flammable state and be distributed with sufficient contiguity to ensure spatial propagation of the burn. Initiation of a savanna fire requires an ignition source which, in the absence of human populations, is almost always lightning; thunderstorms occurring towards the end of the dry season, which produce abundant lightning but little rainfall, are especially potent ignition sources. However, where human populations occur, ignition rates are dramatically increased by the activities of people.

Flammability of fuels is influenced primarily by their moisture content; when this is too high, heat is dissipated in evaporation of moisture from adjacent fuels, rather than in volatilization of flammable gases. There is also some evidence to suggest that higher contents of some mineral nutrients can reduce fuel flammability (Philpot 1970). Graminoid layers provide ideal fuels. They are composed primarily of living and standing dead leaf tissues which dry quickly, especially under dry-season conditions. They are well aerated, allowing a free flow of oxygen to the burning sites, and are usually relatively low in mineral nutrient content. Except in zones of very low rainfall, or after a recent fire or defoliation, the fuels are sufficiently continuously distributed to enable fire propagation. However, because of lower fuel loads, savanna fires do not achieve the high intensities that can occur in some forest fires, and consequently their spread is very sensitive to minor discontinuities in fuel or conditions that affect fire propagation. Among the latter, wind speed and slope are of primary importance. Wind greatly increases the lateral heat transfer that is essential to maintain a fire, and savanna fires usually burn rapidly downwind to form elongate burned strips, with later slow lateral spread in a cross-wind or upwind direction. Fires burning upslope also progress more readily than those on flat surfaces as a result of upward convective heat transfer, while fire movement downslope is inhibited owing to convective heat transfer being directed away from adjacent fuels (Van Wagner 1988). As a result of both slope and wind effects, savanna fires tend to propagate most readily on extensive flat terrain where dry-season trade winds are consistent and fire propagation rates are unaffected by abrupt changes in slope. In contrast, savanna fires propagate much less consistently in topographically broken terrain, and in areas where savanna and forest form a landscape mosaic, transitions from one to the other frequently occur at abrupt changes in slope (Figure 7.12).

Information on the frequency of savanna fires is rich in anecdote but often lacks reliable data. This relates both to the absence of continuous and consistent observations of savanna fires in any region, and to the very patchy burn patterns and the difficulties of recording these accurately. In one attempt to overcome these difficulties, Hutchinson (1977) used fire maps prepared by forestry personnel over an 11-year period to calculate an average fire recurrence interval of 18 years for the Mountain Pine Ridge savannas of Belize, with approximately two-thirds of the fires being lightning-initiated. This is a relatively isolated savanna, and the presence of human populations would almost certainly increase fire frequency dramatically. Where savanna fire is

Figure 7.12 A boundary between savanna and dry forest, Guri, south-eastern Venezuela; view looking west. Savanna changes to forest beyond the ridge crest, where westward-moving fires lose intensity and move more slowly downslope

used intentionally, such as to improve the forage for livestock, it becomes a virtual annual dry-season event in areas where rainfall is sufficient to produce an effective fuel load in one growing season (Cahoon *et al.* 1992).

Savanna plant response to fire

The basal meristems of grasses and other graminoids are rarely damaged by the low-intensity fires that are characteristic of savannas, and redevelopment of leaf tissue from these meristems usually begins within days of a savanna fire (Figure 7.4). The flowering of many graminoid species is also stimulated by fire, and mass flowering of the savanna herb layer is common in the weeks following a burn. The quantities of nutrients released as ash during a burn are modest, so these are normally retained by adsorptive processes in the topsoil of savannas, after being mobilized following rainfall (Kellman *et al.* 1985), and are reabsorbed by the rapidly regrowing herb layer (Coutinho 1990). Redevelopment of phytomass in the graminoid layer is usually rapid. In the Central Llanos of Venezuela, at a rainfall of approximately 1,250 mm.yr^{-1}, 230–730 g.m^{-2} accumulates in the first year after a fire, with asymptotic levels of 1,200 g.m^{-2} achieved within four years of a burn (Sarmiento 1984).

Figure 7.13 Resprouting after fire by *Miconia albicans*, a common Neotropical savanna shrub. The stems of this shrub are killed by even a low-intensity fire, but it resprouts prolifically from dormant buds located at the soil surface around the root collar

Savanna trees and shrubs are potentially much more vulnerable to fire as a consequence of having exposed meristems, but the existence of populations of these life forms in savannas indicates that at least some species can tolerate periodic burning. Three characteristics appear to be critical to fire tolerance in these plants: (i) the ability to store carbohydrate reserves in root tissues; (ii) for larger plants, the development of a sufficiently thick bark to protect cambial tissue from lethally high temperatures during a fire; and (iii) the ability to resprout adventitiously from cambial tissue that survives a fire. Successful establishment of tree and shrub seedlings in savannas requires a fire-free interval that is sufficiently long to enable these to grow to a fire-tolerant size. For *Miconia albicans* and *Clidemia sericea*, two common Neotropical savanna shrubs, this interval was estimated to be 13 and 7 years respectively (Miyanishi and Kellman 1986a). Fire occurring after these intervals destroyed above-ground tissues in both species, but both were able to resprout adventitiously from root collars (Figure 7.13), drawing on stored carbohydrate reserves (Miyanishi and Kellman 1986b). In both species, accumulation of adequate reserves required at least two fire-free years, but beyond this, minimum stable populations could be achieved at a wide range of fire recurrence intervals (Miyanishi and Kellman 1988). Although comparable studies have not been conducted on other savanna trees and shrubs, it seems probable that these also are recruited episodically during prolonged fire-free intervals, as these populations often exhibit even-sized structures (Frost and Robertson 1987), and increase recruitment when fires are suppressed (see below). As successful recruitment may also require a sequence of 'wet' dry seasons (Medina and Silva 1990), it is probably not surprising that tree and shrub populations in savannas normally are found only at low densities.

The above-ground tissues of larger savanna woody plants are not normally destroyed by fire; some or all of the foliage and small branches may be killed, but the stem and larger branches survive and resprout adventitiously (Figure 7.14). The times and fire-free intervals necessary to achieve this stage of fire tolerance are unknown, but probably are long, given the poor soil conditions and slow growth rates of trees in these situations. This provides a further bottleneck to successful savanna tree recruitment that helps to explain the low densities of these trees' populations in savannas. The characteristic gnarled branching pattern of many savanna trees can probably be attributed primarily to multiple adventitious sprouting episodes after savanna fires and an absence of inter-tree competition for light; many of these trees assume a 'normal' shape when growing on the edges of gallery forests.

Fire at savanna–forest boundaries

Abrupt boundaries between savanna and forest are characteristic wherever these two communities come into contact. This occurs at the edges of the gallery forests that occur in many savannas (Figure 2.7), and where savanna and dry forest form landscape mosaics (Figure 7.12). The abruptness of these boundaries can be attributed to the intolerance of fire by most forest plants, and the inability of graminoids to survive beneath a closed tree canopy; the boundary represents a point at which most fires cease to burn, and is often associated with abrupt changes in slope (Figure 7.12).

The failure of most savanna fires to enter forests, despite the large fuel loads that exist in these communities, can be attributed to both characteristics of the fuel and a protected forest micro-environment. In contrast to the ready flammability of savanna graminoid layers, forest fuels are composed of compact and poorly aerated litter layers that contain relatively high nutrient contents. These dry much more slowly and ignite much less readily than savanna fuels, even when exposed to savanna conditions (Biddulph 1997), and when protected by a forest canopy may remain in a non-flammable state for much of the time. For example, Uhl et al. (1988c) were unable to initiate burns in closed evergreen forest in Venezuela, even after 41 rainless days, and estimate that atmospheric humidities sufficiently low to allow fuel ignition (<65 per cent relative humidity) may occur less frequently than 1 day each year in the forest understorey. Recently created tropical forest boundaries tend to 'close' rapidly owing to foliage development by sub-canopy plants (Williams-Linera 1990), and at long-established boundaries populations of fire-tolerant savanna trees often form a deep and

Figure 7.14 The common Neotropical savanna tree *Byrsonima crassifolia* resprouting epicormically after being damaged in a savanna fire

highly insulative boundary zone (Kellman *et al.* 1994). Thus only where forest canopies are opened artificially, such as by logging, are fire incursions frequent and potentially threatening to the forest community (Uhl and Buschbacher 1985). Elsewhere, most savanna fires tend to die out at the existing forest–savanna boundary and effectively preserve its abruptness.

Savanna fires may occasionally enter adjacent forests during especially dry periods, but these intrusions do not normally destroy the forest–savanna

Figure 7.15 Remnant tree stumps from a former forest cover, Gran Sabana savannas, Venezuela. Older persons in the indigenous Pemon population of the savanna report that this and many other forests in the area were destroyed by slowly burning root mat fires in a very dry period during 1929–30, some of which burned for more than a year

boundary; most fire intrusions are narrow tongues that consume litter layers and destroy smaller understorey plants, but do not destroy canopy trees, although these may be scarred (Kellman and Meave 1997). However, two conditions appear to promote retreat of forest boundaries in response to fire: (i) where savanna fire frequencies are increased greatly, repeated fire-scarring of canopy trees may eventually cause their death, and the opening of the forest understorey to graminoid invasion (Kellman *et al.* 1994); (ii) when forests contain deep root mats, slowly burning fires may enter these during severe droughts, and destroy large areas of forest (Figure 7.15).

Savanna response to fire suppression

Observations in several fire-protected savanna areas throughout the tropics have shown that fire cessation promotes the invasion of woody plant species (Menault 1977; San José and Fariñas 1983; Swaine *et al.* 1992). The initial stages of the process usually involve an increase in the populations of woody savanna plants; these may eventually form a closed thicket (Figure 7.16). Beneath these, the graminoid layer is gradually excluded and some forest tree seedlings begin to establish. These seedlings represent a subset of the forest

Figure 7.16 A thicket of woody savanna plants and young forest trees, representing a forest invasion nucleus in pine savannas in Belize, where fires have been suppressed for several decades

flora that are able to disperse into the savanna and are capable of establishing in the modified, but still stressful, conditions created by the savanna 'nurse' trees (Kellman 1985a; Kellman and Miyanishi 1982). On the nutrient-poor soils of Neotropical savannas, slow growth appears to be especially effective at enabling these seedlings to avoid nutrient exhaustion (Kellman 1989).

This process presumably happens with some frequency at forest–savanna boundaries, at places that may have avoided fire for several decades (e.g. Mariotti and Peterschmitt 1994). It must also occasionally have happened more extensively, to enable widespread forest re-establishment in savanna areas, such as that which took place in the late Holocene (see Chapter 2). A critical stage in the process is likely to be the point of canopy closure by woody invaders. At this stage, graminoids are excluded, fuel characteristics change, and a more protected microclimate reduces fuel drying rates; a much reduced fire probability can be expected beyond this point. The time necessary for this to happen, relative to the fire recurrence interval for the savanna, is likely to be critical to the completion of the process. Where canopy closure is slow, as it is likely to be on soils of low fertility, a reoccurrence of fire is probable, returning the site to a savanna state. In contrast, where closure requires less time than the fire recurrence interval, transition to forest becomes much more probable (Kellman 1984).

Figure 7.17 Destruction of *Pinus oocarpa* in mountain pine savannas, Belize, where fire suppression for several decades led to a large accumulation of fuels that promoted a severe wildfire

Fire management in savannas

The natural prevalence of fire in tropical savannas, and its importance to the functioning of these systems, implies that any attempt to manage savanna-covered areas, whether for livestock rearing, forestry, watershed protection or conservation, is unlikely to be successful without intentional manipulation of this phenomenon. Complete fire suppression in a nature reserve, even when practicable, is likely to ensure encroachment by woody plants. When fire suppression is used to ensure tree recruitment in pine savannas, but not followed by a prescription burning programme, fuels accumulate and complete loss of the pine stand in a subsequent wildfire usually follows (Figure 7.17). Similar approaches to watershed protection in savanna areas are likely to produce low-frequency fires of high intensity that result in catastrophic runoff and erosion. Many examples of effective fire management in tropical savannas now exist (e.g. Van Wilgen *et al.* 1990); unfortunately, many policy-makers are reluctant to go beyond fire suppression, which is unlikely to be a wise management tactic in these fire-prone systems.

DERIVED SAVANNAS

While there is much evidence for the antiquity of tropical savannas, there is also evidence that a proportion of these may have been derived more recently, as a result of human activities. For example, in recently derived savannas, tree stumps may still exist (Figure 7.15). In others, soil carbon isotope signatures may indicate a relatively recent origin from forest (e.g. Schwartz *et al.* 1986), and a savanna flora composed of relatively few weedy grasses, rather than a richer and more endemic flora, is suggestive of a recently derived savanna. We may distinguish two sorts of derived savannas: those produced by recent expansion of a pre-existing savanna, and savannas newly created in otherwise forest-covered landscapes.

The resistance of forest to savanna fire incursions makes savanna expansion due to fire spread an unlikely occurrence under normal fire recurrence intervals. However, an increased frequency of fires can lead to a gradual degradation of forest edges, and in exceptional circumstances can lead to sudden large savanna extensions (Figure 7.15). Most of the savannas of continental South-East Asia are considered to have been derived from the expansion of limited 'core' areas of savanna into dry forests, under increased fire frequencies induced by human populations (Blasco 1983; Stott 1990). Savanna expansion can also be speeded by human activities that open a forest canopy in areas adjacent to savannas, making forest fuels more flammable. Logging of forest adjacent to savannas (or pastures) can have this effect, and the creation of shifting cultivators' fields at forest edges often results in the transformation of the abandoned site to savanna owing to fire incursion (Figure 7.18). A form of savanna expansion caused indirectly by people is that produced by an over-concentration of elephants induced by human population increases in Africa (Laws 1970).

Creation of entirely new savanna in a forested landscape normally requires an extensive canopy opening that persists for sufficient time to permit entry of fire. In the absence of human activities, hurricanes provide the only mechanism of extensive forest canopy opening that occurs with any frequency. While some post-hurricane fires have been reported (e.g. Wolffsohn 1967), and may have initiated many of the eastern coastal savannas of Central America, rapid forest recovery from hurricane damage is the more normal pattern (e.g. Vandermeer *et al.* 1995). However, the presence of human populations, especially those practising shifting agriculture, results in a much more pervasive form of forest canopy opening, and creation of derived savannas.

The best-documented, and apparently most widespread, example of savanna creation by this means comes from the moister areas of South-East Asia, and is associated with the population biology of one especially aggressive weedy grass: *Imperata cylindrica*. *Imperata* is a weed of shifting cultivators' fields throughout this region, and difficulty in controlling this species is a

Figure 7.18 An abandoned shifting cultivator's field that was created in forest imme-diately adjacent to savanna, Gran Sabana, Venezuela. The early second-growth vege-tation, which contains many grasses and sedges, makes this site vulnerable to incursions of fires originating in the adjacent savanna for several years after site abandonment

primary reason for field abandonment (Kellman 1969). This and other weeds are normally suppressed by fast-growing pioneer trees that develop from a seed bank after fields are abandoned (Kellman 1970), but when this seed bank is depleted owing to prolonged field use, *Imperata* may become domi-nant. While protection of these *Imperata* grasslands from fire for extended time periods normally allows tree reinvasion, this grass possesses the flam-mability characteristics of grasses generally, and the sites remain fire-prone. In the presence of human populations, a high ignition frequency of these grasslands is assured, either intentionally or accidentally, which results in the elimination of any establishing trees and may lead to the gradual expansion of the grasslands and their coalescence with others (Seavoy 1975). Most of the extensive *Imperata cylindrica* grasslands of the South-East Asian archi-pelago have probably originated in this way.

In other regions, less aggressive weedy grass species can have a similar effect, although the speed of expansion of the derived grasslands may be less dramatic. Creation of an isolated system of grasslands in montane forest in Peru by this means has been described by Scott (1978). Most of the grass-lands of India and Sri Lanka are considered to be derived (Pemadasa 1990). The persistence of most derived savannas depends on a sufficiently high fire frequency, and this is assured in the presence of human populations. It is also facilitated by expansion of the patches of derived savanna, as most ignition derives from spread from adjacent savanna, rather than independent ignition. Isolated small derived savannas may be expected to have a low probability of persisting if human populations vacate the region.

A DYNAMIC INTERPRETATION OF THE 'SAVANNA ANOMALY'

Savannas represent a plant community type in which trees fail to establish a closed canopy, and thus allow the existence of a graminoid layer beneath them. Explanations for the quasi-stable balance between the two life forms have usually been sought in conditions which are thought to inhibit tree development (e.g. Sarmiento and Monasterio 1975) or in sharing of a critical moisture resource (Walter 1973). However, the very fluid relationship between forest and savanna that has been revealed by the palaeoecological record, and shorter-term observations of savanna response to variations in fire and grazing frequencies, suggests that a much more dynamic interpretation of savannas' existence is required.

We propose that the interaction between graminoid and tree layers in savannas goes well beyond simple competition for moisture, and involves the graminoid layer acting as an 'attractor' of two other phenomena that inhibit tree seedling recruitment: fire and grazers. As long as a graminoid layer sufficient to carry fire or attract grazers can persist, tree seedling recruitment will be severely limited. However, an increase in tree populations to the point of canopy closure will effectively eliminate the graminoid layer and result in a fundamental switch of the community to a stable forest state that, in the absence of forest clearing, will resist reversion to savanna. We therefore propose that areas in which the rate of tree growth and canopy closure is slow relative to the return time of events inhibiting tree seedling recruitment will be inherently predisposed to savanna persistence once such a community forms initially. In contrast, areas without such growth limits will tend to return to forest following disturbances. Environmental conditions that may be expected to slow tree growth rates are moisture deficits in all parts of the tropics, and problematic soils, especially in the Neotropics (Kellman 1984).

This proposal does not *require* that savannas exist at sites where these conditions prevail, but simply states that their *probability of existence* increases on these sites. Nor does the proposal suggest that any single factor is of paramount importance in 'explaining' the existence of savannas, but envisages a suite of potential factors that may interact with fire and grazing to maintain savannas once these form. As such, the proposal appears to accommodate the multiple partial associations between savannas and certain environmental conditions that were discussed earlier. Because a dynamic equilibrium is proposed, a change in savanna frequency and extent is to be expected when any of the interacting variables changes. Such changes are most likely to occur in the frequencies of fire and grazing, rather than in soil or climate conditions, and the great increase in extent of savannas at the expense of forest in the past century is entirely compatible with a dramatic increase in fires due to increasing human populations.

8

TRADITIONAL TROPICAL AGRICULTURAL SYSTEMS

Human populations throughout the world depend almost completely on agriculture to supply their food. Agriculture also supplies a range of other important non-food items, such as fibres, to human societies. The well-being of human populations in the tropics depends, in the first instance, on the existence of successful forms of agriculture in these environments. In most tropical countries, human populations are fed primarily by local agriculture. This provides a degree of autonomy to these societies, but also makes the provision of an adequate human diet primarily a local responsibility. Successful food agriculture also permits the development of cash cropping, and produces a surplus that permits the development of non-agricultural activities.

'Successful' agriculture is often taken to be synonymous with 'productive' agriculture – agriculture that yields large quantities of products per hectare and can thus support large numbers of people, directly or indirectly. While high productivity is an important component of successful agriculture, we choose to define this term more broadly, requiring not only that it be productive, but that it provide a nutritionally adequate diet, and be sustainable both economically and environmentally. By environmental sustainability, we deem that the agriculture must be capable of being practised indefinitely without ongoing deterioration in the environmental matrix in which it exists. This is especially critical in relation to the soil resource on which agriculture depends most directly.

In this chapter we examine various forms of agriculture that exist, or have existed, in tropical environments, without external subsidies of energy or materials. We term this 'traditional agriculture'. In Chapter 9 we go on to discuss the prospects for intensifying agricultural production in the tropics without precipitating environmental degradation. Most traditional agriculture in the tropics is focused on the production of food for human consumption, so we begin by sketching human dietary needs. We then go on to review the origins of agriculture in the tropics and the major forms of traditional agriculture in these regions.

HUMAN DIETARY NEEDS

Humans are heterotrophic organisms that rely on the consumption of other organisms to meet their dietary needs. However, they are omnivorous, and can make use of a broad range of plant and animal tissues as food. Basic human dietary needs can be categorized into: (i) an energy source; (ii) protein; (iii) a variety of vitamins derived from organic sources; and (iv) some inorganic mineral nutrients, notably calcium, iron and phosphorus. Most diets that contain adequate quantities of protein and sources of energy also contain sufficient vitamins and minerals, so most attention in human nutrition has been focused on protein and energy needs of the human body, and the ability of various foods to fulfil these.

Energy for sustaining human body processes can be derived from three sorts of tissues: (i) plant carbohydrates; (ii) fats and oils from plants and animals; and (iii) plant and animal protein. However, energy extraction from protein is an inefficient process, and the human body resorts to this source only when others are unavailable (Chrispeels and Sadava 1977). Moreover, not all plant carbohydrates are available to humans as an energy source. The human digestive system can make use of sugars and starches (a form of plant energy storage tissue), but is unable to utilize cellulose, which makes up cell walls and much of the bulk of plants. Human energy requirements are usually measured in kilojoules (kJ, = 0.24 Calories) per person per day, and have been estimated for persons of various age, sex and level of activity (Table 8.1). Energetic requirements of the human body increase with maturation, and vary with sex and level of activity. While much of the energy consumed is used in basal metabolism, strenuous physical activity requires additional energy intake, although little extra protein. In the absence of an adequate energy intake, the human body catabolizes its own fat reserves, and ultimately protein, as an energy source, leading to progressive emaciation and ultimately death. Most human populations in the tropics derive the great majority of their energy requirements from plant tissues. These include a wide array of starchy root tissues, sugars, and starches and fats in fruits and seeds. This variety of alternative energy sources provides for a high degree of flexibility in meeting human energy requirements.

Somewhat less flexibility is available in meeting human protein requirements. Proteins are complex organic molecules that are present in all organisms, and have nitrogen as an important constituent (hence the large nitrogen requirements of plants; Chapter 4). The total protein requirement of humans increases rapidly to adult levels during maturation, and is especially high in pregnant and lactating women (Table 8.1). Protein deficiency induces a variety of symptoms, such as bloated abdomens and bleached hair, that are known collectively by their West African name 'kwashiorkor'; children are especially vulnerable to this after weaning.

The basic structural units of proteins are 20 different amino acids, 12 of

Table 8.1 Recommended daily intake of energy and protein for human populations in Africa

Population	Energy (kJ)	Protein (g)
Children		
<1yr	3,445	14
1–3yr	5,710	16
4–6yr	7,685	20
7–9yr	9,200	25
Adolescents		
Male, 10–12yr	10,920	30
Male, 16–19yr	10,835	30
Female, 10–12yr	9,870	29
Female, 16–19yr	8,275	29
Adults		
Male, sedentary	9,700	31
Male, active	10,625	31
Male, very active	12,475	31
Female, sedentary	7,100	24
Female, active	7,895	24
Female, very active	9,280	24
Female, pregnant	+1,470	33
Female, lactating	+2,310	41

Source: Latham 1979: Table 1

which the human body can synthesize from simpler constituents, leaving eight which must be ingested. These have been termed the eight 'essential' amino acids. For this reason, the human body requires not only sufficient total protein intake, but that this include sufficient quantities of each of the eight essential amino acids. Deficiency of one or more of these can induce protein malnutrition. Only hens' eggs contain the eight essential amino acids in the correct proportions for human diets, and egg protein is usually used as a datum against which to rank other protein sources (Table 8.2). In general, animal products rank highly as human protein sources, while most plant proteins are deficient in one or more essential amino acid. Clearly, this presents potential nutritional difficulties to populations whose diet is primarily composed of plant products. It requires either that an appropriate blend of plant products be consumed, or that plant protein be supplemented by small quantities of animal protein. In general, the latter procedure is considered to be more easily achieved (Chrispeels and Sadava 1977).

Tropical crop types

While a large number of different plant species are used as sources of food in the tropics, most populations (except pastoralists) rely on one or more of a

Table 8.2 Protein content and content of the eight essential amino acids in common tropical foods

Food	Isoleucine	Leucine	Lysine	Methionine	Phenylalanine	Threonine	Tryptophan	Valine	Protein %	Protein Score
Egg	393	551	436	210	358	320	93	428	12.4	100
Beef	301	507	556	169	275	287	70	313	17.7	69
Chicken	334	460	497	157	250	248	64	318	20.0	64
Pork	356	563	625	188	288	319	85	388	11.9	71
Fish	299	480	569	179	245	286	70	382	18.8	70
Cow's milk	295	596	487	157	336	278	88	362	3.5	71
Cheese	339	661	553	188	337	257	77	494	18.0	77
Rice	238	514	237	145	322	244	78	344	7.5	57
Maize	230	783	167	120	305	225	44	303	9.5	41
Sorghum	245	832	126	87	306	189	76	313	10.1	31
Millet	256	598	214	154	301	241	122	345	9.7	53
Cassava	175	247	259	83	156	165	72	209	1.6	42
Sw. potato	230	340	214	106	241	236	107	283	1.3	53
Yam	234	404	256	100	300	225	80	291	2.4	56
Taro	219	460	241	84	316	257	88	382	1.8	53
Soybean	284	486	399	79	309	241	80	300	38.0	66
Peanut	211	400	221	71	311	163	65	261	25.6	51
Bean	262	476	450	66	326	248	63	287	22.1	63
Chickpea	277	468	428	65	358	235	54	284	20.1	54
Coconut	244	419	220	120	283	212	68	339	6.6	55
Pumpkin	341	625	397	125	391	319	81	391	4.0	81
Tomato	112	169	175	37	112	138	52	131	1.1	27

Source: FAO 1970

Note: Amino acid contents are expressed as mg.g⁻¹ of protein nitrogen. Hens' eggs are considered to have a perfect blend of amino acids, and other foods are scored relative to this as a protein source

Table 8.3 Energy and protein contents and yields of major tropical food crops

Crop	Average tropical yield ($t.ha^{-1}$)	Edible energy value ($MJ.kg^{-1}$)	Edible energy yield ($MJ.ha^{-1} \times 10^3$)	Protein content (%)	Protein yield ($kg.ha^{-1}$)	Crops per year*
Starchy crops						
Cassava	9.63	6.3	50.3	1.6	154	1
Sw. potato	5.86	4.8	24.8	1.6	94	2
Yams	7.00	4.4	26.2	2.0	140	1
Banana	13.00	5.4	41.4	1.1	143	1
Cereals						
Padi rice	2.69	14.8	27.9	7.5	202	2
Maize	1.59	15.2	24.2	9.5	146	2
Sorghum	0.87	14.9	11.7	10.5	91	2
Millet	0.65	15.0	9.8	10.5	68	2
Legumes						
Soybean	1.34	18.0	24.1	38.0	509	2
Peanut	0.89	22.9	20.4	25.5	227	2
Beans				22.0	132	2
Chickpeas				20.0	132	2

Source: Norman *et al.* 1995

Note:
* Potential number of crops, assuming no moisture or temperature limitation

limited number of staple crops for sustenance. The nutritional characteristics and yields of these crops are therefore of major importance in the diets of tropical peoples. The major tropical crops may be grouped into three broad categories: (i) high-bulk starchy crops; (ii) cereals (grass seed); and (iii) legume seeds (also called 'pulses'). The average energy and protein contents and yields of the most common tropical crop species are provided in Table 8.3.

The starchy crops, which are primarily root crops, stand out as providing high energy yields, with that of cassava being especially large (Table 8.3). However, the protein content of these crops is uniformly low. Thus, while they may be capable of a moderate protein yield per hectare, reliance on these crops as a protein source would require the consumption of a very large quantity of them. Other constraints are their relatively slow growth, which limits the potential for multiple cropping, their bulk, which poses problems in transportation, and a limited storage life relative to seed crops. These crops must therefore be viewed primarily as sources of dietary energy, which need to be supplemented with plant or animal protein for an adequate diet to be achieved.

The energy yields of cereal grains generally fall below those of the main starchy crops (Table 8.3), but their protein content is much higher, and all are capable of being potentially double-cropped, and of being easily stored.

Of the major cereals, padi (irrigated) rice stands out for its high yields of both energy and protein, the latter tending to be of moderately high quality (Table 8.2). While the protein content of these crops is qualitatively imperfect (Table 8.2), much less protein supplementation would be needed in the diets of populations that rely on these cereals as their principal foodstuffs.

The outstanding feature of leguminous crops is their very high protein contents, many of which equal, or exceed, those of animal products (Table 8.2). Soybeans have an especially high protein content, and a protein yield per hectare that is more than double that of any other major crop (Table 8.3). Given the relatively high quality of the protein of this crop (Table 8.2), it represents a potentially important source of supplemental protein for human populations in the tropics.

The type of processing applied to a crop before consumption varies widely, and can have a major impact on its quality. In some instances, specialized processing is necessary to remove toxic compounds from the crop. For example, the bitter varieties of cassava contain hydrogen cyanide, which must be leached prior to consumption. In other instances, non-essential processing can reduce the quality of a crop. For example, polishing rice seed removes an outer layer that is high in the vitamin thiamine, and this practice can lead to the nutritional disease beri-beri in populations that rely heavily on a rice diet. A particularly important processing procedure in parts of the Neotropics where maize is a staple food is the soaking of this cereal in an alkaline solution prior to the making of tortillas. This serves to increase the availability of lysine and the vitamin niacin, and to enhance the balance of the essential amino acids (Katz et al. 1974). An especially thorough review of traditional crop-processing techniques is provided by Stahl (1989), to which the reader is referred.

ORIGINS OF AGRICULTURE

Throughout much of prehistory, human populations in the tropics and elsewhere subsisted by hunting and gathering products from natural ecosystems. However, signs of incipient agriculture appear in the archaeological record at about 10,000 years BP in both the New World and Old World tropics, and in the semi-arid subtropical environments adjacent to these regions. The transition from hunting and gathering societies to agricultural ones was once thought to be a relatively sudden 'revolutionary' process, but current data indicate a much more gradual transition between the two, in which agriculture coexisted with hunting and gathering for prolonged time periods, and only gradually replaced it (Harris 1989). The essential feature of agriculture is the propagation and tending of plants, as opposed simply to the collection of their products. This process results in 'domestication' of the plant species cultivated, and has produced varying degrees of change in their genetic structure. The distinctive flora of different tropical regions

Table 8.4 Probable areas of origin of the major tropical crops

Crop	Probable area of original domestication
Starchy crops	
Cassava (*Manihot esculenta*)	N. South America and/or Central America
Sweet potato (*Ipomoea batatas*)	N. South America or Central America
Yams: (*Dioscorea alata*)	South-East Asia
(*Dioscorea rotundata*)	West Africa
(*Dioscorea cayenensis*)	West Africa
Bananas (*Musa* spp.)	South-East Asia/Melanesia
Taro (*Colocasia esculenta*)	South-East Asia
Cereals	
Rice: (*Oryza sativa*)	South and South-East Asia
(*Oryza glaberrima*)	West Africa
Maize (*Zea mays*)	Southern Mexico
Sorghum (*Sorghum bicolor*)	Sub-Saharan Africa
Millet (*Pennisetum glaucum*)	Sahel region of Africa
Legumes	
Soybean (*Glycine max*)	China
Peanut (*Arachis hypogaea*)	N. South America
Beans (*Phaseolus vulgaris*)	Andean South America and Central America
Chickpeas (*Cicer arietinum*)	Mediterranean
Fruits and nuts	
Avocado (*Persea americana*)	Central America
Cacao (*Theobroma cacao*)	N. South America or Central America
Cashew (*Anacardium occidentale*)	N. South America
Citrus (*Citrus* spp.)	China and South-East Asia
Coconut (*Cocos nucifera*)	Pan-tropical
Coffee (*Coffea arabica*)	Ethiopia
Mango (*Mangifera indica*)	India
Pineapple (*Ananas sativus*)	N. South America
Other crops	
Cottons (*Gossypium* spp.)	Various
Rubber (*Hevea brasiliensis*)	Amazonia
Sugarcane (*Saccharum officinarum*)	New Guinea

offered unique suites of potential domesticates, resulting in very different groups of crop species being domesticated in these places (Table 8.4). These appear to have been brought into domestication at different times throughout the past 10,000 years, with some species, such as rubber, having been domesticated only in the past century.

The incipient agricultural context in which crop domestication took place is unknown and has been the subject of much debate (see Harris and Hillman 1989). Relatively sedentary populations are often assumed to be necessary for the experimentation that would be needed in incipient agriculture, and appear to have existed in at least some places, such as the Basin of Mexico (Niederberger 1979). However, Groube (1989) suggests that promotion of wild plants by forest management techniques may have been the main precursor to agriculture in New Guinea, and rudimentary forms of forest 'caching' of root crops in forest micro-sites have been described among contemporary hunters and gatherers in the Philippines (Griffin 1989). Neither would require sedentism. It therefore seems quite probable that a variety of different forms of incipient agricultural experimentation were conducted in different regions, varying with the habitats present and potential domesticates available.

There has also been much debate about the factors that encouraged the transition from hunting and gathering to agriculture. While agriculture can support much higher population densities than hunting and gathering, Cohen (1977) has argued that it supplies a nutritionally inferior diet and should therefore be avoided by human societies. On this basis, he suggests that an expanding human population and a deteriorating hunting and gathering resource may have driven the relatively synchronous worldwide move to agriculture after 10,000 BP. Palaeopathological studies of human skeletal remains generally confirm a decreased quality and length of life during the transition from hunting and gathering to agriculture (Cohen and Armelagos 1984), and appear to support the Cohen hypothesis. A contrasting interpretation of the development of agriculture in tropical rain forests is provided by Piperno (1989), who suggests that the food energy sources available in these forests, rather than protein supplies, are likely to be limiting on human populations. Piperno argues that these energy sources are only sporadically available in space and time in tropical forests, a conclusion supported by Milton's (1984) study of nutrition among an Amerindian hunting and gathering society in Amazonia. In the rain forests of Panama, increased human populations appear to have followed, rather than preceded, the development of agriculture; this appeared in the area at about 5,000 BP using maize, a crop that had been domesticated in the semi-arid highlands of southern Mexico, and later diffused southwards (Piperno 1989). The contradictions between these alternative interpretations may reside more in the spatial scale used by each author than in fundamental differences in the underlying process. Cohen's (1977) hypothesis envisages a 'macro' spatial scale and assumes the coexistence of alternative life support systems in the same region, between which choices are made. Piperno's (1989) work refers to a small spatial region in which no agricultural alternative existed prior to its introduction from an external source.

The adoption of agricultural techniques involving the propagation of, and care for, certain preferred plant species resulted in a fundamental shift in the sorts of natural selection operating on these species. Preferred plants were not only propagated, but almost certainly promoted by being freed of competitors, and otherwise favoured by rudimentary forms of irrigation and fertilization. The genotypes responding best, in terms of crop yields, to these treatments were probably the ones selected for further propagation, setting in motion an evolutionary process of selection for yield, an increasing dependence on the favourable conditions provided by humans, and a decreasing ability to survive in their absence. This selection of crop genotypes showing high yield responses to large inputs continues to the present (see Chapter 9), and it is therefore not surprising that most crop plants, relative to their wild relatives, are competitively inefficient, nutrient-demanding, moisture-requiring and intolerant of soil toxic conditions, such as high aluminium levels.

However, accompanying this selection for dependence on human technologies was also selection by early farmers for a remarkable diversity of varieties within cultivated crop species. For example, approximately 50 varieties ('land races') of maize have been recognized in Mexico, where this crop was domesticated, and a further 30 in Peru, where secondary diversification took place (Wilkes 1989). This varietal diversity not only has produced great ecological flexibility among crops, allowing them to be grown in a great variety of environmental conditions, but also has provided potential flexibility in response by farmers to such conditions as droughts and pest outbreaks. Farmers can spread risks by planting locally a range of varieties of differing susceptibility to these potential catastrophes (Clawson 1984). Unfortunately, much of this varietal diversity, which is preserved in the fields of traditional farmers, is in danger of being lost under the impetus to adopt uniform new high-yielding varieties of the same crop. We will return to this issue in Chapter 9.

The process of domestication, operating synchronously in different parts of the tropics and adjacent regions, provided suites of crop species for local use. Many of these crops were later diffused elsewhere, promoting the development of agriculture there, or being incorporated into existing agricultural systems. While cultural preferences have often preserved local crops as the staple in different regions, and adopted others only as ancillary, this has not been universally the case. For example, sweet potato (*Ipomoea batatas*), a Neotropical domesticate, has become the principal food crop of New Guinea highlanders during the past five hundred years, displacing other crops whose identities remain unknown (Golson 1989). The widespread diffusion of tropical crops over the past five hundred years has ensured that most tropical farmers now have available to them a diverse array of potential crop types with which to fashion, and improve upon, their agricultural systems.

The adoption of agriculture, whether a product of chance or necessity, introduced a radically new sort of ecosystem to the tropics, especially the forested regions. High-diversity vegetation, made up of large woody plants that are able to tolerate low soil fertilities and intense biotic interactions, began to be displaced by low-diversity herbaceous crop vegetation whose species were competitively inefficient, nutrient-demanding and productive of readily consumable tissues. The low phytomass and ephemeral life cycles of these crops precluded the development of effective nutrient and soil conservation mechanisms in these new communities, and chemical protectants to herbivory were probably selected against during domestication. Intrusion of these new and very different 'agro-ecosystems' into tropical landscapes necessarily precipitated a suite of 'sustainability' problems. The competitive inefficiency of crops meant that these could be sustained only with the expenditure of large amounts of energy to protect them from more aggressive competitors ('weeds'), especially in aseasonal environments of year-long growth. Infestations of crops by pests would be probable if agricultural activity persisted for long, as these plants lacked the defences provided by toxic chemicals or spatial isolation in natural populations. Loss of soil fertility would be expected, owing both to removal of nutrients in crops and to the relative inefficiency of herbaceous vegetation at conserving nutrients. Finally, replacement of forest by less protective herbaceous vegetation set the scene for an increase in soil erosion rates. The success of tropical agriculture continues to depend upon the development of techniques with which to cope effectively with these fundamental agricultural problems of weeds, pests, soil fertility and erosion.

SHIFTING CULTIVATION

Shifting cultivation (also called 'swidden farming') represents the most widely practised form of agriculture that does not depend on external subsidies of energy or materials for its successful continuance. Today it is concentrated in tropical latitudes, where it is practised by some 500 million persons on 300–500 million ha (Goldammer 1993). However, it was worldwide in extent prior to the development of more intensive forms of agriculture, and can be considered the ancestral agricultural form in most regions. It is practised across the full range of moisture conditions found in the tropics, from regions covered by evergreen tropical forest to those supporting only dry thorn woodland. Its continuance in many tropical environments, including those subject to human population pressure, can be taken as an indication of the difficulty of developing more intensive forms of agriculture in these regions. Shifting cultivation's central feature is its avoidance of the problems posed in sustaining a simplified crop community at a site (i.e. weed competition, pest infestation, declining soil fertility) by abandoning sites when these problems emerge locally, rather than by attempting to control them and continue the raising of crops. As such, it can be characterized as a system of 'passive avoidance' of agricultural problems (Kellman 1974b).

Shifting cultivation systems involve clearance of fields in areas of natural (i.e. non-planted) vegetation, normally woodland, for use for several years (often two to three). This is followed by its abandonment, often for several decades, during which time a volunteer plant community re-establishes itself

Figure 8.1 The landscape of shifting cultivation near Kavanayen, southern Venezuela. In this area the concentration of human population around a mission settlement, and growth in the size of this population, is leading to rapid consumption of the original forest and expansion of derived savanna

at the site to form a 'fallow' phase. The cycle is completed when the regrown fallow vegetation is once again cleared for agriculture. Clearly such a system of agriculture requires large areas for each farmer, as an individual site may be used for only 2–3 years during a period of 20–30 years. Landscapes used for shifting cultivation are thus characterized by low population densities and, from the air, appear as a patchwork of a few fields and remnant patches of the original forest, and many abandoned fields at various stages of fallow vegetation regrowth (Figure 8.1).

The cropping period

Site clearance for a field normally takes place early in the dry season; felled debris is left to dry throughout this period and ignited before the wet-season rains begin (Figure 8.2). The labour requirements for felling may vary a great deal depending upon the type of vegetation cleared; Conklin (1957) estimated that clearing of woody second-growth vegetation required 150 hours of human labour per hectare among the Hanunoo of the Philippines, while the same operation in mature forest required at least 350 h.ha⁻¹. A hot burn which consumes most plant debris is usually essential for several reasons, and

Figure 8.2 A new shifting cultivator's field after felling and burning, Belize. Large quantities of ash have been released during the burn, and appear as a patchy white residue on the soil surface

a number of reburns may take place if the first is incomplete. Most importantly, a complete burn releases most of the non-volatile mineral nutrients stored in above-ground plant tissues as ash, representing carbonates, phosphates and silicates of nutrient cations. In this form they are relatively soluble and enter the soil solution of upper soil horizons when the rains begin. Here, cations such as Ca^{2+} and Mg^{2+} displace H^+ ions in the exchange complex and raise soil pH (Nye and Greenland 1960). This has three immediate beneficial effects: (i) soluble aluminium levels are decreased; (ii) the cation exchange capacity (CEC) of the variable-charge clay mineral fraction is raised; and (iii) populations of nitrifying micro-organisms increase. The first process ensures that aluminium toxicity effects on crops are minimized, while the second provides an enlarged exchange capacity that can adsorb at least some of the cations released by the burn. Increased populations of nitrifiers promote increased decomposition of organically bound nitrogen in surface soils, and the release of nitrate into the soil solution. The two major nutrient elements that do not experience a net transfer from biomass to soil after a burn are nitrogen and sulphur, both of which tend to be volatilized during the process. A further potential benefit of an intense burn is soil sterilization, which may be especially important when fallows are of short duration and there may be a carry-over of pests and weed seed from previous site usage. Uhl *et al.* (1981)

Figure 8.3 Maize seedlings establishing in a recently created shifting cultivator's field, Belize. At this time, the field is usually free of weeds and other pests, and soil physical and chemical conditions are very favourable for crop growth

found that viable seeds in the upper 5 cm of soil dropped from 752 per square metre under forest to 157 per square metre after the cutting and burning of forest in Venezuela.

Provided that a well-developed fallow has been cleared and well burned, conditions are usually ideal for crop growth at the commencement of the rainy season, when the first crops are planted (Figure 8.3). Soils are free of weeds and other pests, have an enlarged CEC and lower Al levels, and have received a large flush of soluble nutrients from the fallow burn. Surface soil organic matter levels characteristic of the fallow vegetation that was cleared persist temporarily, contributing to the CEC and imparting good soil structure and high infiltration capacities. Thus, while the removal of the vegetation canopy and litter layer potentially will result in some increased rates of soil erosion, these will be much lower than would develop after prolonged site usage. The highest yields achieved in a shifting cultivation cycle are almost always during the first crop (Figure 8.4).

A wide variety of annual or rapidly maturing perennial crops are grown by shifting cultivators, with considerable regional variability in crop complexes. While some of this variability reflects cultural preferences, a remarkable willingness to adopt new crops is often exhibited by these farmers, and there has been a widespread dissemination of crops throughout the tropics

213

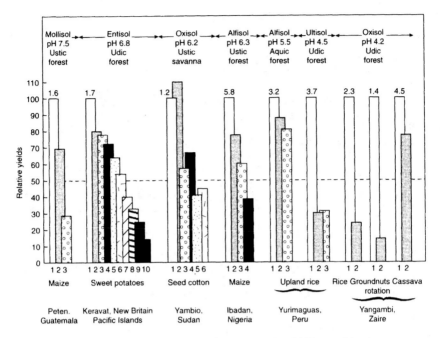

Figure 8.4 Yield decline with continued cropping in shifting cultivation systems. Figures above the bars indicate first-year absolute yields (t.ha⁻¹). Numbers below the bars indicate years of cropping

Source: Norman *et al.* 1995: Figure 1.2

during the past five hundred years. Thus, cassava (*Manihot esculenta*; Figure 8.5), which was domesticated in the Neotropics and is a common crop of shifting cultivators there, is now widely planted throughout the African and Asian tropics, where its tolerance of low soil fertility and high aluminium levels makes it especially valued on the poorest soils (Norman *et al.* 1995). Much has sometimes been made of the diversity of crops planted by shifting cultivators, but most rely upon a few staple crop species to supply most of their food needs. A variety of other edible, or otherwise useful, species are often planted in shifting cultivators' fields but these are almost always supplemental to the few principal crops. Among these secondary crops may be longer-lived perennials, including trees, that mature slowly but are capable of persisting at the site for several years after it is abandoned, and are harvested sporadically. However, several varieties of the principal crop species are often planted in the same field; these may mature at different times, or be tolerant of different extreme conditions such as drought or pests. This has usually been interpreted as a means of risk-spreading and harvest security by the cultivator (e.g. Clawson 1984).

Figure 8.5 Cassava (*Manihot esculenta*) planted in a shifting cultivator's field, southern Venezuela

Problems of weed infestation are usually minimal during the first year of cropping and little effort needs to be expended in controlling weeds. However, in subsequent cropping cycles weed populations rapidly accumulate in fields, and increasing effort must be expended in weeding. Uhl and Murphy (1981) estimate that approximately 36 per cent of the human energy expended by shifting cultivators on in-field activities during a two-year cropping period in southern Venezuela is required for weed control (Table 8.5). This compares with 21 per cent for cutting, burning and other forms of site preparation, and 43 per cent for planting and harvesting. Accompanying the weed infestations may be a buildup of other pest populations, but little energy is usually expended in controlling these, except in deterrence of birds and mammals that may consume fruit, seed or root crops prior to harvest.

More gradual changes in soil conditions may be expected during the cropping period. Mineralization of soil organic matter proceeds and, in the absence of large litter inputs, leads to a decline in soil organic matter content, and less nitrification in the surface soil. Nutrients added to the surface soil during the burning of cut fallows may be lost by absorption and removal in crops, in leaching, and in erosion. Montagnini and Jordan (1983) estimate that nutrient removal in cassava harvests over 2 years represents only 1–4 per cent of the nutrients held in forest biomass prior to site clearing, but Nye and Greenland (1960) estimate much larger proportionate losses in crops

Table 8.5 Estimated human energy inputs to cassava production for a 1-ha shifting cultivator's field over a two-crop cycle in southern Venezuela

Activity	Energy consumption (kJ × 10³)	% of activity type
Field activities		
Cutting forest	203.7	16.9
Burning	9.7	0.8
Gathering stem stocks	42.8	3.6
Planting	177.7	14.8
Cleaning and weeding	426.7	35.5
Harvesting	342.6	28.5
Total field activities	1,203.2	100.0
Food preparation		
Peeling	396.6	45.3
Grating and soaking	228.4	26.1
Squeezing and sifting	66.3	7.6
Fire making and baking	183.7	21.0
Total food preparation	875.0	100.0
Travelling	468.8	100.0

Source: Uhl and Murphy 1981

(4–52 per cent) for a 3-year cropping period following a 17- to 18-year fallow. Large losses due to leaching during the cropping period are often assumed in the non-technical literature, but there are few reliable data in support of such a process. Nye and Greenland (1960) argued that leaching would be driven primarily by the NO_3^- anion concentration in soil solution, and Uhl and Jordan (1984) have recorded elevated levels of this anion, plus Ca^{2+}, Mg^{2+} and K^+, in soil solutions at 40 cm depth beneath a shifting cultivator's field over a 2-year cropping period. However, it is difficult to translate these concentrations into net losses without an accurate soil moisture budget and information on solution concentrations at greater soil depths (see Kellman and Roulet 1990). Significant ion adsorption in deeper horizons seems probable, including that of anions, as these layers are unlikely to have experienced an increase in pH after burning, and so retain anion exchange capacity. Sequential measurements of available soil nutrients during cropping periods usually yield equivocal results (Nye and Greenland 1960), with any trends often falling within measurement errors.

There are also few data on erosion rates under systems of shifting cultivation, most efforts having been devoted to this process under more intensive forms of land use (see Chapter 9). Losses of mineral sediments by surface erosion have been estimated in a hypothetical shifting cultivation cycle of 2 years of upland rice and 15 years of fallow in Mindanao, using data from (unreplicated) erosion plots located on 25 per cent slopes in various fields and fallows (Kellman 1969). This produced an estimated gross sediment loss of 110 g.m^{-2} over the 17-year period. Losses from beneath mature forest for the same period were estimated at 69 g.m^{-2}, leading to an estimate of 139 cultivation cycles (2,363 years) being needed to thin the soil by 1 cm. Even if a less protective crop cover, such as maize, was assumed, 1 cm of net soil loss would still require 1,343 years. These are extremely low rates of accelerated erosion, and are comparable to estimated rates of weathering of geological materials in the tropics (e.g. 1.8 cm per 1,000 yr; Lewis et al. 1987). They suggest that erosion is unlikely to be a serious problem under such a low-intensity system of shifting cultivation. However, the same study showed dramatically increased erosion rates if cropping periods were prolonged; for example, losses on a 12-year rice field were over 80 times those of a new field, probably reflecting the lower soil organic matter content and infiltration capacities that develop under prolonged field usage (Kellman 1969).

Reasons for field abandonment

Crop yields on shifting cultivators' fields almost always show a monotonic decrease with time (Figure 8.4), and field abandonment is a rational choice when greater yields for the same labour input can be achieved by moving elsewhere. Declining yields are often attributed exclusively to deteriorating soil conditions, but there are few convincing data in support of this contention. In contrast, both the rapid accumulation of herbaceous weed phytomass (Table 8.6) and the large expenditure of human energy needed to control these (Table 8.5) indicate weed infestation to be the primary cause of declines in yield, and the major incentive to field abandonment. Preparation of a new field in high forest requires less than two-thirds of the human effort expended on weeding (Table 8.5), and considerably less still where second-growth forest is cleared. Provided that new terrain is available for field preparation, it makes eminent sense to prepare a new field, rather than to expend increasing effort on weed control.

This does not preclude the likelihood of soil fertility becoming limiting on crop growth eventually, in situations where weed competition is contained; positive responses by crops to fertilizer applications on permanently farmed fields clearly indicate that this is so (see Chapter 9). The role of animal pests in crop declines in shifting cultivators' fields has received little attention. Montagnini and Jordan (1983) found that 7–14 per cent of phosphorus absorbed by cassava was consumed by sucking and chewing insects in the first year of cultivation in Amazonia, and concluded that these pests had some effect on yields. However, insect population densities, and their proportionate phosphorus uptake, actually declined during the second year, suggesting that insect pest effects may be episodic, in contrast to the monotonic increase in weeds that is the norm. In the same study, root-boring insects were found to cause less than 2 per cent damage to the cassava root crop.

Table 8.6 Changes in the weed community of a shifting cultivator's field over a two-crop cycle

Weeding	Months	Forbs and grasses		Woody plants	
		Density	Phytomass	Density	Phytomass
1	10	6.1 ± 2.7	7.1 ± 5.4	12.7 ± 6.0	16.6 ± 6.5
2	16	9.7 ± 5.6	55.2 ± 24.9	2.2 ± 0.9	8.5 ± 4.7
3	21	72.6 ± 30.8	41.7 ± 16.3	3.4 ± 0.7	3.4 ± 2.3
4	26	28.2 ± 16.2	24.2 ± 12.5	2.2 ± 1.2	5.0 ± 4.1
5	31	53.1 ± 22.6	27.5 ± 16.7	1.8 ± 0.8	3.2 ± 2.5

Source: Uhl *et al.* 1982: Table 2

Note: Weeds were measured at the time of five weedings in 27 randomly located 1.5-m^2 plots, and are expressed as mean dry weight (g) per plot ± 95% confidence interval

Although neither soil fertility decline nor insect pest accumulation seem to be the primary cause of field abandonment in shifting cultivation, these exist as latent problems to prolonged field usage. We will return to potential strategies for coping with these problems in Chapter 9.

The fallow period

An abandoned shifting cultivator's field, embedded in a matrix of woodland, has some characteristics of a very large natural forest gap, but also some important differences. Competition by large forest trees is temporarily removed and a competitive advantage given to light-demanding species, including the fast-growing pioneer trees that are found sporadically in the tree-fall gaps of mature forest. However, the much larger size of the gap, and the length of time over which it is maintained in an open state, allows many other light-demanding herbaceous species not normally found in forest gaps to locate and expand their populations in these large disturbances. These become the weed populations of the field. A further important difference from natural gaps is the severity of disturbance in fields relative to that in natural forest gaps; this further shifts the competitive advantage away from mature forest trees and towards pioneers.

The vegetation of the earliest stages of the fallow period is essentially the weed vegetation of the final crop cycle. This normally comprises three distinct components:

1 Populations of herbs, both graminoids and forbs, that make up most of the weed phytomass and probably compete most directly with crops.
2 A population of stump sprouts ('coppice'), from trees and other woody plants of the previous fallow, that have been able to survive being repeatedly cut back during weeding operations in the cropping period.

Figure 8.6 Stump-sprouting from a felled tree in an abandoned shifting cultivator's field, Belize. In many areas this 'coppice' is the primary source of woody plants during the fallow period

3 A population of fast-growing pioneer trees that derive primarily from the soil seed bank.

The 'mix' of these three components varies both regionally and with the intensity and longevity of field usage prior to abandonment.

In some regions, burning eliminates the stump sprout component (Kellman 1970; Uhl *et al.* 1981), while in others this forms a prominent component of the early fallow (Figure 8.6). Ewel (1980) has suggested that resprouting is more common in drier forests than in moister forests, and it is probable that prolonged shifting cultivation in a region could select for stump sprouters among the tree flora. Prolonged usage of fields, involving repeated weedings, can be expected to deplete the resprout population and reduce the population of pioneer tree seedlings developing from the seed bank. Uhl (1987) found that stump sprout densities in Amazonia declined from 0.43 per square metre prior to the first weeding of a field to 0.07 per square metre at the time of site abandonment 3 years later, and Kellman (1970) found a negative correlation between pioneer tree abundance in the early stages of fallow in Mindanao and the number of times the site had been weeded (Table 8.7). Prolonged usage of a field may therefore be expected to favour herbaceous weeds, at the expense of woody plants, in the early stages of fallows.

Figure 8.7 A simplified stand dominated by a species of the Neotropical pioneer genus *Cecropia* developing on an abandoned shifting cultivator's field, Belize

Table 8.7 Changes in the population of pioneer tree species in fields and young fallows (<3 yr) in Mindanao, relative to the number of weedings or clearings that the site was estimated to have been subjected to subsequent to clearance from forest

Estimated number of weedings/clearings	Species	Summed frequency
1	9	59
1	7	29
2	16	69
2	8	39
2	8	38
6	6	17
18	4	5
25	4	6

Source: Kellman 1970: Table 31

Note: Relative abundance of the population is expressed as the summed frequency of occurrence of individual pioneer tree species

Where usage of a field has not been prolonged, as is typical in shifting-cultivation systems with low population densities, there is usually rapid growth of woody plants deriving from stump sprouts or the seed bank. These overtop and suppress the herbaceous weed populations within a few years of field abandonment, and can create a micro-environment approaching that of mature forest within a few years of canopy closure (Kellman 1970). However, the floristic composition of this early woody regrowth is usually much simpler than that of mature forest, and sites are often dominated by a single tree species (Figure 8.7). Shorter-lived pioneer trees may die within a decade and release slower growing pioneers that established concurrently, or slower-growing shade-tolerant tree species deriving from stump sprouts. Also, the seed of other, slower-growing forest trees may disperse to the site if a seed source and dispersers are available, and contribute to a gradual re-establishment of dominance by the regional forest flora. These sequential changes in the fallow vegetation are usually characterized as a form of 'secondary succession'.

During the early stages of growth of these secondary forests, leaf phytomass develops rapidly, leading to high productivity, and the accumulation of woody phytomass is especially rapid during the first 15–20 years of succession (Brown and Lugo 1990). Litterfall also increases rapidly, resulting in large returns of organic matter to the surface soil, and an increase in soil organic matter content, cation exchange capacity and infiltration capacities there. Large quantities of nutrients are reabsorbed in the developing phytomass and circulated through the topsoil in litterfall. For example, Uhl and Jordan (1984) have found that a 5-year woody fallow was made up of the following proportions of nutrients contained in the original forest phytomass: N, 15 per cent; P, 23 per cent; K, 39 per cent; Ca, 48 per cent; Mg, 45 per cent. While a return to topsoil conditions that are equivalent to those of mature forest may require 80–100 years, these rapid early changes beneath woody fallows ensure that soil conditions recover much of the way towards those within a few decades. Nye and Greenland (1960) conclude that under a fixed cycle of cropping periods and fallows, both organic matter and nitrogen reserves in the topsoil may be expected to equilibrate at some new level that is below that of mature forest, but nevertheless sustainable. The degree to which other soil nutrients can achieve a similar new equilibrium level in such a system is unknown, but in theory will depend on the rates of nutrient loss in crops, leaching and erosion, relative to the inputs in weathering and from atmospheric sources. Recent work on the nutrient retentiveness of tropical soils (see Chapter 4) suggests that earlier assumptions of rapid leaching losses were probably in error, but we still lack reliable estimates of the rate of nutrient loss by this, and the erosional pathway.

The favourable effects of the fallow described above are largely dependent on the establishment of dense tree populations on abandoned fields. In the absence of this, not only will nutrient reaccumulation be far slower, but weed and other pest populations will tend to persist at sites and carry over into the next cropping period. Where aggressive grasses, such as *Imperata cylindrica*, are prominent in the weed community, and a high fire frequency prevails, fallows can rapidly become derived savannas composed of these weedy species. These are of much lower agricultural value as fallows.

Figure 8.8 Shifting cultivation in gallery forests along the Rio Yurani, southern Venezuela. When fields are located at the forest–savanna boundary, this activity can encourage an expansion of savanna into the gallery forest (cf. Figure 7.18)

The beneficial effects of a woody fallow results in woodland being sought after by shifting cultivators in regions where both forest and savanna exist. This is true in savanna regions, where shifting cultivation is often concentrated in gallery forests (Figure 8.8), and in areas where secondary grasslands are being expanded by human activities. However, the absence of forest, or, in many cases, a shrinkage of forest, may force the use of savanna. This is especially true in Africa, where shifting cultivation is practised in many drier savanna areas. Clearing of a grassland fallow is often a labour-intensive operation requiring removal of grass root systems and much surface soil disturbance. This is especially so when weedy grasses have not been replaced by others during the fallow succession. A grass fallow accumulates many fewer nutrients in phytomass, and adds less organic matter to the surface soil, than a woody fallow. As a result, on clearance there is only a small flush of nutrients in ash, and surface soils are of lower organic matter content, of lower CEC and more erodible. As a result, crop yield declines are more likely to be driven by declining soil fertility in these areas than by weed competition (Nye and Greenland 1960). However, partially offsetting these difficulties is the fact that many drier African savannas are underlain by alfisols, of higher inherent fertility than the oxisols and ultisols that are characteristic of moister forest-covered areas.

The stability and sustainability of shifting cultivation systems

Shifting cultivation has usually elicited two contrasting reactions from external observers. Among European colonial administrators and their post-colonial successors, shifting cultivation has usually been seen as a primitive and wasteful form of land use that consumes large areas of forest in the support of few persons. The pejorative term 'slash-and-burn' agriculture echoes this view of the system. An alternative, and recently more widespread, view conceives of shifting cultivation as a benign system of sustainable forest use that can persist without external subsidies of energy or materials. While empirical data exist to support elements of either view, both treat the issues of stability and sustainability axiomatically, and therefore can add little to a rational evaluation of the question.

Shifting cultivation is energy-efficient, but no more so than other forms of traditional agriculture. Uhl and Murphy (1981) have calculated an energy conversion efficiency ratio (energy output in crops : energy input in human labour) of 13.9 for shifting cultivation in Amazonia, which is almost identical to the ratio of 13.8 for traditional wet rice cultivation in the Philippines calculated by Freedman (1980). However, both ratios are well above those calculated by Freedman for more intensive forms of agriculture in the Philippines (2.9–6.6), indicating the inherent energy efficiencies of systems that make little use of fossil fuels.

Most shifting cultivation systems are probably non-destructive of environmental resources if practised at a low intensity and in relatively flat terrain; in steep terrain erosion rates are so great that even infrequent site usage for crops is likely to cause much accelerated soil loss (Figure 8.9). However, the population densities supportable at this intensity are sufficiently low as to make it questionable whether alternative uses, such as forestry, could not support more people. Low-intensity shifting cultivation also has other potential socioeconomic drawbacks. A highly dispersed system of agricultural land usage contains inherent logistical difficulties. Travel time to fields may be considerable (cf. Table 8.5), and the absence of a road network will limit the development of cash cropping. It is also difficult and costly to provide social services, such as schools and clinics, to a highly dispersed population, and providing these at a few central places may lead to over-concentration of populations and destabilization of the agricultural system (Figure 8.1). A shifting cultivation system commits its practitioners to a labour-intensive lifestyle; when combined with the logistical difficulties of developing cash cropping and of providing social services, this sets relatively low ceilings on the material standard of living that they can expect to achieve.

While shifting cultivation may be a sustainable form of land use at low, and stable, population densities, the historical record indicates that most human populations, including those of shifting cultivators, have been subject to continuous increases throughout prehistory, with rates of growth increasing

Figure 8.9 Signs of severe surface soil erosion, represented by rills and roots exposed by sheetwash, in a shifting cultivator's field recently created on a very steep slope, Oaxaca, Mexico

dramatically in the past century (see Chapter 9). This has required most shifting cultivation systems to cope with increasing human populations, and we may ask what the response to these demographic pressures has been. For much of prehistory, and in the presence of a low global human population, the response appears to have been to migrate to new, unoccupied territory and to continue the existing economic activity (Cohen 1977). In the absence of this migration 'escape valve', the options for shifting cultivators become: (i) intensify the existing system, or (ii) develop alternative, more productive forms of agriculture.

The inevitability of the second process was argued by Boserup (1965), who saw over-population as a stimulus to the adoption of more productive forms of agriculture. This argument was based on the assumption that lower labour inputs per person are required in shifting cultivation than in more intensive forms of agriculture, and on the assumption that significant latent productivity existed in the environments used by shifting cultivators. We shall see in the next section that the labour inputs needed in shifting cultivation are indeed often lower than those needed in other, more intensive forms of agriculture. We will also see that there have been examples of alternative forms of agriculture being developed under population pressure. However, these developments have often been concentrated on especially favourable

Figure 8.10 A landscape being used intensively for shifting cultivation in Belize. The forested terrain in the background is a forest reserve which is unavailable to the shifting cultivators, whose increasing population has forced a progressive shortening of fallow periods

soils, and the long-term sustainability of many of these remains questionable. Elsewhere, the persistence of shifting cultivation in the face of population pressure suggests much less cause for optimism about the inevitability of technological progress than is embedded in Boserup's hypothesis. Grigg (1979) has provided a critique of other aspects of the Boserup hypothesis.

The most common response to population pressure among shifting cultivators is to decrease fallow lengths and increase field sizes, while continuing the same cropping cycle. This probably reflects the immediacy of crop yield declines with prolonged field use (Figure 8.4), relative to the more gradual negative effects on yields of decreasing fallow length. Most long-fallow shifting cultivation systems could probably be intensified to a new, shorter fallow system while preserving the system's stability (Kellman 1969). However, while the effects of continual fallow shortening may be sufficiently gradual as to be almost imperceptible in one farmer's lifetime, they are nevertheless real. Continuous fallow shortening will result in the failure to achieve any new equilibrium level of organic matter or nitrogen in the soil, and the reaccumulation of fewer nutrients in a shrinking fallow phytomass will probably lead to long-term site defertilization. Moreover, increased land use intensity is likely to have negative consequences at the level of the local landscape (Figure 8.10). Late successional species may go regionally extinct, as may game animals, which can provide shifting cultivators with an important protein source (Linares 1976; Redford and Robinson 1991). The benefits for pest control of spatial isolation of fields will decline, and under very short fallows, weed and other pests may carry over between cultivation periods, leading to more rapid crop infestations. The point at which woody fallows no longer re-establish, and are replaced by derived savanna, usually represents the effective collapse of the system. Using a simple empirical model, Trenbath (1989) has estimated the maximum human population density that can be sustained by a shifting cultivation system to be about 50 per square kilometre, a figure that is close to other estimates of the maximum carrying capacity of the system (UNESCO 1978).

While many examples now exist of shifting cultivation systems in incipient states of collapse owing to population growth (e.g. Figures 8.1 and 8.10), some historical data suggest that similar events have occurred in prehistory. Reasons for the sudden cessation of the building of monumental structures in the Mayan region of southern Mexico and northern Central America in the ninth century AD has been an enigma. While these societies seem to have made some use of more intensive agricultural systems (see below), they appear to have been fundamentally based on shifting cultivation. A collapse of this system has frequently been cited as a potentially important contributor to the societal collapse (e.g. Gourou 1958), but convincing evidence for this process has come only recently. Deevey *et al.* (1979) have conducted an intensive palaeodemographic survey near a Mayan ceremonial centre in the Peten area of Guatemala, in conjunction with sedimentological and palynological analyses of local lakes. The palynological record shows a progressive replacement of forest by secondary vegetation and grassland throughout the period of Mayan occupancy. Accompanying this were signs of accelerating soil erosion in the local watersheds. The demographic record showed a remarkably long period of steady exponential population growth from the first millennium BC to the ninth century AD. However, with a doubling time of 408 years, this population growth was probably imperceptible to an individual inhabitant. Deevey *et al.* (1979) suggest that the gradualness with which population grew and environmental conditions deteriorated precluded the development of density-dependent responses regulating population size, and set the scene for a crisis in carrying capacity. Evidence of decreasing human stature at burial sites during the Late Classic period at the Mayan centre of Tikal was reported by Haviland (1967), who tentatively attributed this to increased nutritional stress. A comparable pattern of deforestation and soil degradation prior to the societal collapse at Copán, another Mayan centre, has been inferred by Abrams and Rue (1989). Here, cessation of monumental construction was followed by a progressive decline in the human population over the following three centuries, presumably as a result of migration away from the centre.

In conclusion, shifting cultivation systems appear to be sustainable at low and stable population densities, although they contain inherent logistical difficulties that make an improved standard of living difficult to maintain. However, there is also ample evidence from contemporary and historical situations to indicate that the systems become unsustainable beyond relatively low threshold densities in the human population. Beyond these thresholds, not only does the productivity of the system decline, but there may be severe environmental deterioration at the regional scale, involving loss of plant and animal diversity, and increased flooding and erosion. Much of this environmental deterioration may be sufficiently gradual as to be imperceptible to the shifting cultivator who is responding to proximal cues of decreased crop yields and increased weed control problems. More optimistically, we may hope that the rapidity of recent population growth may lead to a greater awareness of system deterioration among these farmers. The technical options that are available for developing a more productive system will be discussed in Chapter 9.

INTENSIFIED TRADITIONAL AGRICULTURAL SYSTEMS

Although various forms of shifting cultivation remain the common form of agriculture throughout large areas of the tropics, other, more intensive forms of traditional agriculture exist, or have existed in the past. Most of these are capable of producing much higher outputs of agricultural products per hectare, and are thus capable of supporting far higher population densities than can be supported by shifting cultivation. However, as we shall see, many require much higher labour inputs and often result in lower crop yields per labour-hour than are possible in shifting cultivation. In most instances, these intensified forms of agriculture appear to have been preceded by forms of shifting cultivation, and their development is associated, in the archaeological record, with signs of increased human populations. This appears to support Boserup's (1965) hypothesis that the demands of increased populations stimulated the search for alternative, and more productive, forms of agriculture. However, the temporary existence of many of these systems suggests that they may have lacked long-term sustainability.

Home-site gardens

Throughout the tropics, home-sites that are relatively permanent (including those of many shifting cultivators) are often surrounded by small areas of intensively managed agricultural land (Landauer and Brazil 1990). These have been called home gardens, kitchen gardens, door-yard gardens or home-site gardens. Small numbers of a great variety of useful plant species are often grown here, including trees for fruit and firewood, specialized vegetables, medicinal plants and ornamentals. Small populations of domesticated animals, such as chickens and pigs, are usually also raised in these places, feeding on household refuse and crop residues. Home-site gardens are often sites of agricultural experimentation with new crops, and are especially advantageous for agriculture for reasons of both soil fertility and labour availability.

Home-sites act as sinks for nutrients derived from the entire farming area. These are transferred from outlying areas in crops, or by large animals such as cattle and horses, which graze in the larger farmed area but are usually kept around the home-site at night. Here, organic refuse and manure are deposited. The proximity of home-site gardens to dwelling sites allows intensive hand care to be applied to crops raised in this area by family members of all ages. Weeds and other pests are continually removed, vertebrate frugivores kept away from crops, individual crop plants may be watered, and crop maturation is carefully monitored. These intensive management procedures, when combined with the concentration of soil nutrients, result in these places being highly productive. However, while home-site gardens supply a variety of products, and are important in dietary supplementation (Marten 1990), their small area rarely allows them to provide the staple crops of the household, which must be grown in larger, and more distant, fields.

Raised fields in wetlands

In recent decades, archaeological research using aerial photography has found signs of earlier agricultural activities in tropical wetlands, especially those in the Neotropics (Darch 1983). Excavations in these areas have revealed evidence of systems of shallow canals in the wetlands, with the spoil removed from these deposited as intervening platforms on which intensive crop-growing appears to have taken place. Dating of these features has revealed that most were developed relatively late in the archaeological record (e.g. 200 BC–AD 1500), and their existence in areas of dense prehistoric human populations, such as near Mayan ceremonial centres, has led to the assumption that most were developed in response to increasing human population pressure (Turner and Denevan 1985). Despite being widespread in prehistory, very few of these raised-field systems were in use at the time of European contact, five hundred years ago. Two exceptions to this were the 'chinampas' of the Valley of Mexico, which were a primary source of food to the Aztec capital at Tenochtitlán, and raised-field systems employed in New Guinea.

The environmental advantages of a wetland for agriculture are the relatively high concentrations of nutrients present in mineral and organic sediments, and the availability of a moisture supply. There may also be significant quantities of nitrogen fixed by free-living organisms in the waters of canals (Vasey 1985). The main disadvantage of these sites is the anaerobic conditions that prevail in the soils for all or part of the year, which inhibit the development of most plants' roots. Construction of platforms above water table level from excavated sediments provides a fertile and aerobic substrate

Figure 8.11 An experimental raised field ('chinampa') created in a wetland in Veracruz, Mexico. The crop in the foreground is taro (*Colocasia esculenta*)

for crops, whose roots may have continued access to water through capillarity (Figure 8.11). Moreover, ongoing clearance of canals, and use of canal water for irrigation, ensures continued accretion of nutrients to the raised field.

Maize yields of 3,000 kg.ha^{-1} have been estimated for contemporary chinampas in the Valley of Mexico (Sanders *et al.* 1979), a figure that is approximately double that of dryland maize for the tropics generally (Table 8.3). Turner and Denevan (1985) estimate that a population density of about 540 per square kilometre could be supported by a raised-field system of agriculture, a figure over ten times the maximum density supportable by shifting cultivation. These high yields in raised fields appear to come at the cost of very large labour inputs. For example, Bayliss-Smith (1985) calculates that preparation of raised fields for taro cultivation in New Guinea requires 2,700–5,100 person-hours.ha^{-1}.yr^{-1}, while raising the same crop by shifting cultivation requires only 700–1,200 person-hours.ha^{-1}.yr^{-1}.

Reasons for the abandonment of most raised-field systems, despite their high potential productivity, remain obscure. A decrease in population pressure and reversion to a less labour-intensive form of agriculture remains one possibility. Conversely, raised-field systems may contain some potential instabilities that encourage their failure. For example, periodic droughts may lower wetland water tables to the point at which only dryland crop yields are possible. In semi-arid environments an accumulation of salts may occur in raised fields, while these and all other permanently located agricultural activities remain vulnerable to periodic pest outbreaks, which may prove catastrophic. Finally, human populations working these systems may remain especially vulnerable to water-borne diseases and to diseases such as malaria, whose vectors require standing water in which to breed (Chapter 5).

Dryland terraces

Today, dryland terracing is widely used as a means of erosion control in severely degraded agricultural landscapes (see Chapter 11). Whether these systems have often been established in the past in the absence of erosional crises is unclear. Most agricultural societies (including those of the contemporary industrialized world) have been remarkably insensitive to signs of incipient soil erosion, making it unlikely that terracing for protective purposes was widespread in prehistory. The other benefits of terracing involve deeper soil profiles for plant crop rooting, and possibly increased moisture retention, benefits which may not be apparent to a farmer except in very rocky terrain (Figure 8.12). The proportional increase in benefits provided by dryland terracing is also likely to be low relative to those provided by other techniques of intensification, while labour costs in construction remain high. This may account for the relative scarcity of dryland terracing as a means of intensifying agricultural production in the tropics.

Irrigation

The augmentation of water available to crops has taken several forms in the tropics, depending upon the environmental conditions and the crops available. In its most rudimentary form, irrigation involves 'recessional' agriculture on floodplains, which are used temporarily for crops in the early dry season, after floods recede but while soil remains moist. Such seasonally flooded sites appear to have been the habitat of the progenitors of modern domesticated rices, and collection of wild rices at these places still occurs (White 1989). In addition to providing moisture, flooding usually introduces fresh sediments and their associated nutrients, and provides a harsh physical stress that may mimic the pest-regulatory properties of a temperate winter. Recessional floodplain agriculture continues to be used in some areas, such as north-eastern Nigeria, where bunding has been employed in areas of vertisolic soils to improve water retention for sorghum crops (Connah 1985).

In parts of the semi-arid tropics, especially those with nearby mountains that provide perennial streams, rudimentary irrigation systems have been employed throughout prehistory to enable the growing of moisture-requiring crops in an otherwise inhospitable environment (e.g. Smith 1985). These systems, which used rudimentary dams or shallow wells, were especially

Figure 8.12 An ancient dryland terrace created on shallow limestone soils near the Mayan ceremonial centre of Caracol, Belize

vulnerable to sedimentation and salinization in these dry climates and so were often impermanent. However, others were more robust, and their use has continued until the present (Zimmerer 1995).

In moister parts of the tropics, irrigation has been used most commonly to provide optimum growing conditions for crops whose yields are maximal under very wet or flooded conditions. Taro and wetland (padi) rice have been the two principal crops grown in these systems. Taro (*Colocasia esculenta*), a starchy root crop staple in the Pacific Islands, was raised both in dryland fields and in irrigated 'pondfields' (Kirch 1985). However, it is the cultivation of rice in flooded fields that represents the most successful and sustainable form of irrigated agriculture that has so far appeared in the tropics.

Two species of rice (*Oryza* spp.) were domesticated independently in Asia and Africa from semi-aquatic ancestors. In West Africa, *Oryza glaberrima* appears to have been domesticated from about 3500 BP, and possesses relatively low varietal diversity (Chang 1989; Norman *et al.* 1995). *Oryza sativa* appears to have been domesticated in South and South-East Asia from at least 5500 BP, and has diversified into three major races: *indica*, *sinica* and *javanica*. Within each race, further ecological diversification has taken place along a flooding gradient between unflooded conditions and floodwater depths >1 m. Chang (1989) considers the semi-aquatic varieties most primitive, with upland and deep-flooding varieties being more recently derived

231

from these. Today, rice is cultivated throughout the Asian region both as a dryland crop, principally in shifting cultivation systems, and as an irrigated crop. Its cultivation elsewhere in the tropics has been principally as an 'industrial' crop, rather than by farmers using low-input techniques.

Approximately three-quarters of the world's rice crop is produced using shallow irrigation flooding (Roger and Ladha 1992), in which form this species outperforms other tropical cereals in terms of energy and protein yield per hectare (Table 8.3). It is the staple crop of most of the large human populations of South, South-East and East Asia, and feeds more than half of the world's population (Roger and Ladha 1992). Chang (1989) considers that flooded rice was originally raised in floodplains as a flood recession crop, with bunding and control of irrigation water being a later development. Seavoy (1973) has described such an intensification procedure still taking place in Kalimantan, Indonesia. Under low population densities, rice is raised by a form of shifting cultivation in seasonally flooded marshes, where a large sedge forms the fallow, and is cut and burned before cropping. Under increasing population pressure bunds are gradually added to fields, together with irrigation and drainage canals, and the fallow period is abandoned.

The success of irrigated rice as an intensified form of tropical agriculture can be attributed both to the adaptability of *Oryza sativa* and to the unique flooding conditions under which it is grown. Unlike most other irrigated crops, which require abundant moisture but aerated soil, padi rice can be grown in fields kept permanently flooded. Where rainfall is sufficient, flooding may be induced simply by creating bunds and promoting low soil water permeability by 'puddling' of the soil. More reliable flooding is provided by introducing irrigation water to systems of bunded terraces (Figure 8.13). Rice plants tolerate flooding because they possess specialized 'aerenchyma' tissue that allows oxygen diffusion down stems to roots. However, few other tropical plant species are tolerant of flooding, which greatly reduces potential weed problems in flooded fields. Semi-permanent flooding is also an atypical environment in tropical landscapes, and a smaller potential fauna of soil-inhabiting pests may be expected. Where wet rice cultivation is alternated seasonally with dryland crops, the extreme seasonal habitat alteration may be especially effective at limiting pest buildup.

Flooding also introduces certain soil fertility benefits that go beyond the addition of nutrients in flood water and sediment. In acid soils, pH is increased, and phosphorus becomes more readily available. However, the single most important change is a significant increase in nitrogen fixation by free-living or symbiotic nitrogen-fixing organisms in flooded fields. A compilation of the results of experiments on nitrogen balances in flooded rice soils shows an average positive balance per crop of about 30 kg N.ha^{-1} in fields without inorganic nitrogen fertilizer additions (Roger and Ladha 1992). This has allowed many traditional wet rice agricultural systems to be sustained for many centuries without additional nitrogen fertilization (Norman *et al.* 1995).

Figure 8.13 Irrigated terraces used for rice production along the steep slopes of the Ayung River, Bali, Indonesia

While padi rice has enjoyed spectacular success as a productive and sustainable form of intensified tropical agriculture in the Asian region, its adoption in other parts of the tropics has been limited to large industrial farming operations, rather than small-scale farming. The failure of small-scale traditional farmers to adopt this agricultural system probably relates to the specialized agronomic techniques that it requires. In contrast, other, less atypical crops can be incorporated more readily into their existing farming systems, without major changes in management technique.

9

INTENSIFIED TROPICAL AGRICULTURE

In this chapter we review means of achieving forms of tropical agriculture that not only are more productive than those that exist traditionally, but also are capable of providing high-quality diets and of being sustained both economically and environmentally. We will argue that most recent changes in tropical agriculture have been driven by a need to feed a burgeoning human population, and have therefore primarily consisted of attempts to increase productivity, rather than attempts to improve dietary quality or sustainability. These recent changes have been largely successful in meeting the immediate needs of growing numbers of people, but they have left unaddressed longer-term issues of quality of life and sustainability of the agricultural systems. We will pay special attention to these latter aspects of agricultural intensification, not because we seek to belittle the major achievements of recent efforts at increasing agricultural productivity, but because we fear that a single-minded preoccupation with productivity will leave tropical peoples vulnerable to reliance on economically unsustainable forms of agriculture and a deteriorating environmental resource.

The fundamental problems that must be overcome in achieving successful forms of tropical agriculture are largely those that are dealt with in shifting cultivation by field abandonment: (i) intense interactions between crops and other biological populations such as weeds and insect pests, and (ii) declining soil fertility under prolonged site usage. Intensified agricultural usage also introduces a new problem that rarely emerges in shifting cultivation: accelerated soil erosion. In the sections that follow, we explore the solutions to these agricultural problems that to date have been employed with varying degrees of success, plus others that show theoretical promise, or have proven to be successful in special circumstances. Many of the latter two groups draw on analogies with processes that operate in natural tropical ecosystems, or represent techniques used traditionally by subsistence farmers. The practicability of applying these techniques is likely to vary greatly from place to place, depending on the environmental substrate and local socioeconomic conditions. Consequently, the techniques reviewed can be taken as a set of ingredients that are available, and *potentially* usable, for the development of intensified and sustainable agricultural systems. Whether these are *actually* usable will depend on environmental and socioeconomic contexts.

Any treatment of agricultural intensification in the tropics requires some discussion of human demography in these regions, as this has traditionally driven the demand for agricultural products. Consequently, we begin by reviewing the recent history of human birth rates and death rates in the tropics, the two parameters that determine population size.

THE DEMOGRAPHIC CONTEXT

Unlike other species, whose populations are regulated within upper and lower thresholds by various density-dependent and density-independent processes (see Chapter 5), the human species population appears to have undergone a slow but steady increase (albeit at varying rates and with temporary setbacks) since its emergence in the Pleistocene. The world's population at the initiation of agriculture (8000–10,000 BC) is estimated to have been between 5 and 10 million (Durand 1977). This number is estimated to have grown to 255–330 million by AD 0, with little further growth during the ensuing 1,000 years. By AD 1500, world population is estimated at between 427 and 540 million (Durand 1977), after which growth rates accelerated. It is about to exceed 6,000 million (6 billion) persons (Figure 9.1). This monotonic increase in the human population can be attributed to two unique features of this species:

1 Humans have the capacity to migrate to, occupy and support themselves in almost all terrestrial habitats on the earth's surface. This adaptability has allowed humans to co-opt an increasing resource base.
2 Within these habitats, humans have developed the capacity to support increasing numbers of people through technological innovation, most notably that involved in the domestication of plants and animals.

The ability to expand the available food supply served to remove starvation as a major control on human populations, although this agent may have operated with sporadic effectiveness. In the absence of chronic starvation, parasitic disease probably emerged as the primary controller, but the continuing growth of the human population throughout prehistory suggests that disease and other agents of regulation were insufficient to stabilize the global human population.

A steadily growing biological population reflects one in which average birth rates continually exceed average death rates; long-term stability is achieved only when the two equalize. Maximum birth rates are set by the reproductive biology of the organism involved. Time to achieve reproductive maturity, numbers of offspring per litter, and the gestation period are especially important in determining this rate. In almost all animal species, these reproductive traits result in many more offspring being produced by females than the

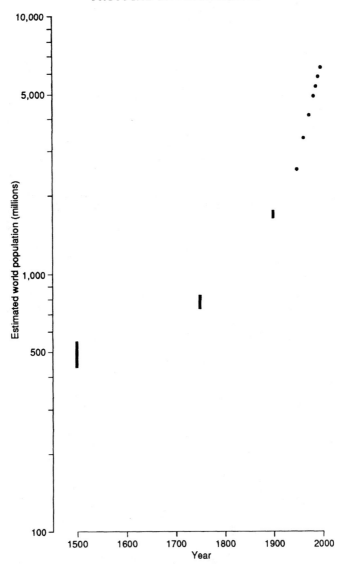

Figure 9.1 Growth of the human population of the world since AD 1500. The three earliest totals are subject to large errors, and a range for estimates is provided. Note that a logarithmic scale is used; the upward convexity of the trend line indicates that rates of growth have been accelerating throughout the 500-year period

Data sources: Durand 1977; El-Badry 1991

minimum of two needed to replace the parents upon their death. The result is a variable, but usually large, intrinsic potential to increase the population's size (r_{max}; see Chapter 5). In humans, for example, if one assumes single births to a woman at 2-year intervals between the ages of 20 and 40 years, 10 offspring result. If all survived to maturity, this would result in a fivefold growth in the human population each generation. In the presence of these high reproductive potentials, population regulation must involve equally high death rates, which are usually age-dependent and especially high in the youngest members of the population. In short, most species produce large numbers of offspring, very few of whom reach reproductive maturity.

In the human species, a slow but steady population growth throughout prehistory reflects death rates that, on average, were somewhat below birth rates; in each generation there tended to be slightly larger numbers of reproductive adults maturing than existed in the previous generation. This positive population balance presumably reflected primarily humans' ability to avoid starvation as a major source of mortality; the fact that the growth was slow relative to potential rates indicated that this and other population regulators were still operating. Among the latter, human diseases, especially parasitic diseases, are likely to have been the principal agents of regulation, but ones that clearly were not able to stabilize the human population. Since the mid-nineteenth century, and especially in the past fifty years, human death rates have been reduced much further as a result of steady improvements in disease control and agricultural productivity. This initiated an unprecedented period of accelerated growth in the human population which is still continuing (Figure 9.1).

In economically developed countries, death rate declines generally began earlier than elsewhere and increased only gradually, setting in motion a prolonged period of steady but not explosive population growth. Throughout this period, birth rates gradually fell in response to socioeconomic changes such as increased urbanization and increased educational levels, which led to a desire for fewer offspring, and technical developments which provided the means to achieve this. The result was a prolonged period of moderate population growth leading to a period of stable population, which many developed countries have now entered (Bos *et al.* 1994). In most economically less developed countries, including almost all of those in the tropics, the whole process of demographic transition has been greatly compressed, especially as a result of dramatic increases in disease control since 1945. Disease control measures have primarily involved a greatly increased ability to suppress disease epidemics as a result of environmental sanitation, immunization programmes and control of the insect vectors, especially with the use of insecticides. These measures, which can be applied relatively cheaply and effectively by governments, have had an especially dramatic effect on death rates since the mid-twentieth century (Figure 9.2). While birth rates in tropical countries have also begun to fall, the decrease has been much less dramatic than that

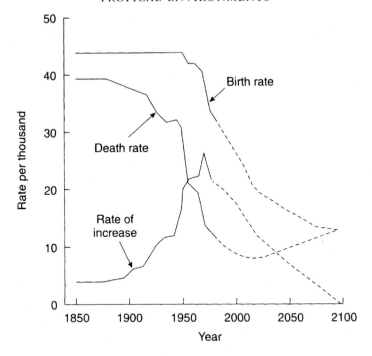

Figure 9.2 Birth rate, death rate and rate of population increase in less developed countries between 1850 and 1980, with projections to the year 2100

Source: Coale 1983: Figure 1

of death rates; this has resulted in a large cohort of people in tropical countries, who otherwise would have died before maturity, reaching reproductive age and contributing further to population expansion.

The rapidity with which these changes have taken place has diminished the prospects that the birth-limiting processes that operated in developed countries will be effective in rapidly stabilizing the population. A 150-year period of demographic transition, such as occurred in developed countries, would witness a population explosion in the tropics likely to be limited ultimately by starvation. Thus, a rapid decrease in the birth rates in tropical countries in the next few decades is essential if this drastic means of population regulation is to be avoided. Fortunately, there are increasing signs of this happening (Bongaarts 1994), although equalization of birth and death rates is not projected to occur until after 2100, when the world population will be approximately 11 billion (Bos *et al.* 1994). Moreover, even if birth rates were reduced immediately to a replacement level of two surviving children per woman, there would remain a considerable demographic momentum effect resulting from a young population containing many pre-reproductive

238

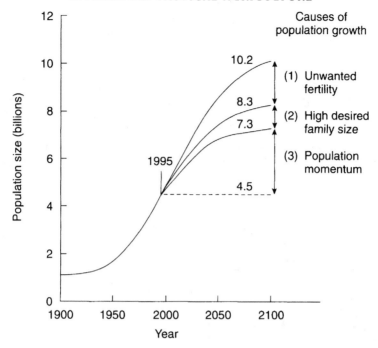

Figure 9.3 Alternative projections of the size of the human population in the developing world between 1995 and 2100. The 'population momentum' component reflects the effects of the large population not yet of child-bearing age, and assumes reproduction only at the replacement level of two children per woman. The 'desired family size' and 'unwanted fertility' component projections are based on responses to an extensive survey programme conducted in Africa, Asia and Latin America

Source: Bongaarts 1994: Figure 4

individuals, which will lead to substantial further population growth for several decades (Figure 9.3).

The unprecedented explosion in the human population of tropical countries during the past half-century has introduced a fundamental distortion to agricultural developments in these regions. It has required that attempts at agricultural change focus almost exclusively on how to produce more food rather than on the quality of agricultural products and the sustainability of the agricultural endeavour. Increased production, combined with food imports from elsewhere, has been remarkably successful in preventing mass starvation in tropical countries. Much of this success can be attributed to intensive breeding programmes which have developed high-yielding varieties of a few staple crops. Success has also come from applying the 'industrial' inputs so successful in temperate agriculture: inorganic fertilizers, pesticides and mechanization. However, these successes should not obscure the fact that they have bought only a temporary delay in a food supply crisis, and this at the expense of conserving and developing other components of sustainable tropical agriculture.

Much of the material covered in the remainder of this chapter addresses the prospects for developing forms of agriculture that are both productive and sustainable. We will argue that the prospects for developing these are good, provided that human populations can be stabilized and eventually reduced in many tropical regions. In the absence of stabilization, improvements in the quality of tropical agriculture will represent only temporary delays in the progression to an inevitable food supply crisis (Kendall and Pimentel 1994). Such a crisis would result not only in mass starvation, but also in severe environmental degradation and diminished environmental carrying capacity, so the desirability of rapid movement to a stable or even shrunken human population in many parts of the tropics is obvious.

INDUSTRIAL AGRICULTURE IN THE TROPICS

In the tropics, forms of agriculture that rely on large external inputs of energy, materials and technological knowledge originated during the era of dominance of most tropical countries by colonial powers. Colonial agriculture focused primarily on cash crops for export, such as sugar, coffee and rubber, rather than on staple food crops, and was usually practised on large-scale 'plantation' farm units. Although cheap local labour was essential for these plantations, throughout the colonial period inputs of inorganic fertilizers and mechanization were increasingly used, as were the fossil fuel inputs that are embedded in these. Pesticides did not become a major input until the past half-century. Crop improvement programmes during the colonial era were focused on selective breeding of the cash crop cultivars that responded well to the prevailing input recipes. These agricultural techniques were essentially those that had proven successful in temperate agriculture since the mid-nineteenth century, when industrial inputs to the agricultural enterprise became increasingly common.

Cash crop plantation agriculture continues in many tropical countries, but, since the mid-twentieth century, in the face of a rapidly expanding human population, there has been an increasing preoccupation with applying comparable technologies to staple food crops. These efforts have involved intensive breeding programmes at several agricultural research centres, such as the International Rice Research Institute in the Philippines, where the existing varietal diversity of tropical crops has been drawn upon to derive hybrid genotypes that show large yields in the presence of high inputs of inorganic fertilizers, pesticides and irrigation water. These developments have usually been concentrated in the most productive agricultural areas, where various forms of intensive traditional agriculture were already being practised, and have come to be known as the 'Green Revolution'. By far the most successful of these programmes involved irrigated rice, and it is largely as a result of developments with this crop that the large and expanding human population of tropical Asia has continued to be fed successfully.

While these developments have succeeded so far in preventing mass starvation in most of the tropics, it is doubtful whether the productivity increases recorded can continue to be sustained by further application of the same techniques, which makes rapid stabilization of the human population of utmost concern. Moreover, there is increasing cause for concern about whether many of these industrial systems are sustainable economically or environmentally over long time periods, and whether they form a model on which to base further changes in tropical agriculture (e.g. Fearnside 1987).

Employment of industrial inputs by tropical farmers makes them increasingly vulnerable to external conditions over which they have little control. All industrial inputs involve a large fossil energy component (Pimentel et al. 1990). Some estimates of energy inputs to several tropical crops grown under industrial conditions are provided in Table 9.1; the very high energy inputs required for nitrogen fertilization are especially notable. Freedman (1980) has estimated that 'green-revolution' padi rice production requires more than 16 times the energy input of traditional production, with almost all of the difference being accounted for by fossil fuels (Table 9.2). While the green-revolution system can be sustained at present fossil fuel costs despite a low output : input energy ratio (2.8, as against 13.9 for the traditional system), a rise of these costs as global fossil fuel reserves are depleted will make the system increasingly uneconomic. Similar arguments apply to potentially increased costs of inorganic fertilizers (Fearnside 1987). Adoption of these systems has often led to substantial social dislocation, as it is primarily the wealthier farmers who can afford the large costs of inputs and have prospered from using these, resulting in a growing gap between rich and poor farmers (Marten 1986).

Whether growing cash or food crops, most forms of industrial agriculture tend to be simplified monocultures. Also, rapidly maturing crops are frequently grown continuously, using two to three crops per year. These simplified systems are more easily mechanized and provide uniform and predictable responses to the large inputs used. Unfortunately, the simplicity of these systems, when combined with their spatial permanence, makes them especially vulnerable to pests. Increased use of chemical pesticides can result both in the development of pesticide-resistant strains and in the destruction of their natural enemies (e.g. Barducci 1972). For example, the brown planthopper is a potential pest of rice in South-East Asia, but in the absence of insecticide use it is controlled by natural enemies (Cook and Perfect 1989). However, heavy insecticide use associated with green-revolution technology in the 1970s precipitated major planthopper outbreaks (Joyce 1988). Pest infestation often results in a spiral of increasing pesticide use (Brown and Marten 1986) which may affect not only other agricultural activities, such

Table 9.1 Energy inputs (kcal.ha^{-1}) involved in producing three tropical crops. These crops were grown using industrial inputs, but also labour inputs far larger than are normal in temperate agriculture

Item	Rice (Philippines)		Watermelons	Bananas
	Wet season	Dry season	(India)	(Taiwan)
Labour (814.4 h)	153,107	153,107	150,400	582,800
Machinery	81,000	81,000	54,000	1,782,000
Fuel	1,327,312	1,327,312	684,840	1,774,040
Inorganic N	485,100	1,293,600	1,716,000	7,276,500
Inorganic P	—	—	75,000	357,500
Inorganic K	—	—	—	1,440,000
Herbicides	69,937	26,073	—	125,540
Other pesticides	255,664	69,937	8,691	1,468,520
Irrigation	227,090	454,180	836,656	314,940
Seeds	352,000	415,600	14,400	—
Transportation	NA	NA	14,906	76,843
Plastic covers	—	—	—	1,482,000
TOTAL	2,951,210	3,820,809	3,554,893	16,680,683
Crop yield	9,540,864	12,324,600	1,832,760	22,888,000
Output : input ratio	3.23	3.23	0.52	1.37

Source: Pimentel 1980. Labour energy inputs, which were not given by Pimentel, have been estimated, using the average of 188 kcal per work hour assumed by Uhl and Murphy 1981 and work hours provided by Pimentel 1980

as fish-rearing in rice paddies (Guerrero 1990; Litsinger 1993), but also non-agricultural systems. A major resurgence of malaria in Central America and India in the 1970s was closely correlated with the level of insecticide use in these areas and is thought to have resulted from the development of insecticide resistance among *Anopheles* mosquitoes (Chapin and Wasserstrom 1981). Other negative impacts on human health have been linked to industrial farming activities through direct intoxication with pesticides (Loevinsohn 1987), and concern exists about the potential contamination of water by fertilizer residues, especially nitrate (Conway and Pretty 1988).

In sum, in agricultural systems employing large industrial inputs, high productivity is bought at the expense of a reliance on external factors over which the farmer has little control, a loss of flexibility and ability to spread risk, and potentially severe environmental degradation. Indigenous 'folk' knowledge about successful low-input management techniques is usually rapidly lost when industrial techniques are adopted. Finally, agricultural systems employing mechanization and tillage frequently produce much bare soil and induce accelerated rates of soil erosion. All of these conditions call into question the long-term viability of these systems as means of improving upon tropical agriculture, whatever their immediate successes in feeding an exploding population. They are probably better seen as providing some

Table 9.2 Comparison of energy inputs and outputs (kcal.ha^{-1}) for traditional and green-revolution padi rice production in the Philippines. Energy inputs in human labour are included within the activities listed

Item	Traditional	Green revolution
Seeding	27,000	31,700
Land preparation	109,650	496,340
Fertilization	9,360	1,954,627
Weed control	—	435,395
Insect control	—	65,365
Irrigation	3,600	1,183,980
Harvesting	131,500	167,847
Threshing	46,800	139,430
Seeds, drying, transportation	—	1,020,000
TOTAL	327,910	5,494,684
Energy yield	4,537,500	15,972,000
Output : input ratio	13.84	2.91

Source: Freedman 1980

ingredients to be used selectively when developing more sustainable forms of agriculture.

CROP BREEDING PROGRAMMES

Throughout prehistory, crops have been selected primarily for high yields. Much of the selection took place in agricultural milieux in which potential problems for plant growth were minimized (competition, soil infertility, moisture supply, pests and diseases). Even in shifting cultivation, these problems are usually reduced by field abandonment. This tradition of selection for yield in an 'undemanding' habitat has intensified since industrial inputs have been used to minimize constraints on crop growth. Intensification of agricultural production in the absence of access to these industrial inputs today confronts crops with problems that were previously avoided. Crops are grown more continuously and face declining soil fertility levels, or are grown on poorer soils that previously were avoided, and interactions with weeds and other pests intensify. Under these circumstances, crop traits that impart an ability to cope with these problems begin to assume significance, and selection for these, as well as for yields, becomes important. Selection for more 'multipurpose' crops probably goes on in most traditional agricultural systems undergoing gradual intensification in land use. This is likely to be a rather slow process because genetic lines are not isolated in farmers' fields.

In recent decades, more attention has been focused on the intentional breeding of 'non-yield' traits into tropical crops, as the limitations to some industrial inputs have become more obvious. This is especially true for the breeding of resistance to insect pests and diseases (Maxwell and Jennings 1980). These breeding programmes usually involve mass screening for resistance among the large germ plasm collections of crops and crop-relatives from traditional agricultural systems that are maintained at international research centres. This is followed by combining the most resistant varieties with other high-yielding varieties (e.g. Bellotti and Kawano 1980; Pathak and Saxena 1980). The speed of genetic change can be much more rapid in these programmes than under farmers' selection in agricultural landscapes because pure lines are screened and fertilization in subsequent crossings is tightly controlled.

An extension of these sorts of breeding programmes to include selection for other traits, such as tolerance of aluminium and low soil fertility, will be necessary if farmers without access to high-quality soils or industrial inputs are to benefit. Breeding for tolerance of these conditions is likely to prove a more tractable problem than breeding for pest resistance, as pests can continue to evolve and breach crop defences. This requires that the breeding programme be a continuing process. Research on the genetic engineering of crop plants is still in its infancy, and has so far been concentrated on temperate crops (Gasser and Fraley 1989). The technology has been applied to rice and soybeans, and there is no doubt that it can be applied to other tropical crops. A more fundamental problem in relation to the tropics is whether small farmers there are likely to be able to afford the purchase price of seed of such engineered crops, whose cost is likely to be very high. While these crops will probably have an impact on large-scale agricultural enterprises in the tropics, it seems unlikely that they will have much impact on small-scale farming.

Preservation of crop genetic diversity

The adoption by farmers of genetically improved crop varieties that have been bred at central research institutes poses several potential problems regarding sustainability. If adoption of the new varieties is accompanied by abandonment of indigenous crop varieties, there may be considerable erosion of crop genetic diversity in rural landscapes, which previously enabled some spreading of risk. Paradoxically, the loss of traditional varietal diversity destroys the very ingredients used in breeding the new varieties, and requires that henceforth these must be stored in centralized germ plasm banks. A further problem is the possibility that the new varieties may become regional monocultures and so be far more susceptible to major pest outbreaks. For these reasons, Altieri and Merrick (1987) have recommended that every effort should be made to preserve elements of traditional farming systems, including the local crop varieties, in rural development projects. A somewhat more optimistic view is presented by Brush (1989), who argues that many traditional farmers continue to grow traditional varieties in more marginal habitats, even while adopting genetically improved crop varieties.

A much more technological view is taken by Evans (1993), who argues that loss of spatial crop varietal diversity can be compensated for by temporal diversity, created by the continuous rapid breeding and release of new varieties by agricultural research institutes. While this may be technically true, it ignores a larger sociopolitical and logistical problem associated with entrusting diversity maintenance to centralized institutions. To maintain diversity at the regional scale, new breeds would need to be diffused at very high rates through regional farming systems. This may be feasible in developed countries but is quite impractical for much of the tropics. It would also place central control of germ plasm resources in a few centralized agencies. To date most of these have been public agencies, which release new varieties at relatively low cost, but the co-option of this activity by private for-profit agencies would effectively exclude most tropical farmers from access to this critical resource.

WEEDS

Weed infestations are the immediate cause of field abandonment in most shifting cultivation systems (Chapter 8), and it is therefore not surprising that these become a major problem when more permanent forms of field usage are attempted. For example, an experimental continuous-cropping system in Amazonia collapsed after seven crops under 'conventional' chemical weed control, but could be continued successfully with 'full' chemical weed control (Sanchez and Benites 1987). However, the cost of the full treatment was economically unsustainable: US$225 per hectare as against US$25 per hectare for the conventional treatment. Intermingling of weeds with crops can have some benefits to farmers, such as providing emergency food sources or helping to diversify predator complexes, but these benefits are usually outweighed by their negative impact on crop yields. Consequently we will concentrate upon these, and how they may best be avoided or reduced.

The negative effect of weeds on crops is primarily competitive. Some weeds, such as *Imperata cylindrica*, also have an allelopathic (chemically toxic) effect (Eussen 1979), while others may act as alternative hosts for other crop pests. Crops differ in their susceptibility to weed competition. The effects on crop yields of not weeding experimental maize/beans/squash polycultures in Mexico are shown in Table 9.3. These data indicate that in this system squash is the most susceptible to weed competition, and maize the least so. Most weeds of permanently farmed tropical fields are annual or perennial forbs and graminoids. In traditional agricultural systems these are controlled by hand-weeding and, in some instances, by use of simple animal-drawn ploughs. In padi rice systems, alternation between flooded and non-flooded conditions helps to suppress weed infestation (William and Chiang 1980). In shifting cultivation, ultimate weed control is achieved by shading from larger woody plants in fallows. Although there is a long-term trend towards increased

Table 9.3 Effects of weeds on crop yields (kg.ha^{-1}) in two experimental maize/beans/
squash polycultures in southern Mexico

	Site C-34			CSAT site		
	Weeded	*Unweeded*	*% red.*	*Weeded*	*Unweeded*	*% red.*
Maize	1,720	1,286	−25.2	262	176	−32.8
Beans	110	75	−31.8	210	128	−39.0
Squash	80	9.6	−88.0	0	0	—

Source: Amador and Gliessman 1990

homogenization of tropical weed floras on permanently farmed land (Kellman 1980), the weed community that develops in any field is likely to be quite idiosyncratic, depending on the local habitat conditions and the history of agricultural usage there. This compositional variability means that prescriptions for weed control will frequently have to be quite site-specific. Despite this, certain generalizations about the functioning of weed populations and communities can be developed that can provide guidance in the search for site-specific control programmes.

Industrial means of weed control in temperate zones have conventionally involved mechanized soil tillage at frequent intervals and, increasingly in recent decades, the use of chemical herbicides. These techniques have proven to be less successful in the tropics, especially in wetter areas where weed growth is continuous. Here, use of herbicides results in rapid changes in weed populations towards herbicide-resistant species, and has resulted in the need to apply increasingly diverse herbicide mixtures. For example, over a 2-year experiment in Amazonia, use of the standard herbicide Metachlor resulted in a rapid change in the weed community to one dominated by the aggressive grass *Rottboelia exaltata*, which was resistant to this herbicide (Mt Pleasant *et al.* 1990). While *Rottboelia* and other species could be controlled by using a variety of herbicides in combination, the high cost relative to hand-weeding (US$185 as against US$50 per hectare) led the authors to conclude that herbicide use was probably uneconomical. Biological control of especially noxious weed species, which are often exotics, by the introduction of host-specific predatory organisms has been developed primarily for temperate-area weeds and much less for those of the tropics. These programmes usually require a prolonged period of research development and, when applied in the tropics, have met with variable success (Swarbrick 1987).

After a weeding operation, weed populations re-establish from viable seed or vegetative fragments left in the soil. These carried-over diaspores may be considered a form of infection and can occur at very high densities. For example, the upper 4.2 cm of surface soil in permanent pastures in Belize was found to contain 5,226 viable weed seeds per square metre, while those of cropped fields contained 7,623 per square metre (Kellman 1974a). While weeding will give temporary relief to the crop from weed competition, rapid re-establishment of weeds can be expected if a large infection persists. Consequently, effective weed control requires not only the removing of established weed populations, but also the minimizing of the carried-over infection between weeding operations. Vegetative fragments normally lose viability rapidly if favourable conditions for vegetative regrowth do not exist, and populations of buried viable seeds normally decay at exponential rates in tilled soil (Roberts 1970). Consequently, a relatively rapid diminishment of weed infections can be expected if diaspore replenishment is prevented. However, even a low density of maturing weeds can produce large numbers of new seed, making continuous and frequent weeding operations necessary for effective control.

Even when weeding is frequent, some carry-over infection must be expected, and the most effective means of coping with this is to limit the opportunities for new establishment from this source. Mechanical control is ineffective because it exposes bare soil and the diaspores that it contains to the well-illuminated conditions that are usually important in stimulating seed germination or vegetative sprouting. Consequently, only with very frequent weeding, such as that possible for home-site gardeners, does mechanical control become an effective long-term strategy. It is unlikely to be sustainable economically at a larger scale. The alternative is to limit the opportunities for weed establishment initially, by minimizing the availability of light and and soil resources. This can be achieved by using combinations of rapidly developing crops or other plants that compete little with crops to establish soil covers that are maintained throughout the cropping cycle.

Some crops, such as sweet potato, whose above-ground organs comprise spreading vines, can rapidly establish a ground cover and minimize weed establishment. Other crops, such as maize, are much less effective at establishing ground cover, and monocultures of these crops remain vulnerable to weed infestations. The degree to which an individual crop plant can monopolize the resources needed for growth, and limit their availability to a potential weed competitor, is constrained by its above- and below-ground architecture. Consequently, combining crops of different architectures may be expected to make more effective use of growth resources and limit the opportunities for weed competition (e.g. Table 9.4). However, the success of crop mixtures in suppressing weed development is likely to be very dependent on the *timing* of crop growth relative to that of weeds, with optimum weed control likely to be provided by the maintenance of continuous crop covers at a site. Multiple cropping has other advantages beyond weed control, and will be discussed further in the sections that follow.

There has been increased interest recently in the possibility of combining various 'cover crops' with staple food crops such as maize, as a means of controlling weeds and providing other agricultural benefits. The latter include

Table 9.4 Phytomass (kg.m^{-2}) of major weed species in maize and mung bean planted as monocultures or intercropped in the Philippines

Crop	Purslane	Goosegrass	Jungle ricegrass	Crabgrass	Total
Maize	0.32	0.46	0.76	0.41	2.06
Mung bean	0.24	0.40	0.02	0.76	0.85
Maize + mung bean	0.04	0.14	0.04	0.21	0.28

Source: Bantilan and Harwood 1977

protection of the surface soil from erosion and, when the cover crop is leguminous, the potential addition of nitrogen to the soil. One of the most successful tropical cover crops used so far is velvet bean (*Mucuna* species complex), whose use has been expanding rapidly in recent years (Buckles 1995). This Asian cover crop was introduced to the African and New World tropics within the past century. In Central America, it was introduced as a cover crop in banana plantations and has subsequently been adopted spontaneously by many small farmers, who use it as a cover crop beneath maize. On the Atlantic coast of Honduras, velvet bean is used in combination with dry-season maize crops (Figure 9.4). The cover is maintained in the field at all times, but slashed, without subsequent burning, prior to the planting of

Figure 9.4 A cover crop of *Mucuna* being grown on permanently cropped land, eastern Honduras. On the left, the *Mucuna* has been slashed with a machete in preparation for replanting the site with maize

maize. This is a relatively simple operation from which the velvet bean recovers spontaneously, by which time young maize plants have grown to a height at which they are no longer competing for light with the cover crop. No further weeding is required, and a protective soil cover is preserved on the site at all times. The fertilizing effect of this cover crop will be discussed further below.

An alternative means of providing field cover and weed control is to use organic mulches comprising crop residues or other material imported to the field from elsewhere. Although these mulches have been shown to suppress weeds (Wade and Sanchez 1983), and are used in some traditional agricultural systems (e.g. Allen 1985), they suffer from the need for large farm areas on which to grow the mulches, and the high labour inputs needed to harvest, transport and apply these to fields. The potential for weed control using cover crops grown on-site appears to be much larger.

OTHER PESTS

Included here are a variety of organisms that may affect the yields of crops negatively; these are primarily herbivorous insects, soil nematodes and various micro-organisms causing diseases in crops. Herbivorous insects normally cause the most severe problems, and have been the group upon which most efforts at control have been concentrated. Although these organisms are usually of secondary importance to weeds as pests in shifting cultivators' fields, they have the potential to become severe problems when field usage becomes permanent in the tropics. There is an abundance of potential pest organisms available in most tropical biotas, and year-long conditions for population growth. Crops often have few chemical defences, and, when fertilized, may become even more attractive to herbivores. The crops are almost always grown in much higher concentrations than are characteristic in natural tropical plant communities, where isolation due to low densities appears to provide some protection (see Chapter 5). The tendency for recently domesticated crops, such as rubber, to become severely infested with native diseases when grown at high densities in their native region appears to confirm these predictions.

In natural plant communities in the tropics, outbreaks of herbivores and parasites are contained by a combination of chemical defences, host avoidance, and the existence of multiple biological interactions with 'natural enemies' that suppress population eruptions (Chapter 5). Above, we have discussed crop breeding programmes and the need for greater emphasis on breeding for pest resistance in these. Here, we will focus on the issue of how avoidance and biological interactions may best be used as a means of coping with pest infestation.

Rotation of fields among different crops has long been used in temperate agriculture as a technique providing multiple benefits, including avoidance of pest infestations. Much less attention has been devoted to this technique in the tropics. In padi rice systems, alternation of flooded and dryland cropping periods can reduce pest infestations considerably (Litsinger 1993), but it is less clear whether similar benefits accrue in rain-fed agricultural systems, where a less drastic habitat change is involved. In theory, crop rotation should be most effective against crop-specialized and sedentary pests such as nematodes and soil-borne fungi, and less effective against crop-generalist and more mobile insect, or insect-borne, pests (Litsinger and Moody 1976). One would also expect rotations to be most effective when they are among crops of very different morphology and taxonomy, which may be expected to be susceptible to different pest complexes. Most small-scale tropical farming tends to rely on a limited number of staple crops within a region, which offers limited scope for the development of diverse rotational systems. However, the large number of staple crops that are now available pan-tropically suggests that considerable potential exists for the development of crop rotation as a pest avoidance technique.

Much more research has been focused on the potential for using crop mixtures ('polycultures') in the tropics as a means of controlling pest infestations. The Janzen–Connell predation hypothesis, which was outlined in Chapter 5, predicts that herbivory should increase with increasing population density, with a simplified agricultural monoculture being at greatest risk. At a more general level, there has been a long history in theoretical ecology of equating stability of populations with community diversity; this led to the general prediction that diverse mixed-cropping systems should stabilize pest populations, and promoted much research on the topic. This simple generalization has been increasingly questioned, and there have been theoretical demonstrations that diversity can also destabilize a system (Pimm 1984). In an agricultural system, intercropping of plants that provide alternative hosts to the same pest species could well exacerbate infestation. Consequently, the right *kind* of crop diversity, rather than diversity *per se*, is likely to be necessary for potential pest control by crop mixtures. Clearly, identification of this requires an understanding of the underlying mechanisms that may operate.

In many traditional tropical agricultural systems, crops are grown intermingled, rather than as segregated patches. For example, the traditional cropping system of Central America involves a polyculture of maize beans and squash (Figure 9.5). The persistence of these polycultural cropping systems suggests that some benefits accrue to crops that are grown intermingled with other crops. An extensive review by Risch *et al.* (1983) of 150 studies which examined insect pest abundance in agricultural systems of differing diversity showed that 53 per cent of species were less abundant in more diversified systems, 18 per cent were more abundant, 9 per cent showed no difference and 20 per cent showed a variable response. This indicates an effect of diversity which is predominantly beneficial in pest control, but not universally so. Power (1990) has also assembled a list of examples in which crop disease incidence was lower in more diverse cropping systems. Two general

Figure 9.5 A polyculture of maize, squash and beans being established in a shifting cultivator's field, Belize. At this stage of crop development bean plants remain small, but maize plants are beginning to elongate vertically, while squash vines have begun to spread over the soil surface (where weed seedlings are beginning to emerge)

hypotheses have been advanced to account for why more diverse cropping systems should have fewer pests: the 'resource concentration' hypothesis and the 'natural enemies' hypothesis (Root 1973). The first represents a generalized form of the Janzen–Connell hypothesis, while the second proposes that a diverse community preserves a more diverse array of predators and parasites which serve to regulate the pest populations. Risch *et al.*'s (1983) review of the available data led them to conclude that there was more evidence in support of the 'resource concentration' regulation mechanism among herbivorous insect pests, with diversity being most effective in impeding pest movements in complex systems.

This conclusion does not preclude the 'natural enemies' process being important in certain circumstances. The seriousness of pest outbreaks in areas where broad-spectrum pesticides have been used excessively (Joyce 1988) is strong circumstantial evidence for the pest having been controlled previously by a natural predator complex. Examples also exist of effective predator control being sustained by natural enemy complexes that are maintained within fields or in adjacent non-farmed areas. One of the most effective predator complexes in the tropics is provided by ants (Carroll and Risch 1990). Many of these are generalist predators which attack a range of herbivorous insects, and may also consume smaller seeds, especially those of weedy grasses. Their removal by insecticidal treatment can result in a dramatic resurgence of the crop pests on which they previously preyed (Perfecto 1990). Ants are abundant in the tropics, easily manipulated, and establish stable territories. Their greatest potential as pest control agents appears to lie in long-lived perennial tree crops, where a stable mosaic of ant territories may develop (Leston 1973). While ant biotas offer a generally promising resource for biological pest control, it must be stressed that this is not necessarily always so. For example, in the Neotropics, leaf-cutting ants are often one of the most serious herbivorous pests of agricultural crops (e.g. Blanton and Ewel 1985). In citrus trees in Trinidad, these were found to be controlled by populations of another ant, the *Azteca* fire ant (Jutsum *et al.* 1981).

In the wake of failures by pesticides to provide an effective long-term solution to pest infestations in agriculture generally, there has been increased interest in the development of 'integrated pest management' (IPM) schemes for different agricultural regions (Batra 1982). Several examples of these in tropical areas are provided by Brader (1979). Most IPM schemes seek to combine various biological means of pest control, such as inter-cropping or even the introduction of exotic biological control organisms, with more limited use of pesticides to reduce pest infestations to economically acceptable levels. While many dramatic improvements in pest control have been achieved using IPM techniques, their continued partial reliance on synthetic pesticides raises questions about their long-term value in small-scale tropical agriculture. Here, pesticides remain seductive to farmers as a means of short-term relief from unexpected pest outbreaks, and reversion to their use is a persistent problem (Brader 1979). Advice on pesticide use to technically unsophisticated farmers is often available only from marketing agents, who have little incentive to reduce their use, and the pesticides are often misapplied. Moreover, effective IPM schemes involve considerable research effort both to establish and to maintain; this involves costs which can rarely be afforded in tropical countries. Fortunately, recent IPM schemes are increasingly moving towards developing management stategies that minimize pesticide use and draw upon farmer participation (Litsinger 1993).

Tropical agriculture can encompass a very large number of potential cropping patterns and procedures, any of which can potentially have an augmentative or depressive effect on pest complexes. This makes for a seemingly insurmountable experimental task in identifying ideal agricultural systems for pest control *de novo*. In Chapter 11, we will suggest that an important means of screening this complexity for useful procedures is to use the folk knowledge of indigenous farmers as represented in their traditional farming techniques. These represent the outcomes of prolonged periods of informal experimentation, and can form the basis for further, more formal, experimentation on understanding how these systems work, and how best they can be augmented (e.g. Gliessman *et al.* 1981; Marten 1986). The inherent potential for pest populations to develop to epidemic proportions unpredictably, means that a premium will need to be placed upon the preservation of operational flexibility when developing low-input pest control strategies. Not only will risk-spreading need to be preserved in terms of crop diversity, but alternative cropping procedures must be available in the event of unforeseen pest eruptions. Pest control strategies will also be applied to rural landscapes that are heavily settled. They will therefore need to be co-ordinated regionally, to ensure that measures employed by one farmer are not overwhelmed by pest migrations from adjacent farming units.

SOIL FERTILITY

Plants in natural tropical ecosystems cope with the low fertility of many tropical soils by a combination of biological and non-biological processes that fix nitrogen, suppress acute leaching losses, facilitate root absorption and promote nutrient recirculation (Chapter 4). These are sufficiently effective to permit very small inputs from atmospheric sources to compensate for any leakage from the system, and to allow the development of large nutrient sinks in the biomass of forest communities, if protected from extraneous disturbance (Kellman 1989). Key elements in these processes of nutrient conservation involve high soil organic matter contents and the preservation of aggregated soil structure, large investments by plants in below-ground tissues, and, often, slow growth. Few of these properties occur naturally in agricultural systems. Soil organic matter normally declines rapidly after natural vegetation is cleared (Chapter 8). Crop plants have been selected for rapid growth and high yields of edible tissues in the presence of high nutrient concentrations, rather than for their nutrient-conserving properties. Most critically, a proportion of the nutrients absorbed by crops is removed from fields, rather than re-entering the decomposer cycle on site.

Estimates of the quantities of nutrients removed in several tropical crops are provided in Table 9.5, together with estimates of nutrient inputs that may be expected from atmospheric sources and soil weathering. The latter input could be expected to be accessible to plants only in more recent soils, where significant weathering continues in the rooting zone. These data show that the crop removals of calcium and magnesium can probably be offset by net inputs on those tropical soils that contain weatherable minerals in the upper solum, including alluvial and recent volcanic soils, vertisols and those on limestone terrain. Potassium inputs are also probably sufficient to offset losses in basic food crops (provided only single-cropping is employed) but would be insufficient to meet the large losses of this element in bananas and cacao. However, removals of nitrogen and phosphorus in crops can be expected to be far in excess of any net input available from weathering or atmospheric sources, identifying these two elements as those most likely to become rapidly exhausted in permanently farmed fields. When further

Table 9.5 Estimates of nutrient removals in different crop types and nutrient inputs from weathering and atmospheric sources

	N	P	K	Ca	Mg
Crop removal					
Maize	23.9	4.3	4.3	0.3	1.6
Rice	32.3	8.6	9.4	2.2	3.8
Peanut	34.3	3.1	5.5	1.1	NA
Cassava	6.4	0.9	9.5	1.6	NA
Yams	7.1	0.6	7.4	0.1	NA
Bananas	8.6	1.6	21.5	0.2	NA
Cacao beans	13.5	3.4	11.2	NA	NA
pods	11.2	1.1	24.7	NA	NA
Inputs					
Weathering	—	0.2	7.6	7.6	3.4
Atmospheric	0.3	0.2	3.7	2.2	0.4
Total	0.3	0.4	11.3	9.8	3.8

Sources: Crop removals calculated using tropical yield averages from Norman *et al.* 1995, and nutrient concentrations quoted by Nye and Greenland 1960, except for cacao, where yields of Nye and Greenland 1960 used. Weathering input estimates from Lewis *et al.* 1987, and atmospheric inputs from Kellman and Carty 1986

Notes: All data in kg.ha^{-1}.yr^{-1}, and assume only a single crop grown per year

potential nutrient losses, due to accelerated leaching and erosion in fields, are added to those of crop removals, it is clear that soil fertility is very likely to become limiting to permanent agricultural production eventually, even on relatively fertile tropical soils. On the more common oxisols and ultisols, this limitation is likely to develop very rapidly.

Industrial solutions to this fertility problem have involved large inputs of inorganic fertilizers, principally N, P and K. Greenwood (1981, 1982) has argued that this solution remains economically profitable in developing countries, providing a mean 'value for cost' ratio (increased value of crop divided by fertilizer cost) of 4.8. However, he acknowledges that many tropical farmers do not have the financial resources to purchase fertilizers. Those intensified traditional tropical agricultural systems that have successfully coped with the problem of declining soil fertility have done so either by importing nutrients from elsewhere (home-site gardens), or by using the high nutrient influxes that occur naturally in floodplains and wetlands. Shifting-cultivation systems avoid the problem by early site abandonment. Nutrient exports in crops represent an inevitable 'cost' of permanent agriculture. The magnitude of this cost will vary with the yields that are considered acceptable, and these will vary from place to place depending upon the local cost of fertilizers, and the increased yields that can be expected with their application. For most tropical farmers, coping with the soil fertility dilemma will require the adoption of techniques that maximize nutrient recycling and minimize nutrient losses, while reducing or eliminating the need for external fertilizer imports to the farm.

Nitrogen sources

More nitrogen is lost in crop removal than any other mineral nutrient (Table 9.5), and the energy cost of producing this element in fertilizer is several orders of magnitude greater than that of producing phosphorus and potassium (Lockeretz 1980). As a result, it represents one of the largest energetic inputs to industrial agriculture (Tables 9.1 and 9.2), making its use both costly at present and likely to become increasingly more so as fossil fuels are depleted. Fortunately, the tropics contains large numbers of organisms capable of fixing atmospheric nitrogen, including both free-living organisms and symbionts of higher plants, primarily species of the family Leguminosae. The incorporation of these organisms into tropical agricultural systems offers considerable potential to free farmers from the need to apply nitrogen fertilizers (Mulongoy et al. 1990).

The existence of nitrogen fixation in irrigated rice padis has been largely responsible for the success and sustainability of this intensive form of traditional agriculture (Chapter 8). However, a large number of other N-fixers are capable of performing this function in non-flooded soils. Plant species capable of symbiotic fixation include a wide range of life forms, from herbaceous leguminous crops, cover crops, and a variety of shrubs and trees. Table 9.6 includes estimates of the N-fixing capabilities of a variety of these plants. Clearly there is a great deal of variability both among and within species; the latter probably reflect differences in the availability of appropriate rhizobial inocula in the soil, which can be quite variable and may need augmentation in agricultural systems (Peoples and Craswell 1992). Crops capable of nitrogen fixation are potentially independent of the need for N fertilization. Although higher yields may be achieved if this is provided, increased nitrogen fixation is normally stimulated by low availability of this element in the soil (Peoples and Craswell 1992), providing a strong incentive not to fertilize leguminous crops. Use of N-fixers to supply nitrogen to other crop plants involves a much less direct benefit, and requires that this crop make use of nitrogen that is first fixed, but later released into the soil. To have net utility, it also requires that the additional benefits provided to the crop by nitrogen fixation exceed potential negative effects caused by competition for light and other soil resources. Leguminous crops that yield less, but leave more nitrogen in the soil for subsequent crops, are likely to be the most suitable N-fixers for incorporation into low-input agricultural systems (Giller et al. 1994).

Leguminous trees planted as rows ('alley crops') within fields often suppress, rather than stimulate, yields of crops in nearby rows (Lal 1991), suggesting that there is no net benefit to the crop. Conversely, herbaceous cover crops

Table 9.6 Estimates of nitrogen fixation (kg.ha^{-1}, per crop or year) by a range of tropical plants

Plant	N fixation (kg.ha^{-1})
Crops	
Beans	3–57
Peanuts	37–206
Soybeans	0–237
Chickpea	37–97
Pigeon pea	68–88
Cover crops	
Mucuna pruriens	250–350
Pueraria phaseoloides	9–23
Centrosema pubescens	150
Stylosanthes spp.	2–84
Trees and shrubs	
Acacia albida	20
Acacia mearnsii	200
Allocasuarina littoralis	220
Casuarina equisetifolia	60–110
Erythrina poeppigiana	60
Inga jinicuil	35–50
Leucaena leucocephala	100–500

Sources: Data from various sources assembled by Peoples and Craswell 1992 and Norman *et al.* 1995, and from Triomphe 1996

beneath tree plantations such as rubber can stimulate yields (Peoples and Craswell 1992). When synchronously established polycultures of short-lived leguminous and non-leguminous crops are grown together, there is probably little direct transfer of nitrogen to the non-fixer (Norman *et al.* 1995), and any benefits accrue only to the succeeding crop. However, where the N-fixer is planted in rotation, considerable benefits may accrue to the following crop. For example, in the maize and velvet bean cover-crop system used in northern Honduras (Figure 9.4), corn crops have been grown for the past 16 years without fertilization, and have provided yields approximately double those achieved without the *Mucuna* (Triomphe 1996; Triomphe pers. comm.). Experiments with this species in Brazil have also shown it to have a capacity to suppress nematodes (Lathwell 1990). Where tree crops, such as cacao, are grown beneath leguminous shade trees (Figure 9.6), nitrogen fixed by the latter is potentially available to the crop when being recycled as litter. For example, Aranguren *et al.* (1982) have concluded that the output of nitrogen in the harvest from a Venezuelan cacao plantation was more than compensated for by nitrogen inputs in the litter from shade trees. There would thus appear to be many opportunities for the incorporation of N-fixing plants

Figure 9.6 A cacao plantation, Dominican Republic. The small understorey cacao trees are being cultivated beneath a cover of leguminous shade trees

into a variety of tropical agricultural systems as crops, cover-crops, components of rotations and shade trees. Giller *et al.* (1994) conclude that biological nitrogen fixation will be unable to sustain the very high crop yields that have been achieved by fertilization with inorganic nitrogen, but that with suitable experimentation the process can probably lead to at least a doubling of the crop yields achieved by most traditional farming practices. A potential problem may arise if reliance is placed on one, or only a few, N-fixing plants, as these may become virtual regional monocultures, and so be especially susceptible to pest and disease outbreaks.

Other nutrients

For nutrients other than nitrogen that are likely to become limiting on crop production (primarily phosphorus and potassium), a biological fixation source does not exist, which provides for a far less tractable problem. Only in two situations are there likely to be sufficient natural inputs of these sorts of elements. (i) In floodplains of rivers that carry large sediment loads from actively eroding mountain ranges, sufficient quantities of nutrients may be added. (ii) In the mountain ranges themselves, active erosion may serve to preserve a shallow solum with an abundance of weatherable minerals within the rooting zone of plants. However, under agricultural usage, problems may arise in both habitats. Use of floodplains for agriculture usually involves water control measures, and these often reduce sediment input to fields; when an upstream dam has been established, such as the High Aswan Dam on the Nile River, sediment inputs may effectively cease altogether. Conversely, agriculture in steeply sloping terrain is likely to produce such accelerated rates of erosion that soil thinness becomes limiting to agricultural production.

Tropical soils often contain appreciable quantities of occluded phosphorus – that is complexed with iron and aluminium oxides – and further 'fixation' of phosphorus in fertilizer added to these soils is a persistent problem in tropical agriculture. This occluded phosphorus represents a reservoir that may become slowly accessible to plants, especially through mycorrhizal activity. With the exception of padi rice, most major tropical crops contain vesicular-arbuscular mycorrhizas (VAM) (Howeler *et al.* 1987), and inoculation can significantly increase P uptake and reduce fertilizer needs (Table 9.7). The effectiveness of different VAM species varies, so introduction of appropriate inocula, and management of soil to preserve these, can be an effective strategy to ameliorate phosphorus deficiencies (Howeler *et al.* 1987). Pesticides can depress VAM populations, providing a further reason to avoid their use. Although mycorrhizal infections can greatly facilitate access to otherwise unavailable P reserves in tropical soils, they provide no net addition to these. Consequently, under prolonged cropping of tropical soils that lack inputs of P or other nutrients from weatherable minerals, artificial fertilization is likely to become a necessary condition of further usage.

Some of the most complete studies on the potential for fertilizer usage on infertile oxisols and ultisols have been carried out at Yurimaguas in Amazonian Peru, leading to what has been termed the 'Yurimaguas Technology' (Sanchez *et al.* 1982). This work has shown that high crop yields can be maintained on these soils, using rotations of rice, corn, soybeans and peanuts, and large fertilizer inputs of N, P, K and Mg, combined with small inputs of the micro-nutrients Cu, Zn, Bo and Mo. A benefit : cost analysis indicated that these

Table 9.7 Effects of inoculation with native vesicular-arbuscular mycorrhizal fungi on plant growth in a Colombian ultisol, without fertilization

Plant species	Non-inoculated	VAM-inoculated
Cassava	0.34	4.33
Beans	1.11	3.08
Maize	1.19	4.84
Rice	3.79	3.83
Cowpea	0.96	2.60
Stylosanthes guianensis	0.08	1.25
Andropogon gayanus	0.15	1.26

Source: Howeler *et al.* 1987

Note: Data are above-ground phytomass (g) per experimental pot inputs were economically sustainable on the basis of prevailing prices for fertilizers and crops in the area. The economic feasibility of this technology has been questioned by Fearnside (1987) on several grounds, however. He points out that development of the technology required large and continuing research costs in terms of frequent soil analyses to guide fertilizer applications, and questions whether these resources would be either available to, or afford-able by, farmers. He also concludes that both fertilizer cost and crop prices were indirectly subsidized and suggests that fertilizer costs will almost certainly rise as finite world supplies become depleted. On this basis, Fernside questions the economic feasibility of the technology, and points out that few local farm-ers were adopting the techniques. Subsequent research at Yurimaguas has emphasized lower-input solutions (Sanchez and Benites 1987). This has shown that acceptable crop yields were sustainable over a 3-year period with less fertilization, but the system collapsed unless uneconomically high herbicide treatments were used to suppress weeds. This system was thus promoted as a 'transitional system' to some other, presumably more viable, agricultural system. Insofar as this low-input system could be sustained for no longer than a shifting cultivator's cropping period, it hardly appears to be a sustainable solution.

From this work, it is difficult to conclude that large fertilizer inputs repre-sent a viable solution to the problem of how to sustain intensive cropping indefinitely on low-fertility tropical oxisols and ultisols. Where the necessary research infrastructure is available and affordable, and cash crops of sufficient value are produced, a heavy-fertilization strategy may be temporarily successful, especially when practised on very large management units. For example, in the Cerrado region of Brazil, a large expansion of intensive crop production (principally soybeans, upland rice and maize) has taken place over the past three decades on large farms (Klink et al. 1993). This uses heavy inputs of lime to raise pH and suppress aluminium toxicity, and large appli-cations of phosphorus fertilizer (Villachia et al. 1990; Goedert 1983). Both materials are relatively cheaply produced and available in Brazil, and their use is subsidized by government agricultural policies. There has been some recent interest in the potential of mechanically ground, but otherwise untreated, rock phosphates as a cheap source of fertilizer in tropical countries that possess deposits of these materials (Hammond et al. 1986). Work on these materials indicates that their major limitation is the slow release of P to crops, but suggests that they may have some potential for use with acid oxisols and ultisols, and in flooded rice agriculture. Hammond et al. (1986) recommend that they be used as a supplement to fast-release P fertilizer sources, but provide no cost–benefit analysis of their potential when used as the sole source of phosphorus. To the very large numbers of small tropical farmers, in countries without fertilizer reserves or government policies that subsidize fertilization, there would appear to be few alternatives to accepting cropping systems that offer lower crop outputs. However, these crop yields can be sustained by adopting measures that maximize nutrient conservation and recycling in their agricultural systems.

Nutrient retention and recycling

Nutrients are lost to agricultural systems in crop exports and in system 'leakage' in the form of leaching and erosion. The most obvious way to minimize losses is clearly to remove as few nutrients as possible in crop removal from fields, or to return unused residues to fields. For example, return of cacao pods to fields after beans are removed would drastically reduce the nutrient drain from cacao plantations (Table 9.5). Up to one-third of the nitrogen removed in coffee beans can be returned in the form of processing waste (Bornemisza 1982). These forms of recycling activity are clearly quite labour-intensive.

The magnitude of nutrient losses by way of eroded sediments from different forms of agricultural use is relatively well documented, and will be treated in the following section. The magnitude of losses by way of leaching is, in general, very difficult to measure, and many of our conclusions about how best to minimize such losses are based on inference. Losses by this pathway will reflect the combined effects of the ambient concentration of nutrient ions in the soil solution, and the flux of soil water through the solum. Consequently, processes that minimize both may be expected to reduce leakage. Bare soil in agricultural fields may be expected to experience maximum leakage for several reasons. No plant uptake is occurring, and in most tropical soils a limited exchange capacity exists to otherwise immobilize nutrient ions. Moreover, agricultural soils will often have lost soil organic matter, which may be expected to lead to lower cation exchange capacity (CEC), and a less well-developed macro-pore structure. Consequently, any cropping system that minimizes soil exposure, even if cover is provided only by a weed community, may be expected to minimize leaching losses.

In addition to providing effective soil covers, polycultures of crops are often assumed to exploit soil moisture and nutrient resources more effectively than monocultures, and in so doing suppress leaching losses. This effect has recently been demonstrated experimentally by Tilman et al. (1996). Tree crops, including rubber or other crops requiring a covering tree stratum, such as cacao and coffee (Figure 9.6), have often been assumed to be very effective at nutrient retention for the same reasons that natural forest communities are: soil organic matter and CEC remain high, soil structure remains well developed, and a permanent root system is deployed. Santana and Cabala-Rosand (1982) used lysimeters buried in soil beneath a cacao plantation in Brazil to estimate leaching losses. While acknowledging the difficulty of quantifying these losses, they concluded that they were probably minor.

In traditional farms, home-site gardens represent places where nutrients are concentrated and reused, representing effective nutrient retention at the farm-unit scale. However, this concentration comes at the expense of nutrient removal from surrounding areas of the farm, and unless these areas are very extensive, they may eventually suffer nutrient depletion. A fully integrated farm-scale recycling system would require that nutrients in organic residues be returned to their source areas in the form of compost, manure or night-soil. Traditional Chinese agriculture probably goes further than most other agricultural systems in such farm-scale recycling, with fish cultured in ponds and small-animal rearing being incorporated into the nutrient recycling process (e.g. Guo and Bradshaw 1993). As in the case of crop waste return to fields, such a system may be expected to require very large labour inputs.

SOIL EROSION

The gradual loss of surface soil materials is a normal process in undisturbed ecosystems, and this loss is usually compensated for by continual weathering of soil parent materials at the base of the solum. Erosion is achieved primarily by the action of moving water, but wind erosion may be added to this in semi-arid tropical environments. In the more humid parts of the tropics, basal weathering rates often exceed surface erosion beneath tropical forest, except in very steep terrain, leading to a progressive thickening of the weathered mantle and remoteness of the weathering zone from plant roots (Chapter 4). Under these circumstances, it could be argued that, in theory, accelerated rates of soil erosion may be beneficial, allowing plant roots more ready access to the weathering zone and the nutrients being released there. However, the massive watershed-level readjustments in sedimentation and hydrology that extensive soil thinning would require make this an impractical solution to soil nutritional problems. Instead, most attention is focused on how to minimize surface soil erosion rates under agricultural usage, and preferably to maintain these at levels comparable to those prevailing under undisturbed vegetation. The materials lost in most erosional processes are surface soil materials. As these normally contain the highest contents of soil organic matter and soil nutrients, loss of this material results in a disproportional loss of the soil resource most important for agriculture (cf. Table 9.8). While here we are concerned primarily with the negative effects of these removals, the deposition of eroded materials elsewhere, especially in river channels, can have large negative effects on the 'downstream' systems (El-Swaify 1993), although occasionally eroded sediments can provide a nutrient source to agricultural systems in these areas.

Accelerated soil erosion occurs on the fields of shifting cultivators (Table 9.11, Figure 9.7), but the periods over which these farmers use their fields are normally too short for the negative effects of this process to become obvious to them. Even under permanent field usage, soil loss due to erosion tends to be a gradual process whose negative effects are cumulative, but seldom so immediately obvious that they promote attempts at remedial

Table 9.8 Enrichment ratios of nutrients in sediments eroded from experimental erosion plots in Nigeria

Treatment	Organic C	N	P	K	Ca	Mg
Bare fallow	2.21	1.54	5.96	1.36	1.52	1.12
Maize-maize (ploughed)	2.75	1.73	5.55	2.02	1.45	1.16
Cowpeas-maize (ploughed)	2.22	1.65	5.83	1.62	1.62	1.23

Source: Lal 1976b: Table III

Note: Ratios are based on the concentration of materials in sediments in comparison to their concentration in the upper 5 cm of surface soils

action, until accumulated sediment loss has become severe. This is especially true where large industrial inputs, especially of inorganic fertilizers, are used to offset the effects of erosion. Pimentel *et al.* (1995) estimate that approximately 10 per cent of all energy inputs to US agriculture are used for this purpose. Small-scale tropical farmers, who do not have access to these large inputs, are especially vulnerable to the negative effects of erosion. This may make them more receptive to the idea of taking remedial measures.

The process of erosion involves the detachment of soil particles from larger soil aggregates, followed by their movement, normally in a downslope (or downwind) direction. Splashing by individual raindrops is the primary mechanism of particle detachment (Lal 1990), with the detached particles subsequently being subject to downslope displacement due to gravity or, more commonly, in water moving across the surface as overland flow. Overland flow may itself cause some particle detachment, and normally concentrates rapidly into a system of rills, down which detached particles are moved in suspension. These may be deposited locally as colluvial 'footslope' deposits, or continue in suspension into the stream system. Two conditions have a major influence on erosion rates: the degree of exposure of the soil surface to raindrop impact, and the infiltration capacity of the surface soil. The susceptibility of a soil to being eroded depends also on the degree of aggregation of soil particles, and the stability of these aggregates. Most highly weathered tropical soils have stable soil particle micro-aggregates as a result of the abundance of iron and aluminium oxides and their dominance by variable-charge clays (Sollins *et al.* 1988). This inherent stability is greatly increased by the presence of soil organic matter. As soil particle aggregation both promotes a large pore structure and rapid infiltration of water, and resists soil particle detachment, high surface soil organic matter content normally serves to suppress erosion. When the soil is also covered by a protective litter layer, as it is in tropical forests, very low soil erosion rates may be expected.

Two other conditions also have an important effect on soil erosion rates: intensity of rainfall, and surface slope. More intense rainfall increases the

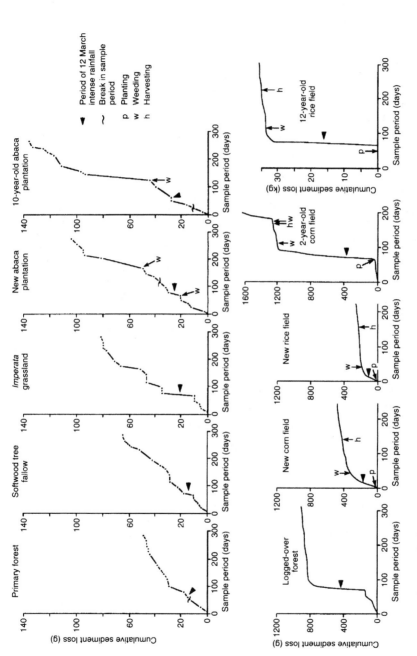

Figure 9.7 Cumulative sediment losses recorded on experimental erosion plots with different covers located in close proximity to each other and maintained for up to 300 days in Mindanao, the Philippines. Plots were 8 m long and 1 m wide and located on slopes of approximately 25%. Note the variable magnitude of the sediment scales

Source: Kellman 1969

Table 9.9 Effects of soil slope on soil losses (t.ha^{-1}.yr^{-1}) from experimental erosion plots in Nigeria subjected to different treatments

Treatment	1% slope	5% slope	10% slope	15% slope
Bare fallow (ploughed)	11.2	156.2	232.6	229.2
Maize (ploughed)	1.6	11.0	7.2	40.7
Maize (ploughed, mulched)	0.0	0.0	0.2	0.0

Source: Lal 1976a

Note: The mulch was rice straw applied at 6 t.ha^{-1} twice each year at the time of crop planting

probability that a soil's infiltration capacity will be exceeded and overland flow will begin. In the tropics, rainfall is primarily convective in origin, and rainfall intensities tend to be higher than in temperate areas (Chapter 3); as a consequence, higher erosion potentials exist. Raindrop size also tends to be larger in convective storms, increasing the kinetic energy available for particle detachment in the soil splash process (Lal 1990). Increases in soil surface slope also tend to increase erosion rates by promoting greater downslope displacement of soil particles in the splash process, and by increasing the velocity of overland flow and hence its capacity to detach and transport soil particles. This slope effect is illustrated in Table 9.9 with data from experimental erosion plots in Nigeria. These show that the effect is most pronounced when soils are exposed, and virtually absent when an effective soil cover is provided in the form of mulch.

Agriculture promotes accelerated rates of soil erosion primarily by increasing the exposure of surface soil to rain splash, and secondarily by reducing the soil's infiltration capacity. The latter effect is usually a result of lowered soil organic matter levels, but where machinery is used, soil compaction may exacerbate the effect. Soil exposure often varies throughout a cropping period, and so also does the field's vulnerability to erosion, and the co-occurrence of intense rainfall and a period of low cover can result in very large erosional losses. An example of this is provided in Figure 9.7, which shows cumulative sediment losses from a series of experimental plots in the Philippines. Most of the year's soil loss in rice and corn fields took place on a single day of intense rainfall that occurred soon after planting of crops. In contrast, permanently vegetated plots showed few signs of increased soil loss during the same rainfall event, and in the abaca plantation (a tree crop) accelerated erosion was more closely linked to weeding activities than rainfall intensities. The single most important means of reducing accelerated soil erosion during agricultural usage of a site is by minimizing soil exposure. Secondary means of erosion control involve promoting high soil infiltration capacities by increasing soil organic matter content, and by avoiding the use of steep slopes.

Table 9.10 Effect of cover by *Mucuna* on runoff and erosion from small (3- to 4-ha) watersheds in Nigeria during one day of intense rainfall

Parameter	*With* Mucuna *cover*	*Without* Mucuna *cover*
Runoff (mm)	4.04	10.01
Duration of surface flow (h)	1.30	2.12
Erosion (kg.ha⁻¹)	76	560

Source: Lal 1990

Coverage of the soil surface by either living or dead plant material dissipates raindrop energy and prevents surface soil splash. However, to be effective, this coverage needs to be close to the soil surface. Non-tillage agriculture, which leaves crop and weed residues in place and avoids mechanical disturbance to the soil surface, is an effective means of reducing erosion. Soil coverage is provided by crops, but can be greatly augmented by the use of cover crops, or dead organic material (mulch) produced by crops and weeds in the field, or imported from elsewhere. Various crops differ in their ability to provide soil coverage, and in the speed with which this coverage develops. For example, sweet potato can provide an extremely effective soil cover, but this develops only gradually. Relay cropping, in which crops are interplanted in sequence, is likely to maximize soil coverage. An alternative, and especially effective, means of suppressing erosion is to preserve a permanent cover crop on a field, into which crops are planted. The effectiveness of *Mucuna* (Figure 9.4) in this role is illustrated in Table 9.10.

Tree crops do not, in themselves, suppress erosion unless a well-developed litter layer, or layer of understorey vegetation, is allowed to develop beneath the trees. In the absence of this, tree canopies can concentrate rainfall into very large drops, which reach terminal velocity in about 8 m and can lead to severe erosion (Brandt 1988). This effect is shown in Table 9.11, where tree plantations without a litter or weed cover exhibit especially large erosional losses. Thus, attempts to combine annual crops with a tree cover, as has become popular in various 'agroforestry' schemes, have the potential to create severe erosion problems unless great care is taken to ensure a complete ground cover. In contrast, combining tree crops with an understorey cover crop or layer of mulch can result in very little erosion. The effectiveness of a mulch at suppressing soil erosion is also shown in Table 9.12, where this treatment to experimental plots reduced erosion rates to values comparable to those beneath a natural grass cover. Clearly, the common practice in the tropics of burning crop and weed residues, rather than using these as a mulch, is likely to promote erosion. Mulches also reduce evaporation from the soil surface, which may be especially beneficial in semi-arid areas. Maintaining a continuous soil cover by means of crops, cover crops or mulches not only reduces erosion but also tends to suppress weeds and enhances on-site nutrient retention in an organic compartment, so multiple benefits accrue from promoting these sorts of techniques.

Table 9.11 Surface erosion (t.ha^{-1}.yr^{-1}) in tropical forest, shifting cultivation and various tree crop systems

Cover type	Min.	Median	Max.
Natural tropical forest (18/27)	0.03	0.3	6.2
Shifting cultivation fallow (6/14)	0.05	0.2	7.4
Tree plantations (14/20)	0.02	0.6	6.2
Multi-storeyed tree gardens (4/4)	0.01	0.1	0.15
Tree crops with cover crop/mulch (9/17)	0.10	0.8	5.6
Shifting cultivation field (7/22)	0.4	2.8	70.0
Agricultural intercropping in young forest plantations (2/6)	0.6	5.2	17.4
Tree crops, clean-weeded (10/17)	1.2	48.0	183.0
Forest plantations, litter removed or burned (7/7)	5.9	53.0	105.0

Source: Wiersum 1984

Note: Data in parentheses represent number of locations and number of treatments, respectively

Table 9.12 Effects of soil cover on erosion (t.ha^{-1}.yr^{-1}) from experimental plots on steeper slopes in Tanzania

Treatment	10% slope	19% slope	22% slope
Natural grass cover	0.08	0.14	0.10
Mulched	0.12	0.19	0.18
Bare fallow	37.8	92.8	88.2

Source: Ngatunga *et al.* 1984

Note: The mulch was applied as straw at 6 t.ha^{-1} twice per year

Sloping terrain is clearly best avoided for agricultural usage, except for such erosion-inhibiting systems as tree crops which provide an effective ground cover. However, population pressures frequently force the use of sloping terrain for annual crops, which precipitates potentially severe erosion. In these circumstances, some form of mechanical slope modification is usually needed to alleviate the problem. In terrain of moderate slopes, ridges and furrows ploughed along the contour can arrest overland flow. In more steeply sloping terrain, the construction of terraces may be necessary, or the creation of barriers behind which eroding sediments can accumulate to form a terrace. These techniques will be more fully examined in Chapter 11.

In the semi-arid tropics, water-driven erosion may be augmented by wind erosion during the dry season. The preservation of soil coverage suppresses wind as well as water erosion, but may be more difficult to maintain continuously in areas with a prolonged dry season and lower phytomass production. Reducing the velocity of wind at the soil surface is the second major technique for suppressing wind erosion. This can be achieved by the preservation of standing crop residues on fields throughout the dry season, and by the planting of windbreaks of trees across the prevailing dry-season wind direction. The beneficial effects of such windbreaks can be felt at distances up to 20 times the height of the trees used (Lal 1990).

THE LABOUR COSTS OF SUSTAINABLE AGRICULTURE

The material discussed in earlier sections illustrates that techniques exist for practising agriculture in tropical environments in an environmentally sustainable way. However, most of the techniques discussed require relatively large inputs of human labour, making it unlikely that they will be attractive to farmers as long as alternatives exist which are economically feasible. These alternatives principally involve mechanization of farm operations. For example, farmers in the Mexican tropics have shown little enthusiasm for the adoption of the raised-field system of their ancestors despite active promotion of these techniques, because although the system is productive it is also extremely labour-demanding. Sustainable techniques are likely to become more attractive when they can be shown to minimize labour; a major attraction of the *Mucuna* cover-crop system (Figure 9.4) is its ability to control weeds, whose control by hand is inordinately labour-intensive (Chapter 8). They are also more likely to be attractive where farming is being practised in terrain so steep, or on farm units so small, that mechanization becomes impractical. Fortunately, these are likely to be places where adoption of sustainable techniques will have the greatest impact on slowing rates of environmental deterioration. In flatter terrain, and on larger farm units, labour-intensive forms of sustainable agriculture are unlikely to become attractive until fossil fuel costs increase appreciably, or the negative effects of high-input agriculture become manifest.

A potentially promising means of reducing human labour inputs into sustainable forms of agriculture is increased use of animal power on farms. Horses, burros, mules, water buffalo and oxen have been important components of traditional agricultural systems, but have tended to be displaced by machines in the past century. Kemp (1987) estimates that draught animals still provide a source of power to the agriculture of developing countries that is approximately equal to that provided by tractors, and that the animal population is utilized at well below its total power capacity. Increased use of animal power can dramatically decrease the human labour inputs needed in various agricultural operations, especially if efficient 'low-technology' animal-drawn implements are used (Kemp 1987). A return to this source of farm power, and elaboration of techniques for making optimum use of it, represents a potentially rewarding area for research in sustainable tropical agriculture.

10

ANIMAL PRODUCTION AND UTILIZATION IN THE TROPICS

Animals make many contributions to tropical societies. One of the most important of these is food, especially meat, eggs and dairy products. Animal protein makes up a much smaller proportion of total dietary protein in most tropical countries than in temperate countries (Figure 10.1). In contrast to the situation in most temperate regions, where dietary protein levels are generally more than adequate, protein–energy malnutrition is a major health problem in the tropics, especially among children (Golden 1993). Animal products are generally ranked more highly than plant foods as a protein source owing to their high content of essential amino acids (Table 8.2), and are a good source of many other nutrients (Southgate 1993). Although meat is not an essential dietary component, supplementation of plant-based diets with small amounts of suitable animal products can make a nutritionally complete diet easier to achieve (Garlick and Reeds 1993). The efficiency with which animals turn plant matter into protein varies among species (Lovell 1979), but in general they are relatively inefficient as protein-producing systems. Their production primarily for this purpose is likely to be desirable only in certain circumstances such as on a small scale to supplement protein quantity and/or quality in a plant-based diet, as a way of producing food from lands which are marginal or unsuitable for crop production such as those in arid and semi-arid regions, and as a method of making use of resources such as crop residues or unused land within cropping systems.

Animals also fulfil numerous other functions in the tropics. Both wild and domestic animals are used to supply products such as hides and fibre. Manure is a valuable source of fertilizer and fuel, and is sometimes used as a building material for dwellings (Figure 10.2). In some societies, livestock are considered as a form of currency, a means of accumulating assets, and may have religious significance. The use of draught animals is still widespread throughout the tropics, and constitutes the most important source of power for working agricultural lands on small farms in many regions. Ninety-five per cent of Thailand's rice production is estimated to depend on draught buffalo and cattle, and most agricultural power in many other Asian countries is also derived from animals (Chantalakhana 1981). In Mali, 95 per cent of

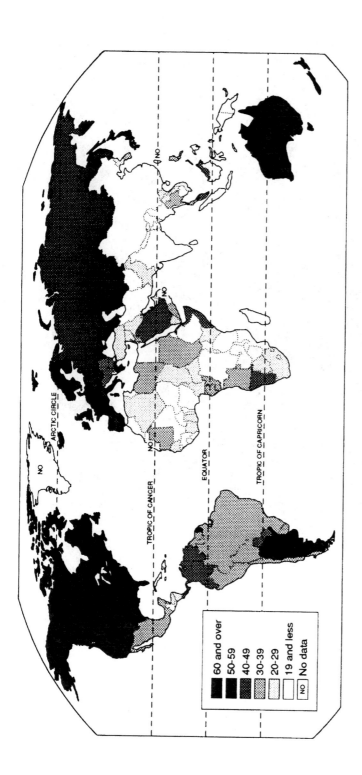

Figure 10.1 World consumption of animal protein as a percentage of total protein, 1986–8. Animal protein makes up a much smaller proportion of total dietary protein in most tropical countries than in temperate countries, and is particularly low in large regions of Africa and Asia

Source: Grigg 1995: Figure 4

Figure 10.2 These Maasai women in Kenya are using a mixture of cattle dung and mud to plaster their home. Domestic animals provide many products in addition to food throughout the tropics

farmers surveyed used draught oxen for crop production activities, primarily field preparation, but also for weeding and seeding (Bartholomew *et al.* 1995). Donkeys are also used for this purpose in Zimbabwe, particularly when there is a shortage of cattle due to drought (Hagmann and Prasad 1995). Animals are also widely used for transport. Animal power is more accessible to most farmers in the tropics than increased mechanization as the animals and

270

necessary equipment can be produced and maintained locally, in contrast to more complicated and expensive farm equipment. Livestock feeds can also be produced locally and are less expensive than other energy sources.

In this chapter we will examine several animal production and utilization systems including husbandry of domestic livestock, aquaculture, and the harvesting, ranching and farming of wild animals. We will discuss the role of each of these activities in providing for human needs, and prospects for conducting them in an environmentally and economically sustainable way.

HUSBANDRY OF DOMESTIC LIVESTOCK

In this section we focus on domestic ungulates which are raised in pastoral systems or in combination with crop agriculture. The majority of the world's domestic livestock is located in the tropics, including about three-quarters of the cattle, sheep and horses, and most of the goats, buffaloes, mules, asses and camels (Table 10.1). However, production per individual from these animals in terms of growth rate and meat and milk production is usually considerably less than in temperate zones (Minson 1980; Thomas 1995). For example, although the tropics contain 75 per cent of the world's cattle, only 44 per cent of the world's beef and veal are produced there (FAO 1995). Several fundamental differences between livestock production systems in these two regions are probably responsible for this discrepancy. Many temperate systems are heavily subsidized and depend on high levels of energy input, while inputs are generally low in tropical systems, pastures are often of poor nutritional value and animals frequently spend more time existing at maintenance levels (Humphreys 1991; Jones 1988; Minson 1980). Animals which have been selected for their adaptation to the thermal stresses imposed by many tropical climates are not highly productive because of a basic antagonism between heat tolerance and high metabolic rates (Collier and Beede 1985). Moreover, the calculation of productivity solely in terms of food products does not take into account contributions such as draught power, transportation, fertilizer, fuels and capital accumulation, which are very important in the tropics. Some of the environments in which animals are raised in the tropics represent extreme conditions for both humans and livestock, and risk avoidance is a more important goal than increasing productivity in some situations.

Many livestock husbandry systems rely on the use of land which is marginal or unsuitable for other, more productive uses. The nature of these areas often presents special problems for the production of domestic livestock and its long-term sustainability. We will discuss these problems, and potential livestock and pasture management strategies which require levels of inputs likely to be practical within the context of existing production systems in the tropics.

Table 10.1 Numbers of domestic animal species (1,000 head) worldwide and in tropical regions

	World	*Africa*	*Central America*	*South America*	*Asia*	*Oceania*
Cattle	1,288,124	192,180	50,060	279,656	410,118	34,049
Sheep	1,086,661	208,845	7,238	94,054	340,102	182,758
Goats	609,488	176,089	12,907	22,819	373,005	1,007
Buffaloes	148,798	3,200	9	1,436	143,640	—
Horses	58,158	4,758	8,455	14,493	16,007	430
Mules	14,952	1,394	3,675	3,451	6,124	—
Asses	43,772	13,408	3,647	4,105	21,325	9
Camels	18,831	13,815	—	—	4,761	—

Source: FAO 1995

Types of livestock used in the tropics

Cattle are the most abundant of the domestic ungulates in the tropics (Table 10.1), in terms of both absolute numbers and biomass, and they are found throughout most tropical regions. For these reasons, we have placed much of the emphasis of the following sections on cattle. However, many of the principles discussed will also apply to other livestock species. Sheep and goats are also abundant in the tropics, where they are used primarily for fibre, meat and milk. They are often raised in conjunction with cattle, particularly in pastoral systems. Other species are found in lower numbers, but are regionally very important. The tropics are home to almost the entire world's population of domestic buffaloes, most of which are found in Asia (Figure 10.3). They are used primarily for draught and meat in South-East Asia, but have been bred for milk production in other areas such as India (Mannetje 1982). Horses, mules and asses are found in smaller numbers, and are used mainly for transportation and draught power. Camels are used almost exclusively in arid and semi-arid regions of Africa and Asia, where they provide milk, meat and transportation.

Most cattle breeds in the tropics are derived from either *Bos indicus, Bos taurus*, or crosses between these. These two species are thought to have originated in Asia from a common ancestor and can interbreed without any subsequent reduction in fertility (Sanders 1980). *B. indicus* (zebu) are humped cattle which are generally more tolerant of conditions encountered in tropical rangelands including high temperatures, reduced availability of water, inadequate quantity and quality of forages, long travel distances and tropical diseases (Homewood and Rodgers 1991). Numerous *B. indicus* breeds have been developed and these are widely distributed throughout the tropics. Although some *B. taurus* breeds are indigenous to tropical regions, most are associated with temperate zones owing to their generally superior production under those conditions (Chenoweth 1994). They have been widely used for

Figure 10.3 Water buffalo (*Bubalus bubalis*) cow and calf being weighed for sale at a livestock market in Sabah, Malaysia. These animals are a very important source of agricultural power in South-East Asia

cross-breeding with *B. indicus* cattle in attempts to develop breeds with improved reproductive traits, growth rates and beef quality, but which retain characteristics more suited to survival in tropical environments.

Dietary requirements of livestock

Domestic ungulates are herbivorous, and meet their dietary requirements by grazing (foraging on herbaceous plants), browsing (foraging on woody plants) or a combination of these. Most livestock species, including cattle, buffalo, sheep and goats, belong to the suborder Ruminantia of the order Artiodactyla. Ruminants have a specialized digestive system, including a four-chambered stomach, which enables them to digest plant fibre high in cellulose by microbial fermentation. Equines such as horses and donkeys, which have a caecal digestive system, are also able to digest these materials through a similar fermentation process. These types of digestive systems provide ungulates with the ability to convert plant materials which are indigestible by humans into food, energy and other products. This can allow productive use to be made of materials such as crop residues and roadside vegetation by animals raised in close association with crop agriculture and human settlements, and of lands which are too dry, steep, or infertile for reliable annual cropping.

Domestic livestock, like humans, need water, energy, protein, vitamins and minerals. Maintenance diets are those which will meet the basic nutritional requirements of animals with no loss or gain in body energy and protein; additional intake of these nutrients is necessary for growth, reproduction or work. Specific dietary needs are related to a variety of factors, including species, breed, age, sex, reproductive status and energy output required. Nutrient requirements of grazing ruminants are discussed in more detail by McDowell (1985).

Dry matter intake (DMI) of ruminants under normal conditions depends mainly on body size, energy within the diet, and rate of digestion or fermentation (McDowell 1985). However, it is also influenced by external conditions, such as the high temperatures characteristic of the tropics. Feed intake in cattle generally decreases rapidly as temperatures rise above approximately 25°C, leading to a consequent reduction in consumption of energy and nutrients (Collier and Beede 1985). The amount of feed which can be consumed is limited by the amount of digestible energy and protein; when these are low, rumen activity decreases and DMI drops significantly (Breman and de Wit 1983; McDowell 1985). This has important implications where forage quality is low, as livestock cannot make up for poor quality by increasing the quantity of feed ingested. The ability to utilize poor forages varies among animal species, with large-bodied herbivores generally being more capable of using low-quality forage than smaller ones (McNaughton *et al.* 1993). Seasonal differences in environmental conditions and forage quality and quantity have important implications for maintaining continuity of the feed supply for livestock throughout the year in the tropics; we will return to this topic later in the chapter.

Livestock production systems

For convenience we have classified livestock production in the tropics into pastoral systems, intensive systems and integrated livestock–crop production systems, although there is often some overlap between these modes of production. Most pastoral livestock production in the tropics is carried out in extensive systems which typically involve very little in the way of inputs or management. A relatively low degree of intensification does occur in some of these systems, usually involving replacement of native grasses with planted forages. Intensive systems, involving very high inputs and management effort, are uncommon in the tropics and will be discussed only briefly in this chapter. Finally, we will discuss the combination of animal husbandry with crop production, which is practised throughout the tropics, especially on small subsistence farms.

ANIMAL PRODUCTION AND UTILIZATION
Pastoral systems in semi-arid and arid regions

Extensive pastoral systems which involve few inputs are a predominant form of land use in semi-arid and arid regions where agricultural crops are limited by inadequate or unpredictable rainfall (Figure 10.4). In Africa, the relative contribution of livestock to agricultural production increases as rainfall decreases; in some countries it is greater than 50 per cent of total agricultural gross domestic production (Leeuw and Rey 1995). These types of systems usually rely on natural growth of native forage species and/or naturalized exotics. Management of these lands is frequently restricted to manipulation of grazing and burning, although supplements to natural forages or water may be provided.

Subsistence pastoralism in which domestic animals provide the majority of food products is the major mode of livestock production in most arid and semi-arid regions. These types of systems tend to require few capital inputs but rely heavily on household labour (Homewood and Rodgers 1991). Dairy products are extensively used, and it is possible to obtain about 2.5 times as much energy from mixed milk and meat production as from meat utilization alone (Western and Finch 1986). Blood is also used, although it provides only a small portion of energy requirements. According to Coughenour et al. (1985), Ngisonyoka pastoralists in north-west Kenya consumed 76 per cent of their food energy from livestock milk, meat and blood; an additional 16 per cent was indirectly derived from pastoral products which were used to purchase other foods. Eighty per cent of food energy derived from livestock was obtained from the milk of camels, cattle, sheep and goats. Meat and blood each contributed about 10 per cent of livestock-derived food energy, mainly when milk was in short supply. The degree of reliance on livestock products varies among areas. In much of East Africa, pastoralists rely very heavily on milk production, which is possible throughout the year because there are two wet seasons, whereas in regions where animals cannot lactate through a longer dry season, alternative food sources are more important (Western and Finch 1986).

Many subsistence pastoralist systems in these regions are nomadic or semi-nomadic, or involve transhumance, in which animals are moved seasonally by herders while the remainder of the human population is sedentary. There are approximately 50–100 million nomads in the world (depending on the definition used), and most of them live within tropical Africa and Asia (Omar 1992). Pastoralists may represent only a small percentage of the total population of a region where they live, but generally occupy a relatively large land area and own a high proportion of the livestock (Macpherson 1995). These types of systems can involve relatively regular seasonal movements based on latitudinal differences, as in much of the West African Sahel, or altitudinal variations, as occurs at some locations in East Africa. Nomadic movements tend to become more irregular in arid zones with increasing unpredictability of rainfall and forage production (Homewood and Rodgers 1991).

Figure 10.4 Maasai cattle herd in semi-arid savanna, northern Tanzania

Extreme environmental conditions which include low, variable and unpredictable rainfall, high heat stress and occasional disturbances such as fire, disease and fluctuations in wildlife populations can present severe problems for livestock survival, reproduction and productivity in arid and semi-arid regions. This results in pastoral management strategies which are aimed primarily at risk avoidance, with productivity being a secondary goal. Livestock herds in these regions often contain a mixture of species with varied feeding preferences, reproductive characteristics, resistance to disease and drought, and production qualities. Cattle, goats, sheep, dromedary camels and donkeys are commonly kept in Africa; buffalo, bactrian camels, horses, pigs and yaks are used in Asia, and llamas and alpacas are herded in South America (Macpherson 1995). Cattle, donkeys and sheep forage primarily on herbaceous plants, while camels are primarily browsers and goats are mixed feeders (Coughenour *et al.* 1985). Camels and goats are more drought-tolerant than sheep or cattle. Small stock are regarded as being more vulnerable to predators, but are usually important for herd reconstitution after severe losses as they are capable of more rapid reproductive response. Herds often have a high proportion of females both for milking and reproduction. Animal numbers are often maximized in the hope that enough will survive periodic heavy losses to ensure long-term survival. Herds may also be divided into different units which are pastured in different geographical areas and habitat types. Highland or wetland pastures may be used as drought refuges

276

(Homewood and Rodgers 1991; Humphreys 1991). Population increase among pastoralists combined with conversion of traditional grazing lands to cropland in some areas is likely to affect the viability of this mode of live-stock husbandry in the future.

There is some overlap between semi-nomadic and transhumance systems, and sedentary livestock production in drylands, as in the latter systems seasonal movement of stock between pastures is often carried out. In some regions, commercial ranches tend to concentrate on lands of relatively higher potential than do subsistence enterprises (Homewood and Rodgers 1991). The productivity and sustainability of nomadic and semi-nomadic pastoralism as compared to sedentary ranching has been controversial. Nomadism is sometimes considered an outdated and inefficient form of land use, and some governments have attempted to replace it with sedentary systems. However, there have been few long-term studies which compare these types of systems to commercial ranching in similar areas, especially over drought periods. Recent studies suggest that when inputs and outputs are considered, tradi-tional pastoralist systems are well suited to the lands where they are carried out, and overall productivity and efficiency appear to be comparable to those obtained in commercial ranching (Homewood and Rodgers 1991; Humphreys 1991; Mace 1991; Western and Finch 1986). Western and Finch (1986) suggest that traditional pastoral management practices are likely to be the optimal mode of production in very dry environments, while conven-tional ranching may be more effective in less seasonal environments with more productive pastures.

Pastoral systems in humid and subhumid regions

Pastoral production from grazing lands established within forested regions is especially common in the Neotropics, where it involves mainly cattle (Figure 10.5). Some pastoral systems in these regions are derived from forested lands which have been cleared specifically for the purpose of livestock production, while others are areas which were used for crop agriculture and later turned into pasture.

The level of management effort and inputs subsequent to pasture estab-lishment is generally higher in these regions than in arid and semi-arid zones, but varies considerably among these systems. Although planted pastures account for a much smaller proportion of tropical livestock production than do natural pastures (Humphreys 1991), most of the sown pastures that do exist are located in humid and subhumid regions. Herbaceous species are often introduced to provide forage in formerly forested areas. Seeding may be the only input into a pasture following forest cutting and burning. In the Amazon region, if the grasses seeded do not establish well, few investments in subse-quent pasture maintenance are deemed worthwhile. Tree re-establishment usually occurs shortly thereafter, leading to pasture abandonment within a few

Figure 10.5 Shrub encroachment in cattle pasture, Belize. Rapid regeneration of woody species is a common problem for pasture maintenance in forested areas

years (Buschbacher *et al.* 1987). Better establishment of forage grasses may result in progressively higher maintenance efforts such as fertilization and pest control. Six species of African grasses – *Panicum maxicum, Brachiaria mutica, Melinis minutiflora, Hyparrhenia rufa, Pennisetum clandestinum* and *Digitaria decumbens* – have been widely introduced in the Neotropics, presumably because they were more nutritious and productive than native forage species. These species have since become naturalized in many areas (Parsons 1972). However, the productivity of some of them has been poorly sustained owing to problems with soil fertility and insect pests (Humphreys 1991; Serrão and Toledo 1990; see below).

Intensive systems

There are few intensive animal husbandry systems in the tropics; temperate-zone practices such as grain feeding, storing fodder as hay, and finishing animals in feedlots are uncommon. The more intensive animal husbandry systems that do exist are mainly dairy operations; these are often located at higher elevations with cooler temperatures. They involve higher inputs in the way of forage improvement and supplementation, pasture fertilization and irrigation. Concentration of animals, particularly in indoor housing, results in the nutrients in manure being deposited in the animal quarters rather than on pasture or cropland, and can increase disease and pollution problems (Humphreys 1991). There are very few intensive operations aimed at meat production in the tropics. Intensification of overall land use in an area as a result of population increase does not necessarily result in intensification of livestock production. Although livestock may be managed so as to increase overall agricultural productivity through contributions of draught power and fertilizer, the animals are still primarily raised on natural pastures, fallows and crop residues, with few inputs (Leeuw and Rey 1995).

Integrated crop–livestock production systems

Integrated crop–livestock systems usually include a wider range of animal species, such as turkeys, chickens, ducks, goats, pigs, cattle and buffalo; these often fulfil several functions. It has been estimated that about 90 per cent of the ruminants in South-East Asia are found on mixed farms less than 5 ha in size; average farm size in some areas may be a fraction of a hectare (Humphreys 1991). In Africa, the proportion of small ruminants increases as farm size declines, except where draught power is heavily used (Leeuw and Rey 1995). Some livestock–crop systems also include an aquaculture component (as discussed later in this chapter).

Livestock and crop production are usually very closely integrated in these types of systems, with crop residues providing the main sources of feed, and animals providing draught power, which is important to the success of crop production. The contribution of animal excreta, which is used as fertilizer and fuel, can also be very important. Nutrients from animal wastes are more readily available to plants than crop residues. A study of three villages in Thailand showed that the area with the highest stocking rate also had the highest crop production (Humphreys 1991). Manure for fertilizer or fuel is often considered valuable enough to have a monetary or trade value. In Ethiopia the sale of sun-dried manure for fuel is worth approximately US$80 per cow per year (1988 figures) (Humphreys 1991). However, using manure for fuel rather than returning it to agricultural lands can result in decreased soil fertility.

Cut-and-carry systems, in which forages are brought to confined animals, and tethering of livestock are more prevalent than pasture production on many small farms in the tropics, particularly in areas of high human population density where space is at a premium and few grazing lands exist (Figure 10.6). Herding and rotational grazing with temporary fencing are also carried out. The feed for animals in these types of systems generally comes from a variety of sources including pasture, fallow land, crop residues and by-products, roadside grasses, garden weeds and kitchen wastes. Ruminants may be fed forages such as rice straw, bean or groundnut hay, green maize leaves and sweet potato leaves. Livestock diets may show considerable seasonal variation in accordance with the availability of various forages (Humphreys

Figure 10.6 Cow tethered at the edge of a rice field in Bali, Indonesia. Various types of confinement feeding are especially common in South-East Asia

1991). Grazing animals are often used as a biological form of weed control, representing a safer and less expensive method than the use of herbicides (Sánchez 1995).

Livestock are also raised in conjunction with crop production on a larger scale in plantation agriculture. Cattle are grazed in coconut plantations to control weeds and facilitate movement through the area during harvest. Coconut yield has been reported to increase at higher stocking rates, probably owing to more efficient nutrient cycling. Yields of tree plantation crops can be positively or negatively affected by development of pasture; the net value of this is likely to vary considerably depending on the forage, plantation and livestock species used, and their interactions at various stages of growth with respect to light, nutrients, soil erosion and damage to the main crop (Humphreys 1991).

The combination of some animal species in close quarters with each other and with humans in integrated systems can create potential health hazards (Chapter 5). New influenza pandemics have arisen in areas of Asia where integrated systems involving aquaculture, swine and domestic birds exist. The disease risk should be taken into consideration when such systems are designed, with pigs and poultry being kept in separate systems (Scholtissek and Naylor 1988; Naylor and Scholtissek 1988).

Pasture and livestock management

Pasture and livestock management involves the manipulation of systems which are grazed or browsed by domestic livestock, including both natural and planted pastures. The focus may be on one or more of the components of these systems, including vegetation, water, soils, animals, and their interactions. We consider that the high inputs necessary to develop very intensive production systems in the tropics are unlikely to be practical in many instances, so we focus on management strategies which are directed at obtaining optimum production within sustainable extensive and semi-intensive systems.

Forage quantity and quality

The ability of livestock to meet their nutritional needs is dependent on both the quantity and the quality of forage available. Herbivore carrying capacity of natural ecosystems appears to be closely related to primary productivity, with the biomass of both wild and domestic herbivores increasing with primary productivity (Oesterheld *et al.* 1992). Forage quality is also very important, and is a limiting factor throughout most of the tropics.

Forage quality and quantity are variable in both space and time throughout the tropics (Chapter 7); however, some generalizations can be made about their suitability for livestock. Herbaceous species support most domestic herbivores in the tropics, so we focus on this group of forages. Biomass of forage produced generally increases with rainfall, while an inverse relationship between protein content and biomass has been observed (Figure 7.11). Therefore dry regions may produce low amounts of forage of relatively good quality, while wetter conditions may produce more forage of increasingly inferior quality (Breman and de Wit 1983). Tropical grasses are often low in nutrient concentrations, with protein, nitrogen and phosphorus frequently limiting for herbivore production, and mineral supplementation is often necessary (Boutton *et al.* 1988; McDowell 1985; McNaughton *et al.* 1993; Thomas 1995). Most native tropical grasses have the C_4 pathway of carbon fixation; these often have a high content of structural elements which are low in digestibility (Humphreys 1991; Thomas 1995). These characteristics may be compounded by decreasing forage quality at high temperatures and with nutrient-deficient soils (Collier and Beede 1985; Minson 1980).

Seasonality is one of the major constraints for livestock production in the tropics, as it affects forage quantity and quality (Chapter 7), with the dry season usually representing a bottleneck. In arid and semi-arid regions, dry-season forage limitations are compounded by unpredictable fluctuations in annual rainfall totals, which tend to be negatively correlated with annual rainfall totals (Chapter 3), making this problem increasingly severe with aridity. Year-to-year variation in forage yields can be substantial, and the area needed to feed one animal in a dry year can be about three times the usual area required (Breman and de Wit 1983; Evans 1982). A high proportion of annual species and rapid turnover of vegetation production also exacerbate seasonal forage limitations in some arid zones (Western and Finch 1986). Insufficient surface water is a problem in many of these areas, too. The impacts of seasonality can also be significant in humid regions. In the humid tropics of Australia, pasture growth rates of >150 kg.ha^{-1}.day^{-1} dry matter during the warm wet season decline to levels of <50 kg.ha^{-1}.day^{-1} during the cooler dry season (Teitzel 1992). Dry spells in humid areas not only decrease pasture growth, but can reduce the capability of forage species to compete with weeds (Serrão and Toledo 1990).

Forage storage for seasonal use is not a common practice in the tropics. The potential benefits are reduced by problems involved in hay production during the rainy season due to the wet field conditions (which delay drying of the forage), the difficulties of storing tropical forages, and the costs of using land for their production. Building up of livestock body reserves may be more efficient than forage storage in some circumstances (Golding 1985; Humphreys 1991). A sequence of feeds which varies seasonally is sometimes utilized; this is usually most feasible in areas where crop residues provide the main feed source for livestock and dry-season feed may be more readily available. Other methods which are used to maximize seasonal continuity of forage availability in the tropics include variation of the stocking rate and pasture improvement; these are discussed in more detail below.

Livestock management

Oesterheld *et al.* (1992) suggest that basic herd management techniques such as regulating animal distribution and veterinary care can increase the carrying capacity of pasture systems for herbivores by about an order of magnitude. Regulation of animal distribution includes management of both the stocking rate (numbers of livestock per unit of land) and livestock movements. These are central to achieving a balance between meeting livestock forage requirements and maintaining pasture. One of the major difficulties encountered is the seasonal differences in carrying capacity. If animal populations are maintained at high enough numbers to make optimum use of forage produced during the wet season, stock deaths or loss of body condition may result in the dry season. Weight losses are common among grazing livestock in the tropics during the dry season, and may represent a substantial portion of wet-season weight gains (Golding 1985). However, cattle may browse more in the dry season, which can improve their protein intake (Lascano 1991).

Various types of stocking methods exist (Humphreys 1991). Ideal management of stocking rates would provide adequate feed per head while ensuring the maintenance of plant cover, and growth and persistence of desirable species. Conversely, in some circumstances poor management of either may lead to overgrazing and reduction of vegetation cover, increase of relatively unpalatable species at the expense of more desirable species, and soil erosion. Managing animal distributions is made more difficult by seasonal variations in forage quantity and quality. During dry periods, the short-term needs of livestock may conflict with the most appropriate course of action for long-term pasture sustainability; usually the immediate welfare of the animals will take precedence (Humphreys 1991). Adjustment of seasonal stocking rates to match the forage available is a method of coping with seasonal imbalances, but this is frequently not an option owing to the supply–demand situation in market economies, and other practical considerations such as shortage of available land in non-market economies.

Climatic variations over longer periods pose an additional management problem, particularly in arid and semi-arid areas, as carrying capacity is not absolute but will vary with the time frame considered, becoming lower as longer periods are considered (Western and Finch 1986). As stated above, the amount of land required to support livestock in semi-arid and arid regions in dry years may be several times the normal area required. Drought patterns show considerable regional variation. For example, in West Africa wet and dry spells often occur in long runs; dry periods often last 10–18 years and include shorter but more severe droughts, while East and Southern Africa tend to be characterized by more short-term fluctuations with overall wet or dry spells typically lasting between 2 and 6 years, and severe droughts of 1–3 years' duration (Homewood and Rodgers 1991).

Regional differences such as those discussed above suggest that management of animal distributions will be most effective when tailored to specific locales and livestock husbandry systems. There is a high degree of uncertainty about the concepts and methods of determining carrying capacity and optimal stocking rates for some systems, particularly extensive pastoral systems, and estimations of these are subject to a high degree of variability, error and subjectivity (Mace 1991). Systems with a high degree of flexibility in both stocking rate and pasture utilization offer the best prospects for meeting short-term needs of livestock while allowing long-term maintenance of pastures. However, population growth and replacement of grazing lands by other uses are likely to make this increasingly difficult in some regions. For example, better pasture lands may be relied on for feed during the dry season or drought periods. However, with population growth and the need to feed higher numbers of people, these areas are increasingly likely to be converted to crop agriculture. Unfortunately, these lands are likely to be at best marginal for crop production, and in subsequent droughts are unlikely to provide much sustenance for human populations in the form of crops, while their ability to provide animal products will also have been lost (Dodd 1994).

Pasture forage improvement

Fire has traditionally been widely used as a management tool both in natural and planted pastures within the tropics as it is inexpensive, universally available and can be used over large areas. Forage improvement is one of the major objectives of burning, although it is also used for other purposes such as pest control (see below). Burning can remove dead vegetation which is of low nutritive value and may interfere with grazing (Boutton *et al.* 1988), and temporarily increase forage productivity and quality (Chapter 7). This often results in increased weight gains in livestock. It can also have negative effects, such as soil erosion if rains of high intensity occur immediately after burning while ground cover is absent. The timing and frequency of burning, and control of pasture utilization by livestock are important to obtain maximum benefits while mitigating negative effects (Humphreys 1991). Fire management is especially important in the savannas of semi-arid regions (Chapter 7), as there are few other practical options available for forage improvement in these areas.

Pasture forages can be improved by seeding grasses and/or legumes, either alone or in combination with existing native species. This may be directed at increasing forage yield or nutritive value, extending the grazing season by partly compensating for seasonal variations, and increasing the lifespan of the pasture. This type of pasture improvement can result in substantial increases in carrying capacity during the dry season (Golding 1985). Seeding of forage species in the tropics is usually done in humid or subhumid zones. Technical inputs developed for these areas are less suitable for dryland pastures (Humphreys 1991), and as variability and unpredictability of rainfall increase, the risk of failure may become too high to justify the increased costs. The failure of seeded forage species can also increase the risk of soil erosion. Any improvement of pasture forage systems which is carried out in these regions is usually limited to small areas for use during times of highest nutrition stress such as the dry season or reproductive periods (Evans 1982).

Although introduced grasses can improve pasture and livestock production, they are often limited by nitrogen deficiency, pests and lack of persistence (Lascano 1991). The use of legumes has the potential to improve both animal productivity and pasture sustainability. Adding legumes to a pasture may be the least expensive way to overcome a protein deficiency (Minson 1980). As well as having a nutritional value, the increased protein in the legumes can result in higher intakes of grass forages. Animal production is almost always higher from pastures with legumes than from pure grass pastures (Evans 1982). The ability of forage legumes to provide nitrogen via biological fixation also has important implications for maintenance of soil fertility in pastures (see below). Although the use of legumes in pastures is considered a relatively low-input technique, establishment costs are sufficiently high for their ability to persist to be crucial (Miller and Stockwell 1991). Tropical C_4 grasses often have a higher growth potential than C_3 legumes, which may create problems in maintaining the desired grass–legume balance (Humphreys 1991). This balance can be manipulated by alteration of grazing system and grazing intensity (Lascano 1991).

Despite the potential benefits of legumes, they are presently not widely used in the tropics. A key factor in the successful use of legumes in tropical pastures is the availability of varieties which are well adapted to local environmental conditions and resistant to pests and diseases. Unsatisfactory results from existing cultivars have been cited as a reason why few legumes have been used in Amazonian pastures (Serrão and Toledo 1990). It has been suggested that selection of legume varieties which are adapted to a wide range of climate, soil and management conditions is desirable (Miller and Stockwell 1991); however, as many of the promising varieties which have been tested appear to be suitable only within a limited variety of conditions, selection of legumes targeted for local conditions may be a more realistic objective in some instances (Thomas 1995).

Soil fertility

Maintenance of natural soil fertility and correction of nutrient deficiencies are key factors in establishment and maintenance of productive and sustainable pasture systems. Soil fertility affects forage productivity, forage quality, species composition and susceptibility to weed infestations, which in turn affect livestock production. Forage biomass produced depends mainly on the absolute availability of the limiting factor (soil moisture or mineral nutrients), while forage quality depends on the availability of water relative to that of nitrogen and phosphorus (Breman and de Wit 1983).

Natural mineral nutrient deficiencies which affect pasture and livestock growth are prevalent in the tropics (Humphreys 1991). Nitrogen and phosphorus are the major limiting nutrients in most tropical pasture systems. Breman and de Wit (1983) concluded that N and P deficiencies are a more serious limitation to plant productivity than low rainfall down to annual rainfall of about 300 mm in the African Sahel. Fertilizers are not widely applied to tropical pastures, usually because of their high cost and low availability. Nitrogen and phosphorus fertilizers are the most common of those used (Evans 1982). Nitrogen is normally the main limiting factor in native grass pastures (Jones 1990), but nitrogen fertilizer is often economically unfeasible to apply and/or unavailable even in situations where it would be biologically efficient. Phosphorus is usually the most limiting nutrient for pastures seeded with legumes, while pastures based on native grasses are less likely to be responsive to the addition of P. Deficiencies of nutrients other than N and P do exist but are less likely to be limiting. In Amazonian pastures it is possible to maintain reasonably satisfactory levels of Ca, Mg, K, Al and pH levels for over 20 years in well-managed systems (Serrão and Toledo 1990).

Pasture development in forested areas by tree felling and burning results in temporary increases in soil fertility, creating favourable conditions for establishment of forage species (see Chapter 4 for a discussion of nutrient cycling). Subsequent maintenance of soil fertility is highly dependent on pasture management. Application of P at 20–30 kg.ha^{-1} every 3 to 5 years allowed pasture productivity to be maintained at medium to high levels on ranches in the Brazilian Amazon. For maximum biological benefit and to be economically feasible, such management practices need to be initiated during the early stages of pasture usage when productivity is still relatively high; these measures may not be economically viable after productivity has declined to low levels (Serrão and Toledo 1990).

The choice of forages for pasture establishment can have a large influence on maintenance of soil fertility. Planting of pastures on infertile soils with nutrient-demanding species such as commercial cultivars of *Panicum maximum*, *Hyparrhenia rufa*, *Pennisetum purpureum* and *Axonopus* spp. may lead to subsequent problems (Serrão and Toledo 1990). The utilization of legumes in pastures can provide a less costly source of nitrogen via biological fixation in addition to their benefits as forage species. The proportion of legumes in a pasture should be maintained at a minimum value of about 20 per cent to maintain soil nitrogen levels and maximize forage and animal production. This value may increase with pasture utilization (Thomas 1995). The selection of P-efficient legumes can reduce the inputs needed on poor soils. The legume genus *Stylosanthes* has been noted for its ability to use soil phosphorus efficiently and grow well under conditions where deficiencies of this nutrient exist. However, although some species can persist under these conditions, the P concentrations in the forage produced may be insufficient for that needed for optimum animal production. Phosphorus supplements in the water or mineral licks have been suggested as an alternative (Jones 1990).

Ingestion of nutrients in forage plants by livestock can result in the export of some of these from the system, generally in the form of milk and meat, which are mainly composed of carbon, hydrogen, oxygen, nitrogen and micronutrients, and some loss of nitrogen from excreta by leaching and volatilization can be expected (Serrão and Toledo 1990; Thomas 1995). The remainder is returned to the pasture system as faeces or urine, and is often redistributed within the range of the livestock. Sites near Kenyan bomas used to enclose cattle, sheep and goats overnight had higher levels of limiting nutrients, particularly N and P, and higher-quality forages, than did control sites in the savanna (Stelfox 1986). However, in a study carried out in Botswana by Tolsma *et al.* (1987), nutrients were transported to boreholes from the surrounding savanna via cattle dung, but the intensity of trampling around the boreholes permitted the survival only of certain grazing-resistant plants. In situations where livestock production is carried out in combination with crop agriculture or home gardens, manure is often used as fertilizer for these areas.

Table 10.2 Stages of productivity in Amazonian pastures

Stage of productivity	Relative biomass of weeds in the pasture (%)	Potential carrying capacity (AU)*	Average length of time after pasture establishment (yr)
High	5-10	1.0–1.5	3–5
Medium	15–25	0.5–1.0	4–7
Low	30–60	0.3–0.5	7–10
Degradation	>80	<0.3	7–15

Source: Serrão and Toledo 1990

Note: *Animal unit (= 450 kg of live weight)

Pests and diseases

Invasion of pastures by undesirable plants is a significant problem in most regions within the tropics. In humid regions where forested lands have been cleared for livestock production, regeneration of woody species frequently decreases the value and useful life of the pasture (Figure 10.5). Productivity in Amazonian pastures was found to be inversely proportional to the abundance of weeds (Table 10.2). The difficulty of controlling these becomes increasingly higher and less cost-effective as they make up a greater proportion of the grazing area. Decline of productivity in the planted forages often takes place several years after pasture establishment in the Amazon region; this has been attributed to a combination of the characteristics of the forage species used and a lack of suitable management practices. It is most problematic with forages which lack vigour and competitiveness in local conditions, particularly those which have high nutrient demands, high climatic and soil specificity, and clumped growth habit. Many of those currently in use were introduced without any evaluation of their suitability for local conditions (Serrão and Toledo 1990). These problems could probably be decreased significantly, although not eliminated entirely, with additional management techniques for pasture establishment and maintenance. Successful establishment of forage species which are adapted to local conditions and competitive with local weeds is likely to be a key factor for maintaining pasture productivity in humid areas. Subsequent management techniques such as careful control of grazing and burning, weed removal and fertilization can prolong the pasture's lifespan (Serrão and Toledo 1990; Teitzel 1992). The effectiveness of various techniques depends on the forage and weed species present.

Invasion of dryland pastures by undesirable species can also be a significant concern. This is encouraged by heavy grazing, which can make conditions more favourable for woody plants (McNaughton et al. 1993). Fire protection in systems where burning is normally a regular process can lead to an increase of fire-sensitive, unpalatable species (Homewood and Rodgers 1991). Burning plays an important role in determining the balance between herbs and woody vegetation and is often used to control encroachment of woody plants. This balance is also affected by herbivory, and fire may help to keep woody growth at a height and in a condition which encourages herbivory by browsing ungulates (McNaughton *et al.* 1993; Van Wilgen *et al.* 1990).

Other pests and diseases affect pasture productivity by damaging forage plants and reducing seed production of desired species. In the Amazon region, this is a major factor contributing to pasture degradation. The fungi *Fusarium roseum* and *Tilletia airesii* decrease seed production in *Panicum maximum*. Insect pests known as pasture spittlebugs (various species) are the main factor decreasing the persistence of *Brachiaria* spp., which are widely planted in pastures because of their adaptability to infertile soils. Numerous other diseases and insect pests including caterpillar and leaf-cutter ants also create problems in this region (Serrão and Toledo 1990).

Diseases which affect domestic animals directly are also a significant concern, and in some cases drastically reduce the potential for livestock production over large regions. Pathogenic trypanosomoses, which are protozoan parasites, have a major impact on livestock production in many regions of the tropics. Among the best known of these are the African trypanosomes *Trypanosoma brucei*, *T. congolense* and *T. vivax*, which cause the disease known as nagana in animals (Boid *et al.* 1996). These species are carried by tsetse flies (*Glossina* spp.), which also act as vectors for trypanosomes that produce sleeping sickness in humans. Tsetse flies are found in a wide range of habitats from forested to semi-arid regions, and limit the use of cattle over an estimated 10 million km^2 of the African continent (Dransfield *et al.* 1991). Other trypanosomes have a wider distribution, such as *T. evansi*, which affects cattle, camels, buffaloes and horses in Africa, Asia and South America. Diseases which affect draught animals can result in decreased agricultural output, as is the case where buffaloes are infected with *T. evansi* in Indonesia, China, Vietnam and the Philippines (Boid *et al.* 1996). Numerous other diseases and parasites also cause considerable losses and decreased production of various species of domestic stock.

The prevalence and severity of livestock diseases are affected by many of the same factors which influence the transmission of human diseases (Chapter 5), including population densities, movements of hosts and vectors, and environmental conditions; their prevention and control are correspondingly complex. For example, considerable effort has been put into trying to overcome trypanosomiasis limitations in Africa, often by attempts to eradicate tsetse flies. However, this problem still persists over wide areas. These eradication attempts can also have potentially severe negative effects such as widespread habitat loss, environmental deterioration and elimination of wildlife. Rogers and Randolph (1988) suggest that one advantage of the still widespread distribution of tsetse flies in Africa is that it has put the affected area into a developmental 'deep-freeze', and prevented many local mistakes in land development from being made on continental proportions; this also leaves a window of opportunity for more sustainable forms of land use to be developed in the future. New low-technology methods of tsetse control which are less expensive and less environmentally damaging than previous eradication attempts may offer more promising solutions to this problem (Dransfield *et al.* 1991).

The lack of veterinary services exacerbates disease problems in the remote areas characteristically associated with extensive production systems. Diseases pose special problems in nomadic and semi-nomadic pastoral systems, and heavy parasite loads in livestock are common. Seasonal movements may remove herds from areas where high levels of parasites have been built up, but it may also increase the risk of exposure to other infections and facilitate transmission of diseases between domestic stock and wild animals (Macpherson 1995). Immunization and treatment of animals are more often provided in sedentary systems, but even in these areas veterinary care may be limited by isolation or the low value of animals. Burning carried out for pasture improvement may destroy dormant or free-living stages of livestock parasites, which accumulate in heavily used pastures (Homewood and Rodgers 1991). Domestic animals which are raised in combination with crops on small-scale farms often receive higher levels of veterinary care, as their owners are much more aware of individual animals and their problems, and more likely to make this type of investment in animals which provide draught power and fertilizer as well as meat and milk (Ellis 1984).

Land degradation

The issue of long-term land degradation in the tropics, and the potential contribution of livestock production to this, has received a considerable amount of attention (Dodd 1994; Mace 1991; Serrão and Toledo 1990; Sinclair and Fryxell 1985). Pasture deterioration due to problems such as soil erosion, nutrient depletion, proliferation of pests and diseases, acidification and salinization has been documented in many areas of the tropics. Some of the issues which are of particular concern are the causes of these changes, their permanence, the possibilities and methods of restoring degraded lands, and the long-term sustainability of livestock production in marginal environments.

Planted pastures in humid and subhumid regions commonly undergo deterioration to the point where livestock production is not economically or ecologically sustainable, especially in the Amazon region. Grazing lands in this region are often established for socioeconomic or political purposes such as occupation of frontier regions, very often without any consideration being given to prospects for long-term productive use of these lands. Little attention has been given to management for sustainability, resulting in degradation and decreased productivity, and encouraging additional conversion of forest to pasture. At least half of the approximately 10 million ha of pastures developed in the Amazon region in the past three decades are estimated to be in an advanced state of degradation (Serrão and Toledo 1990). Abandonment of pastures which reach this stage is an option which is exercised where new lands are available for further development, or where available land is extensive enough to allow fallowing in a system similar to shifting cultivation (Chapter 8).

Although maintenance of highly productive pasture systems over a period of decades is unusual in the humid tropics, examples of these which have been sustained for over 20 years indicate that it is possible with careful management (Serrão and Toledo 1990; Teitzel 1992). Intensification of pasture and livestock management in humid regions can extend the life of existing grazing areas and restore degraded ones to provide economic and social benefits, as well as reduce the occurrence of pasture abandonment and expansion into new forest lands. Practices such as reclamation of pastures through reseeding, fertilization and application of pesticides are becoming more common, but involve high inputs, which limits their application. Lower-input solutions to restoration and maintenance of derived pastures will depend on selection of new grass and legume varieties that have the capacity to establish rapidly and are highly persistent, and that are adapted to nutrient-poor soils, resistant to local pests and diseases, competitive and, for legumes, efficient in nitrogen fixation (Serrão and Toledo 1990).

Land degradation in arid and semi-arid regions is often attributed to over-grazing by domestic livestock. The assessment of land degradation in these environments is difficult owing to their climatic variability (Chapter 3), which alone results in major changes to natural ecosystems among years and makes it difficult to distinguish between natural and livestock-induced changes, and temporary and permanent effects (Dodd 1994). The view often portrayed of vast areas of Africa's grazing lands being overtaken by desert owing to overstocking of domestic animals does not appear to be supported by evidence; although the possibility of long-term damage certainly has not been ruled out, these systems may be more resilient than once thought (Mace 1991). Permanent vegetation changes in areas with extremely high stocking rates such as around water holes or villages has been documented, but there is no evidence to support livestock as a cause of irreversible degradation away from these areas (Dodd 1994). A study of a pastoral system in northern Kenya concluded that it was regulated mainly by abiotic perturbances, rather than grazing pressure (Ellis and Swift 1988). Few other studies have separated the effects of livestock grazing from abiotic factors such as climate or fire, or distinguished between long-term and short-term effects (Dodd 1994).

Ecological and economic sustainability of livestock husbandry will most likely become increasingly difficult to achieve as human populations increase and more pressure is put on arid and semi-arid areas. Future livestock management systems in these regions are most likely to be sustainable if they are designed on the basis of an understanding of the attributes of traditional pastoral practices and build on the best aspects of these existing systems (Coughenour *et al.* 1985; Ellis and Swift 1988; Mace 1991).

UTILIZATION AND PRODUCTION OF WILD ANIMALS

Wild animals have been used as a resource throughout the course of human existence. The first human populations on all continents were hunter-gatherers, and although there are very few groups remaining which rely primarily on this livelihood, most tropical societies continue to use a variety of products from wild animals. The desirability and sustainability of various consumptive uses of wild animals has been the subject of considerable controversy (Baskin 1994; Eltringham 1994; Geist 1988). The populations of numerous wild species have undergone a precipitous decline in the past century, resulting in significant range reductions for many, and extinction or near-extinction of others. Many of these problems have been attributed directly to habitat loss and over-exploitation.

It is our view that, under certain conditions, the production and/or utilization of wild animal species can both help to meet human needs, and be complementary to protection programmes. Management of wildlife in a way which provides tangible benefits to the human inhabitants of an area creates an incentive for conservation of populations and their habitats, and is therefore more likely to receive the support of local people than legislation which restricts or prohibits their use of wildlife without addressing problems such as crop depredation and competition with domestic livestock. There are a wide variety of wildlife utilization and production systems in the tropics (Hudson *et al.* 1989). In this section we focus on traditional and market harvesting of wild animals for subsistence purposes, and wildlife ranching.

Wildlife harvesting

Wildlife harvesting, which includes hunting and associated activities such as trapping and collecting of free-ranging wild animals and their products, provides an important source of food and income throughout the tropics. The contribution of wild animal meat to nutritional requirements is very high in sub-Saharan Africa, with an estimated three-quarters of the population depending heavily on this source of protein (Asibey 1974; Asibey and Child 1990), and it is also an important dietary component in other regions (Marks 1989; Robinson and Redford 1991a). Different societies vary widely in their degree of dependence on wild meat, but it is especially important for groups whose diet lacks protein from other sources, such as the BaKouele and BaNgombe in Congo, whose staple crops are plantains and manioc (cassava) (Wilkie *et al.* 1992). Wild animals are also harvested to supply other products such as skins, fur, feathers and medicines.

A wide variety of species are harvested as subsistence food sources. Most of these are large-bodied mammals, birds and reptiles, generally with body mass of over a kilogram (Robinson 1996), but other taxa such as amphibians, snails and insects are also utilized. Harvesting may be seasonal in intensity where alternative resources are available during part of the year. Hunting of terrestrial mammals and large birds supplies most wild meat used during the wet season at sites in the Brazilian Amazon, while aquatic food sources are preferred during the dry season when these are concentrated by low water levels (Peres 1990). Wild meat is most important as a food source for a group of Kenyan subsistence hunters and trappers in the late dry season and early wet season when the supply of their dietary staple, maize, is low (FitzGibbon et al. 1995).

Other forms of wildlife harvesting such as market hunting, sport hunting and large-scale cropping or culling are also carried out. Market hunting includes sale of wild animals and their parts for food and other purposes such as the pet trade, zoological collections, biomedical research and collectors' items. The transition from subsistence to market hunting is often a gradual one. Many subsistence hunters also sell some of their take to supplement their income, and some depend heavily on this source of income to purchase basic cooking utensils, clothing, medicine and education (Wilkie et al. 1992). The demand for bushmeat in markets often results in its being more expensive than meat from domestic stock, sometimes selling for several times the price of mutton or beef, which may lead to subsistence hunters selling rather than consuming their take (Asibey and Child 1990).

Sustainability of wildlife harvests

In theory, sustainable harvesting should be achievable by limiting offtake of wild populations so that rates of harvest do not exceed rates of production. However, in practice this has often not been the case. Martin (1973) postulated that the first humans to settle the Americas were likely to have undergone explosive population growth as they spread throughout the continent, resulting in over-hunting, which led to Pleistocene megafaunal extinctions. Numerous studies have documented over-harvesting of wild species in more recent times (Bodmer et al. 1994; FitzGibbon et al. 1995; Robinson and Redford 1991a). Fa et al. (1995) estimated that actual harvests of forest mammals were an average of 4.98 and 1.08 times greater than sustainable harvests at two locations in Equatorial Guinea, although not all species were over-harvested. In the extreme case, the harvest rate of one primate species in this study, Cercopithecus pogonias, was 28 times greater than the estimated sustainable harvest.

The pressure on wild animal populations is certain to increase in the future with human population growth. Although much of this growth is likely to occur in urban areas, these often provide large markets for bushmeat harvested by rural dwellers (Asibey and Child 1990; Fa *et al.* 1995; Wilkie *et al.* 1992). The severity of this problem will be exacerbated by other factors such as habitat loss, fragmentation and degradation which decrease the carrying capacity for animal populations, as well as improved human access to those habitats remaining, and increasing replacement of traditional weapons with more effective ones. Continued uncontrolled harvesting under these circumstances will be certain to result in many local extinctions of game species.

Management of wild animal harvests will become essential in most areas if these are to be carried out on a sustainable basis. Development of sustainable harvesting systems requires, as a minimum, some knowledge of biological parameters of the harvested populations, as well as regulation and enforcement of harvesting rates. Factors which influence the impact of harvesting on wildlife populations are species-specific, and include population abundance, dispersion, reproductive rates and ease of capture (FitzGibbon *et al.* 1995; Peres 1990).

Species which occur at low population densities and have low rates of natural increase are usually the least resilient under high rates of harvest. Many of the large-bodied preferred game species have these characteristics. Primates are extremely vulnerable because of their popularity as a food source and for the live animal trade, and because they frequently have low population densities and low reproductive rates. The social organization of some species enables hunters to kill several individuals at one time; this can result in a high proportion of a local population being killed in one encounter (FitzGibbon *et al.* 1995; Peres 1990). Hunting for the zoo and pet trade may also be skewed towards females to obtain their young (Peres 1990). For these reasons, most primate species are unlikely to be suitable candidates for sustainable harvesting in the small fragments which will comprise many of the remaining habitats in the future. Species with high reproductive rates and which tend to occur at high population densities, such as elephant shrews (FitzGibbon *et al.* 1995) or agoutis (Robinson 1996), are more promising candidates for biologically sustainable harvesting systems. However, economic sustainability must also be considered, as the rewards of smaller species relative to the degree of hunting effort may be lower.

A key consideration for sustainable harvesting of any species is establishment of provisions to facilitate recruitment, particularly in areas which are heavily and continually harvested. This may require maintenance of protected areas for source populations. The possibility of harvesting areas on a rotational basis, with fallow areas serving as reproductive refuges, has also been suggested (Fa *et al.* 1995). Regulation of harvest according to sex and age, such as a male-directed hunting system in which higher proportions of males than females are harvested, may allow utilization without seriously affecting recruitment in some species (Bodmer *et al.* 1994), although strongly sex- or age-biased hunting does have the potential for serious negative impacts on the population (Ginsberg and Milner-Gulland 1994).

In the future, maintenance of game populations at sustainable and productive levels will be made more difficult by habitat fragmentation, which generally increases both the intensity and impact of harvesting activities (Robinson 1996). Species which undergo periodic population fluctuations are more likely to become locally extinct (Karr 1982). In some cases it might be possible to undertake reintroductions of species to isolated habitats. Robinson (1996) has suggested that preferred species are often eliminated from forest patches even under light hunting pressure, so that small habitat remnants (<10 km^2) are unlikely to be useful as game reservoirs; patches of intermediate size (10–100 km^2) could probably sustain light harvesting of some species, but preferred species would be susceptible to local extinction even in larger patches (100–1,000 km^2) under heavy harvesting pressure.

The effectiveness of various strategies for maintaining sustainable wildlife harvests will be highly specific depending on the species, available habitat and hunting pressure. Unfortunately, data are frequently not available for harvested populations, and determining levels of harvesting which can be carried out sustainably is difficult owing to the number of both biological and social factors involved, and the changes in these over time. Several indices and models have been developed to provide a first approximation at assessing the impacts of hunting and calculating sustainable harvest rates (Robinson and Redford 1991b, 1994).

The development of biologically sound harvesting strategies is by itself not enough to ensure sustainable offtake rates. The social and economic issues involved in developing sustainable harvesting systems are likely to be more problematic than the biological aspects. Development of sustainable harvesting systems may well necessitate reductions in current offtake rates, which are likely to involve an immediate decrease in economic benefits. This will often be necessary to prevent severe depletion of the resource in the long term, and should result in future benefits which outweigh the short-term costs. However, consideration of the effect of these immediate losses on subsistence users, and potential replacements for these, may be an essential part of any game management strategy (Bodmer et al. 1994).

Illegal harvesting is a serious problem in many areas, and leaves little incentive for conservation of populations, as animals left by one hunter are likely to be taken by the next. Commercial wildlife utilization in particular has historically had devastating consequences for animal populations, and these continue in many regions despite regulation by international agreements such as the Convention on International Trade in Endangered Species. Community-based management programmes are one strategy by which wildlife resources can be maintained for the benefit of local communities, thereby providing a strong incentive for sustainable harvest. Such programmes are beginning to be implemented in Latin America (Bodmer et al. 1994) and Africa (Matzke and Nabane 1996). The success of these types of programmes will depend on how compatible they are with local social and biological conditions, and they are unlikely to be feasible in all situations.

It is unlikely that the economic return from wildlife harvesting in the tropics, particularly from subsistence hunting, will be high enough to justify setting aside large areas exclusively for this purpose. However, in combination with other land uses such as extraction of timber and other plant products, livestock production, watershed protection or biodiversity conservation, it could continue to meet some of the needs of tropical societies. Natural habitats in which wild animals are harvested are unlikely to be as useful for biodiversity conservation as habitats managed specifically for that purpose, owing to both the depletion of game species and the potential for secondary impacts on forest structure if the species hunted are seed dispersers or prey for other animals (Robinson 1996). However, harvesting could contribute to biodiversity conservation where the benefits obtained provide additional justification for maintaining natural habitats, or buffer zones around existing reserves, that would otherwise not be preserved.

Wildlife ranching and farming

Wildlife ranching involves the extensive production of wild species in large, confined areas. In the tropics, it is best-developed in Africa, where there are many species of native ungulates and large tracts of uncultivated land. African game ranching was proposed several decades ago as a form of land use which would be more productive than cattle ranching, so should provide additional protein for human populations and thereby serve as an incentive for conservation of wildlife (Dasmann 1964). The reasoning behind this supposition was that wild ungulates would be able to make more efficient use of marginal lands. It was thought that there was virtually no overlap in the diets of wild species, which would allow more efficient use of mixed vegetation cover than by cattle, which concentrate on preferred grasses. Low water requirements by wild ungulates, which give them the freedom to range more widely than cattle, would allow better use of drylands, and avoid land damage caused by overgrazing of areas near water sources. Preliminary data suggested that savanna lands supported a higher biomass of wild ungulates than of domestic stock (Dasmann 1964).

Wildlife ranching has since been undertaken at numerous locations, mainly in Southern Africa, but also in the eastern and western portions of the continent. A variety of wild ungulate species are utilized, and these are frequently ranched in association with cattle. However, there has been very little research published in the primary literature on wildlife ranching relative to the material available on domestic livestock.

Mixed-species systems can be expected to have the potential for higher production, owing to use of a wider variety of the available forages. The advantages of this may not be as significant as first envisioned, as wild ungulates do not show the almost complete separation in diets originally suggested and there is considerable flexibility and overlap among species: cattle are not exclusively grazers but do browse when necessary, and some domestic species are primarily browsers (Fairall 1984; Macnab 1991). However, a mixture of wild ungulates will normally make wider use of available forages than domestic livestock. African ungulates fall along a continuum with respect to feeding preferences. At the extremes are species which are exclusively grazers such as wildebeeste (*Connochaetes taurinus*) and zebra (*Equus burchelli*), and species which are exclusively browsers such as giraffe (*Giraffa camelopardalis*) (Figure 10.7) and dikdik (*Madoqua kirkii*). Mixed feeders in between these extremes may prefer grazing or browsing. Wild grazers with extensive dietary overlap in terms of species composition may ingest different plant parts (Gwynne and Bell 1968; McNaughton and Georgiadis 1986). The potential advantages of using wild herbivores will depend on the mix of species in relation to the available forage, and are likely to be maximized where a variety of types of feeders exist, such as on a Kenyan game ranch on which the wild ungulates consist of 60 per cent grazers, 21 per cent browsers and 19 per cent mixed feeders (Sommerlatte and Hopcraft 1992). A comparison of cattle and wild ungulates on a game ranch in southern Africa showed that for cattle, available browse utilization was about 7–8 per cent compared to about 20 per cent for the population of 13 game species, of which only two were browsers. The wild ungulates also used a much wider range of herbaceous species (Taylor and Walker 1978). An additional advantage of using wild herbivores in combination with domestic stock is that they may consume plants which would otherwise have to be controlled (Asibey and Child 1990).

Predictable seasonal differences among ungulate species might also offer potential advantages for staggered harvesting according to body condition. A study of Thomson's and Grant's gazelles (*Gazella thomsoni, G. granti*) in Kenya indicated that these two relatively similar species show a pronounced difference in forage use. Thomson's gazelles preferred hilltop grasslands, and Grant's gazelles preferred lower slopes with taller grasses, more forbs and tree cover, although both species shifted in a downslope direction during the dry season. Seasonal differences in timing of forage quality between these areas, with peaks progressively later from the top to the bottom of the catena, were suggested to be responsible for asynchronous body condition patterns measured in the two ungulate species. Body condition in Grant's gazelles improved more slowly in the wet season but lasted longer into the dry season than in Thomson's gazelles (Stelfox and Hudson 1986).

Other arguments which have been presented to support wildlife ranching suggest that wild ungulates are better adapted to local conditions by being more resistant to drought and disease. Many native ungulates are less dependent on regular access to surface water sources than cattle. This allows wild ungulates to extend their foraging range, reduces energy needed for daily

Figure 10.7 Giraffe (*Giraffa camelopardalis*) browsing *Acacia seyal* in a riparian wood-land on a Kenyan game ranch. Giraffes can utilize plant material at heights that are inaccessible to other browsing ungulates, although they frequently forage at lower heights as well

trips to water, and avoids the trampling and soil erosion which occur at many cattle bores. The resistance of native ungulates to disease has frequently been cited as a beneficial characteristic with respect to their potential for game ranching (Fink and Baptist 1992; Hoogesteijn Reul 1979). Native species are more resistant to some infections such as foot-and-mouth disease and trypanosomiasis, but are susceptible to others which infect cattle such as rinderpest, anthrax and helminthiasis. Many infectious diseases can be transmitted between wild and domestic animals in either direction (Kock 1995). Where disease problems exist, development and maintenance of disease-free herds may be difficult in extensive systems, particularly if there is contact with free-ranging wild ungulates which act as reservoirs. This could limit the number of species which can be mixed with cattle, or exported as breeding stock or unprocessed products (Fairall 1989). Management of confined animals under more intensive conditions is likely to increase their susceptibility to diseases and parasites. Zoonotic diseases are also a potential concern where handling or consumption of wild animals are increased (Kock 1995).

Although earlier studies presented data indicating that a higher biomass of wild animals than of domestic animals can be produced by a unit of land, subsequent studies have shown that this is not always the case (Coe et al. 1976; Dasmann 1964; Kock 1995; Macnab 1991; Sharkey 1970; Taylor and Walker 1978). Comparisons of factors that affect production, such as reproductive efficiency and weight gain, have usually been made over short time periods; different conclusions and a different perspective on carrying capacity might well be reached from longer-term data which included the effects of climatic fluctuations. Some studies have concluded that integrated cattle and wildlife ranching with a balance of browsing and grazing species is likely to make best use of rangelands, and may be most profitable economically in some situations (Kreuter and Workman 1994; Taylor and Walker 1978).

Knowledge which could maximize productivity of wildlife ranching systems, such as determination of optimal stocking rates, population structures and herd management techniques, is less well developed than for cattle ranching owing to the variety of species mixtures and shorter time during which they have been assessed. For example, stocking rates estimated by using standard livestock indices do not take into consideration differences in energy requirements per unit of body weight which may exist among species. Seasonal adjustment of stocking rates is more difficult with wild ungulates than with domestic stock, which are easier to move (Skinner 1989).

The ultimate economic benefits of game ranching are likely to depend more heavily on technical and economic factors than on the biological issues. Although wildlife ranching has been suggested to be a low-input form of land use, it is not by any means a no-input activity. Although it requires fewer watering facilities such as boreholes, dams and piping, the basic infrastructure required for wildlife ranching is likely to be comparable to that for commercial cattle production in terms of roads, firebreaks, vehicles and fuel. Fencing needs to be approximately 2.3–2.6 m in height to contain ungulates (Fink and Baptist 1992; Skinner 1989), and may be costly, particularly on large holdings. The availability of meat processing facilities may also be a limitation. Wildlife ranching is generally less labour-intensive than cattle ranching, as wild ungulates are not usually vaccinated, branded or drenched. Cropping of wild ungulates is more expensive than for cattle (Fink and Baptist

1992), and other handling is generally more difficult with wild stock. A range of potential uses can be made of wild animal species in addition to meat, including production of hides and other products, and non-consumptive uses such as viewing and photography; economic benefits will depend on which options are practically and legally feasible. Slaughter requirements and transportation to markets may be problematic in remote areas, with the lack of a marketing infrastructure being cited as a limitation for some pilot projects on game ranching in Kenya (Kock 1995).

The intended contribution of wildlife ranching to alleviating protein malnutrition among rural Africans (Dasmann 1964) has not been convincingly demonstrated. The extensive nature of this activity requires large tracts of land, which are not available to the average individual or family. Novelty value and lower availability than beef can result in wild meat being expensive. In 1990, the average price of venison from ranched ungulates was more than twice that of beef in Kenya (Sommerlatte and Hopcraft 1992). Cultural resistance to particular types of wild meat also exists in some areas, which may limit the potential of wildlife ranching in this respect (Eltringham 1994; Kock 1995).

Dasmann (1964) suggested that wildlife ranching would serve as an incentive for conservation, and the potential value of this form of land use for that purpose has been controversial (Macnab 1991). Natural habitats maintained for wildlife ranching are of more value for conservation purposes than cropland, and viable populations of economically valuable species are likely to be conserved. However, this may involve relatively few species. Only eight of the more than 26 ungulate species in Southern Africa are considered to be serious prospects for ranching, and of these, only four are currently economically viable (Fairall 1989). Although keeping native habitat intact can also benefit non-economically important species of wildlife, large predators and competitors of ranched species are unlikely to be widely tolerated. The development of legal commercial markets for animal products other than meat, such as hides, which may be necessary to make wildlife ranching economically worthwhile in some situations, could also encourage uncontrollable illegal use (Geist 1988; Macnab 1991). Wildlife ranching is likely to be complementary to other wildlife conservation programmes in many situations, but unlikely to be a panacea for wildlife conservation in the tropics, as is sometimes implied in the popular literature.

Although we have focused on game ranching in Africa, numerous additional possibilities for both wildlife ranching and farming elsewhere in the tropics have been suggested, such as the use of derived *Imperata* grasslands in South-East Asia (Chapter 7) for production of several deer species, ranching of capybaras or cameloids in South America (De Vos 1991), and utilization of other taxa such as reptiles and insects (New 1994; Renecker and Hudson 1991; Robinson and Redford 1991a). Wildlife farming involves more intensive production of wild animals, often only a single species. Provision of feed,

Figure 10.8 Butterfly in breeding facility in Amazon rain forest, Ecuador. Breeding and sale of butterflies has the potential to be an economically rewarding activity which utilizes relatively little land

housing and veterinary care, and control of reproduction, often requires a higher initial investment and greater ongoing costs. It seems likely that the potential advantages of production of some species might be lost with intensification, particularly in comparison to domestic stock which have undergone a long period of selection for desirable traits under these conditions. Farming of species which require less space (Figure 10.8) and relatively few inputs is likely to be a more viable option in areas of high human population density where landholdings are small, or where the available land does not support intensive uses of other types.

Wildlife ranching and farming offer alternatives to conventional livestock husbandry which may have the potential to meet human needs in a more productive and sustainable manner than existing land uses in some situations. However, wildlife ranching on a large scale such as that carried out in Africa is unlikely to replace conventional livestock husbandry. Existing data do not demonstrate its benefits as unequivocally as first proposed, and it is unsuitable for the many small landholdings on which most of the rural human populations live. The relative benefits and disadvantages of most systems have not been well documented, and are likely to vary a great deal among sites and species. We suggest that the abilities of various systems to fulfil their intended purposes need careful evaluation with respect to the biological, social and

300

economic considerations prior to more widespread implementation, particularly with respect to their long-term results and sustainability.

AQUACULTURE

Aquaculture, the farming of aquatic organisms, has been practised for about four thousand years (Iwama 1991). It is carried out in freshwater ecosystems, marine waters (where it is also known as mariculture), and the range of brackish waters in between. It accounts for a small but increasing proportion of aquatic food production worldwide, presently contributing about 13 per cent of world aquatic fish production (Figure 10.9). Fishes account for approximately three-quarters of total aquaculture production worldwide, with slightly over half of that from inland systems and the rest from marine systems. The remaining quarter is made up of a variety of other products such as aquatic plants, amphibians and reptiles (UNEP 1993). Culture of aquatic organisms is best-developed in Asia and the Pacific, which have the highest production and greatest consumption per capita of aquaculture products (Summerfelt 1982). This region accounts for about 84 per cent of the world's total production (Guerrero 1990).

The primary nutritive contribution of aquatic organisms in tropical countries lies in their value as a protein source. Fish have a high protein content in relation to most other foods, with an amino acid composition comparable to that of most other types of animal protein (Table 8.2), and have a higher degree of digestibility than most domestic animal tissues (Summerfelt 1982). Although the contribution of aquaculture to protein production is presently relatively small on a global scale (Iwama 1991), it is quite important in some regions. Aquatic organisms are the principal source of animal protein in many Asian diets, contributing up to 80 per cent of the animal protein intake in some countries (Summerfelt 1982).

Numerous types of tropical aquaculture production systems exist and a wide range of aquatic organisms are cultured; most of these are finfish, crustaceans, molluscs or algae. Two of the most widely cultured fishes in the tropics are the freshwater common carp (*Cyprinus carpio*), particularly throughout Asia, and *Tilapia* species, particularly in Africa and the Middle East (Meade 1989; Summerfelt 1982). These species and numerous others are farmed both in monocultures and in polycultures. Polyculture systems are the traditional method of fish farming in Asia and rely on combinations of species which are non-competitive in terms of food habits, and non-cannibalistic. For example, four species of Asian carp (*Hypophthalmichthyes molitrix, Aristichthys nobilis, Cyprinus carpio* and *Ctenopharyngodon idella*) which are cultured together feed on phytoplankton, zooplankton, detritus and benthic organisms, and vascular aquatic plants, respectively (Summerfelt 1982).

Small aquaculture operations are often integrated with the production of crops and domestic livestock, particularly in Asia. Fish farming practised in

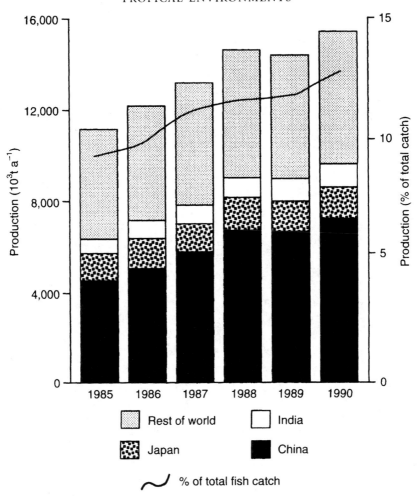

Figure 10.9 Aquaculture production from selected regions and globally between 1985 and 1990 as tonnage and as a percentage of total fish production

Source: UNEP 1993: Figure 3.10

combination with irrigated rice production has a long history in South-East Asia. In Indonesia, rice fields used for raising carp from fry to fingerlings produce 75–100 kg.ha^{-1} (Guerrero 1990). This activity can be practised with minor modifications to rice paddies, such as a trench within the paddy (Stickney 1994). Organic wastes, such as manure from ducks, poultry, cattle or swine, are frequently used as fertilizers. In Thailand, an integrated poultry–swine–aquaculture system produced 4 t of catfish (*Pangasius pangasius*), 8 t of swine and 15,390 chicken eggs in an area of 0.24 ha (Guerrero 1990).

Table 10.3 Efficiency of feed utilization by fish, cattle, chicken and swine

	Feed Composition		Efficiency		
	Protein (%)	Metabolizable energy (Mcal.kg⁻¹)	Weight gain per gram of food (g)	Protein gain per gram of food protein (g)	Protein gain per megacalorie of food energy (g)
Catfish $\{$	30	2.64	0.77	0.41	47.1
	40	2.86	0.91	0.36	50.8
Cattle	11	2.61	0.13	0.15	6.28
Chicken	18	2.60	0.48	0.33	23.0
Swine	16	3.30	0.31	0.20	9.65

Source: Lovell 1979

Future potential of aquaculture production

Aquaculture has the potential to play an increasingly important role in meeting the future food needs of tropical human populations. Fish convert feed into body tissue more efficiently than domestic stock, and produce protein at a lower energy cost than cattle, poultry and swine (Table 10.3). They can therefore provide a more efficient way to supplement human diets with animal protein than by production of other domestic animals. Although most aquatic foods currently consumed by humans are obtained from capture fisheries, there has been growing concern about the sustainability of current yield levels from these owing to overfishing and habitat degradation. It is unlikely that the fishing industry will be able to fill the increasing demand for aquatic foods, and increases in aquaculture production may offer the best prospect for maintaining a secure and sustainable future supply.

Introduction or expansion of aquaculture is probably technically feasible almost anywhere (Thomas 1994), but is subject to the availability of suitable land; for example, coastal areas are often in demand for many other uses, and many of these zones suffer from severe pollution problems. Cultural preferences also influence whether new aquaculture practices or products are adopted (Stickney 1994). Thomas (1994) attributes the limited spread of aquaculture in Africa to the failure of projects for socioeconomic rather than technical reasons. Aquaculture production for market may be limited by the ability to maintain product quality during transportation, as these foods tend to spoil easily.

Increased aquaculture production in the tropics can also be achieved by intensification and diversification of existing systems. Considerably higher densities of organisms per unit of water volume than occur in most natural systems can be maintained by management of cultured species in controlled or partially controlled conditions (Stickney 1994). Productivity of different types of systems is quite variable, and depends on the species utilized and level of intensification. Diversification by the utilization of new species depends largely on obtaining or developing organisms with desirable traits such as ease of reproduction in captivity, lack of cannibalism, rapid growth rates, efficient utilization of available foods, tolerance of hatchery conditions and resistance to disease (Summerfelt 1982).

Extensive aquaculture for production of subsistence foods is often carried out with minimal management and few inputs. At the most basic level, culture systems may involve stocking organisms either captured from the wild or obtained from hatcheries into small natural waterbodies, and relying on existing food sources for growth, and natural reproduction to replace stock that is harvested. Low costs and input for these systems are usually matched by low outputs. Yields vary with the natural fertility of the system, but produce an annual harvest of about 350–400 kg.ha^{-1} (Summerfelt 1982).

Supplemental feeding or pond fertilization usually increases substantially the productivity of small systems. Fertilization with animal or human wastes can increase yield by a factor of five, while provision of high-quality formulated feed can increase the yield of most fishes to 10 times that achieved in fertilized ponds. Yields of 1,000–6,000 kg.ha^{-1} can be found with some fertilization or use of plant wastes (Summerfelt 1982).

Intensive production systems require correspondingly higher inputs of food and energy, and are usually limited to economically valuable species. Most or all of the diet is provided to the cultured organisms, and this must be nutritionally complete (Summerfelt 1982). Feeds usually require at least a small amount of animal protein to meet the amino acid needs of the cultured species (Stickney 1994). Artificial containment systems such as tanks, raceways, pens, cages or silos are often used. Control of water exchange may require water supply and drainage channels, and be carried out with mechanical equipment such as pumps, aerators, and filters. Additional infrastructure such as storage or refrigeration units may also be required. Production levels of 30,000 kg.ha^{-1} of shrimp are possible under intensive culture (Flaherty and Karnjanakesorn 1995). However, these types of production systems often require a substantial capital outlay, which usually puts them outside the realm of subsistence farmers.

Major technological constraints to aquaculture intensification in the tropics include inadequate supplies and quality of young stock, and diseases. Restocking of some species is almost entirely dependent on natural sources, and further work on improvements in artificial spawning and rearing is needed (Guerrero 1990). Aquatic animals are subject to diseases caused by bacteria, viruses, fungi and other parasitic organisms. Together these are responsible for the most significant economic losses within aquaculture production systems (Meyer 1991). Intensification may exacerbate disease problems if it places additional stresses on the cultured organisms such as overcrowding or inadequate dissolved oxygen.

Environmental degradation is one of the most serious factors potentially limiting the sustainability of aquaculture production in the tropics. Feed, oxygen and wastes are all contained or transmitted through water (Colt 1991), so cultured organisms are very vulnerable to effects of environmental deterioration. Pollution control measures and sewage treatment facilities often lag behind human population growth in the tropics, creating severe water pollution problems in many regions; sewage, agricultural chemicals and industrial wastes affect both cultured organisms and the wild stocks which serve as the source for the young of many species (Iwama 1991). Intensification of aquaculture production by supplemental feeding and fertilization also increases the potential for deterioration in water quality. The containment of higher densities of organisms can result in oxygen depletion, release of waste materials such as faeces, food wastes, food additives and therapeutic drugs, and associated effects such as eutrophication. The specific impacts of these are discussed in more detail by Iwama (1991).

Although the culture of marine shrimp has recently increased rapidly in Asia and South America, drastic decreases in production and abandonment of aquaculture ponds due to environmental degradation and disease problems associated with intensive production have already occurred in some areas. Ponds in some areas of Thailand have been abandoned after only 3 or 4 years owing to decreased productivity (Flaherty and Karnjanakesorn 1995). Aquaculture management techniques, such as choice of locations with high flushing rates or continuous replacement of water, artificial aeration, consideration of the feed type and methods of feeding to minimize waste, and control of the release of suspended solids, may help alleviate some of the problems which accompany higher levels of production (Iwama 1991; Summerfelt 1982). However, although improved management can extend the period of viability, it has been suggested that the lifespan of very intensive operations may be much shorter than originally anticipated. Much of the dramatically increased production within the shrimp industry has been achieved through expansion to new areas. This creates two problems: the inevitable shortage of new lands for expansion, and the difficulty of rehabilitation of abandoned areas (Flaherty and Karnjanakesorn 1995).

The environmental impacts of expansion and intensification of aquaculture production are also felt outside the culture operations. Initial expansion of aquaculture operations often leads to habitat destruction in coastal environments, particularly in mangrove forests (Figure 6.8), which are important as nursery areas for many aquatic organisms (Chapter 6). Intensive operations can result in a deterioration of water quality which affects surrounding areas. In Thailand, seepage of brackish water from shrimp farms into ground water supplies and adjacent fields has increased salinity levels beyond tolerable limits for rice production (Flaherty and Karnjanakesorn 1995). Discharges from aquaculture systems can also result in a deterioration of coastal water quality.

The environmental problems involved with development of very intensive aquaculture systems on a large scale, along with the large capital outlay required to develop and maintain many of these systems, suggest that this type of production is unlikely to benefit rural dwellers, and in some cases may result in degradation of their existing agricultural lands. A lesser degree of intensification of small-scale traditional systems is more likely to be sustainable and beneficial as a food source for rural dwellers.

11

LAND USE PLANNING, REHABILITATION AND RESEARCH

As we have emphasized in earlier chapters, the environmental substrate of tropical landscapes is far more heterogeneous than earlier accounts suggested. At local scales, much of this heterogeneity is related to variations in soils and topography, both of which are themselves often closely linked. As a result, most tropical landscapes comprise suites of local environmental site-types that often occur in repetitive and predictable sequences along topographic gradients. These have sometimes been referred to as soil catenas. For example, in river valleys, site types often change from floodplains covered by relatively fertile alluvial soils, through a series of terraces of increasing elevation and age, and decreasing soil fertility, to steeplands in the enclosing hills, where soils may be relatively fertile but thin and subject to high erosion rates (cf. Figure 4.12).

Site types usually differ not only in their potential productivity of materials useful to people, but also in their susceptibility to degradation under use. Consequently, they also differ in the types of land use that can be practised sustainably on them. In general, more productive sites can sustain a wider range of land uses, while the range becomes increasingly restricted as site quality declines. For example, the alluvial soils of floodplains could be used sustainably for forestry, tree crops, pastures, dryland annual crops or irrigated rice culture. Gently sloping lands adjacent to floodplains may be usable sustainably for forestry, tree crops and pastures, while in more steeply sloping terrain only forestry could be practised without environmental degradation. In the steepest terrain even forestry may be inappropriate, and a protective forest cover may be the only sustainable land use.

Despite being potentially usable in a variety of ways, more productive sites will usually tend to be used for purposes offering the greatest returns on investment of labour or other inputs, leaving less rewarding activities to be practised on more marginal sites. For example, forestry will rarely be practised on the most fertile soils, where there are other competing land uses, such as intensive food crop cultivation. A common land use pattern throughout the Asian tropics is for floodplains to be used for irrigated rice culture, while surrounding hilly areas are used for shifting cultivation or tree crops (Figure

307

Figure 11.1 Agricultural land use pattern in an area of recent colonization, Mindanao, southern Philippines. The flat terrain in the foreground is used for irrigated rice-farming, while gentle colluvial slopes in the middle distance are used for permanent dryland crops. The steep limestone terrain in the background is being used for shifting cultivation

11.1). This process sets up an economically rational and environmentally cued pattern of land uses, in which more productive sites are used intensively, while less productive areas are used more extensively. However, such a rational pattern can be distorted by the intrusion of other, non-environmental constraints into the land use system. For example, in eastern Honduras, ownership of most flatlands by large corporations has resulted in these being used for tree crops (bananas and oil palm), while intensive production of most food crops is restricted to adjacent steeplands, to which most of the rural population is confined. Not surprisingly, this distortion of the environmentally rational land use pattern has resulted in severe environmental degradation in the steeplands.

While a less distorted pattern of land use develops in many tropical landscapes, these are not necessarily all sustainable, and one of the most urgent needs in applied tropical environmental science is the development of means of evaluating the stability and sustainability of different land use systems. This is likely to be especially difficult, as environmental degradation often tends to be so gradual as to be imperceptible to an individual in the course of their lifespan. The gradual degradation of the environment surrounding

308

prehistoric Mayan centres represents a classic example of this process (Deevey *et al.* 1979; see Chapter 8). Moreover, where large inputs, such as of fertilizers, are employed, they can effectively mask the signs of environmental degradation associated with soil erosion (Pimentel 1993). For these reasons, careful monitoring of the state of environmental conditions, such as soil properties, may be required for many decades before clear 'signals' of environmental degradation become apparent.

O'Callaghan and Wyseure (1994) suggest simulation modelling of ecosystem properties as a means of projecting environmental deterioration, but this procedure is contingent upon the workings of the system being sufficiently well known to enable reliable simulation. More readily achieved signals of degradation may come from comparisons of ecosystem properties between managed and protected systems, although in these instances it is often difficult to establish whether the managed system has re-equilibrated at some new state, or is subject to ongoing degradation. When available, long-term biological census data may provide signals of system deterioration. For example, Kattan *et al.* (1994) use census data from fragmented cloud forests in Colombia to show that 31 per cent of the original bird fauna has gone locally extinct between 1911 and 1990. Especially useful integrated signals of ecosystem conditions may be provided by the river systems draining managed watersheds, especially if unmanaged control watersheds are available for comparison. Rivers discharge not only water but also sediments, nutrients in solution and water-borne contaminants, and changes in these can provide early symptoms of environmental problems in these areas.

The necessity of evaluating the long-term sustainability of different forms of tropical land use emphasizes the desirability of preserving intact natural ecosystems to serve as controls in comparative studies, and of establishing 'baseline' studies before land use changes take place. It also emphasizes the need to establish ongoing monitoring programmes, not only for such routinely measured phenomena as rainfall and river discharge, but also of river sediment and contaminant loads.

LAND COLONIZATION

Throughout history, migration to unoccupied land has been one of the most common responses to a growing population (Chapters 8 and 9). The phenomenon of land colonization has largely ceased in economically developed countries, and in many it has even been reversed. However, it continues to be widespread in those tropical countries that contain some zones occupied by few people, and other zones with large and rapidly growing populations. One of the most widely publicized examples of recent land colonization in the tropics has been that taking place in Brazilian Amazonia over the past two to three decades, but the phenomenon has been widespread elsewhere in the tropics (Manshard and Morgan 1988). For example, lowland colonization has taken place in all other countries possessing Amazonian territory. In Central America there has been a significant movement of rural populations from the densely settled central highland and west coast regions to the humid tropical eastern coastal zones (Jones 1990). In Indonesia, there has been a major movement of people from the densely settled islands of Java and Bali, to the less densely peopled outer islands, while in the Philippines, movement of people into Mindanao from the densely settled northern islands has occurred throughout most of the present century (Uhlig 1988). Colonization of less distant areas has also occurred in other countries. For example, forested eastern parts of Thailand have been colonized spontaneously in recent decades (Scholz 1988), while in Malaysia, government-planned colonization has taken place in forested areas in the central and eastern parts of the Malayan peninsula (Uhlig 1988). Finally, at very local scales throughout the tropics, there has been a movement of excess population from very densely settled lowlands to more marginal adjacent steeplands.

While a growing rural population in already settled areas, combined with insufficient employment growth in the non-agricultural sector, has been a major impetus to most land colonization in the tropics, other factors have contributed to the process. Colonization of unpopulated border areas for strategic purposes has been encouraged by many governments. Many governments also encourage land colonization in the hope of generating foreign exchange through the exploitation of frontier resources. Government policy may also promote colonization as a means of providing opportunities for particular sectors of the national population, such as landless peasants or ethnic groups. Some land colonization may also be precipitated by emerging market opportunities; in Thailand, increasing world prices for certain dryland crops (sugar, maize and cassava) led to a major recent expansion of agriculture into previously forested upland areas (Scholz 1988).

The degree of government involvement in land colonization in the tropics varies widely. At one extreme are such carefully planned and heavily capitalized schemes as the Jenka Triangle in Malaysia (Wikkramatileke 1972). Here, land use, mainly for tree plantations of rubber and oil palm, was planned in the light of detailed soil surveys, a large infrastructure of roads, housing, clinics, schools and crop processing facilities was provided, and tree planting took place before assignment of land parcels to individual colonists. At the other extreme, colonization may be entirely spontaneous, with settlers often following selective logging of an area, and using the logging trails as a rudimentary transport infrastructure. Under these circumstances, no government involvement may take place save, in some instances, the eventual granting of land titles to the settlers. Between these two extremes are various degrees of government involvement, which often vary widely from country to country. In some instances only a road infrastructure is provided, while in others land may be surveyed, extension services provided and credit made available to farmers. The provision of secure title to land is a perennial problem in land colonization schemes. Although most government policies promote entitlement of individual small farmers, they are frequently poorly educated, lack capital and political influence, and are vulnerable to having their land co-opted by large landowners.

In theory, land colonization offers the prospect of planning for optimum and sustainable use of land from the outset of human occupancy in an area. Unfortunately, realization of this ideal is made difficult by several circumstances. Unoccupied frontier areas are often less productive, or more problematic, environments for human occupancy for the simple reason that earlier human populations have concentrated on the most productive sites. Moreover, there may be limited knowledge about how best to use the land on the part of both colonists and government agencies. Colonists often come from distant areas possessing very different conditions from those of the area colonized, and lack a fund of indigenous folk knowledge developed from prolonged trial and error in the region colonized. Colonists to the Amazonian lowlands often arrive from the dry north-east of Brazil or the Andean highlands, and are likely to be poorly prepared for the problems of agriculture in the humid tropics. A body of folk knowledge may exist among indigenous populations in the areas being colonized, but these groups are often seen as being 'primitive' and potential land use competitors, and their knowledge is rarely made use of. While many government-sponsored land colonization schemes make use of some preliminary resource surveys, these are often of a reconnaissance nature based on remote sensing, and may provide insufficiently detailed information for planning land use at the farm level. For example, the mapping of the Brazilian Amazon with side-looking airborne radar (SLAR) has resulted in a serious underestimation of the proportion of sloping land in this area that is subject to a high risk of erosion (Fearnside 1987). Moreover, even when local environmental patterns are known, agronomists are often unfamiliar with the most appropriate way to manage these. As Moran (1985) has commented, Amazonia is not only a frontier of human settlement, but also a 'knowledge frontier'.

Government policies promoting land colonization are often antithetical to effective land use. The establishment of ownership often requires that 'improvements' be made to the land, which usually promotes excessive clearance of as much forest as possible. Tax incentives may promote land clearance for speculative, rather than productive, purposes; the establishment of large cattle ranches in the Brazilian Amazon represents one especially dramatic example of this process (Hecht 1985). Isolation of areas from markets and the inadequacies of marketing facilities may severely limit the land use options available to the colonist, and encourage exploitative use of forest resources. For example, Brazilian law requires the preservation of 50 per cent of colonized land in forest cover, but the law is frequently ignored (Fearnside 1986), and the forested areas have been subject to heavy exploitation of timber and are degrading rapidly (Uhl and Viera 1989).

As a result of these difficulties, land colonization schemes are often doomed to replicate the misuse of land that is manifest elsewhere in national territories. Traditional land tenure problems of absentee landlords and landless peasants often re-emerge rapidly. Landholdings may become subdivided among offspring, or, where this is prohibited, landless offspring migrate elsewhere to continue the process of colonization. Indigenous populations are often dispossessed of their lands and marginalized socially. Land that should more properly be left untouched is occupied, and inappropriate land uses may be employed. While conservation reserves are now usually designated in colonization schemes, protection of them is rarely enforced, and encroachment on them by expanding colonist populations is virtually guaranteed.

These difficulties call into question the wisdom of promoting land colonization as a national land use policy in tropical countries. These schemes are capable of resettling only a small proportion of most nations' populations and therefore provide only temporary relief from problems of over-population. In the case of Indonesia's transmigration programme, the numbers of people leaving Java have been less than the annual population increase on that island. Also, it has been estimated that the number of actual migrants may be two to three times greater than the number sponsored by the government, which creates potential problems for the planning process (Whitten 1987). In the Jenka Triangle scheme, where land subdivision is prohibited, there are insufficient employment opportunities for the children of original colonists, and many youth are drifting to major urban centres (Bahrin et al. 1988). Temporary relief is bought at the expense of destroying large areas of natural habitat and exposing others to encroachment, and of destroying many of the indigenous cultures that may already exist in the colonized areas. More permanent solutions to the misuse of tropical environments require that problems of over-population, and unequal distribution of land and wealth, be addressed in existing areas of human occupancy, rather than being replicated elsewhere throughout national territories.

In 'frontier' areas of generally less productive habitats, economic developments that are environmentally sustainable are more likely to comprise low-intensity use of the resource base, such as selective forestry and tourism. The incorporation of indigenous populations into these activities would ensure their cultural survival, while providing the opportunities for improvements in material standards of living. It would also achieve the strategic goal of manifest human occupancy of territory, that is beloved by so many national governments.

LAND REHABILITATION

While some rural populations in the tropics are engaged in the land colonization process, most occupy long-settled landscapes. In these places, continual population growth has often placed increasing pressures on the

Figure 11.2 Intensive agricultural land utilization on steep slopes in the Blue Mountains, Jamaica. In an ideal world, terrain such as this would remain permanently forested

land resource and resulted in its progressive deterioration as a result of deforestation and soil erosion, leading to a declining carrying capacity for the human population. Consequently, efforts to prevent further deterioration and ultimately to restore carrying capacity are urgently needed throughout much of the tropics. In this section we will concentrate upon two especially critical problems in this endeavour: the rehabilitation of heavily populated steeplands, and the re-establishment of forest on derived grasslands. The urgency for rehabilitation efforts in these areas relates not only to the need to ensure local land use sustainability and an improved quality of life for local populations, but also to the need for larger-scale hydrological rehabilitation. Most watersheds covered by derived grasslands are subject to large flood risks due to high surface runoff; those whose steeplands are also under heavy agricultural usage also experience high sedimentation rates. Both conditions can promote severe problems for downstream agricultural and urban areas and cause siltation in dams, fishponds and coastal areas.

Heavily populated steeplands

The occupation of steeply sloping terrain by agricultural populations (Figure 11.2) is usually a result of over-population in nearby flatlands, co-option of these by large-scale agricultural enterprises, or both. Initial occupation usually

Figure 11.3 Landslips and earthflows taking place on fields created on excessively steep slopes, eastern Honduras. Tree roots help to stabilize soil mantles in steep terrain, and their removal leaves the soil vulnerable to mass movements when it becomes saturated during the wet season

involves shifting cultivation, whose fallow lengths shorten as populations increase. More frequent cropping produces accelerated rates of sheet and rill erosion, to which all agricultural fields are prone, but also the potential initiation of gullying, and various forms of mass wasting (land slips, earth flows; Figure 11.3). Where farm animals are raised, over-grazing may take place on pastures, which promotes further, accelerated erosion. The process of intensified land use is usually accompanied by almost complete deforestation, with forest remnants being restricted to the most inaccessible sites, such as ravine slopes and mountain-tops (Figure 5.9). Over time, these forest remnants are usually used as sources of fuelwood, which leads to their gradual degradation as tree recruitment is insufficient to sustain tree populations. When these forest remnants are exhausted, crop residues and animal manure may be used as a fuel source, precluding the use of these materials in nutrient recycling or as a source of mulch. While many degrading landscapes may contain sustainable components that can be preserved in rehabilitation efforts, others, which have undergone extreme deterioration (Figure 11.4), may depend entirely on the introduction of new techniques.

Most heavily populated steeplands should, ideally, be returned to the forested state that they originally existed in, but this represents an unrealistic

Figure 11.4 Extreme land degradation in the mountains of central Honduras. In this totally deforested steepland area, erosion and mass wasting on agricultural land is extensive. Some attempts at soil stabilization are being made on fields in the middle distance, using rudimentary terracing

objective except over very long time periods. In the short term, the most immediate needs are usually to slow the rate of degradation by encouraging less damaging forms of land use; much of this involves taking measures to slow rates of soil erosion. However, soil conservation measures are unlikely to be successful unless perceived as beneficial by local inhabitants, rather than as simply fulfilling the needs of an external agency. In the Himalayan foothills, rehabilitation of the steepest watersheds has been achieved by promoting these as sources of irrigation water for adjacent rain-fed farmlands (Dhar 1994). The construction of earthen dams, and protection of watersheds from grazing and tree felling, has allowed a significant extension of the cropping period using irrigation. Livestock is now pen-fed using forage cut in the watersheds, where both improved tree establishment and much increased forage grass production has taken place. An important component of this rehabilitation process has been the development of co-operative village societies to manage the common watershed resource (Dhar 1994).

Replacement of annual crops by tree crops or natural forest covers may be a long-term objective of rehabilitation, but over shorter time periods soil conservation measures on steeply sloping agricultural lands are necessary. Certain cover crops, such as *Mucuna*, may serve to reduce erosion considerably, but

Figure 11.5 Terracing to reduce the rate of soil erosion in steeply sloping fields, central Honduras. Perennial grasses have been established along contour lines and will help to trap eroding sediments and create a low terrace

on steep slopes other, more drastic conservation measures may be needed. The establishment of terracing using stone walls, or contour lines of perennial plants that act as sediment traps, may be needed to reduce surface erosion and redirect runoff water (Figure 11.5). However, Lal (1990) warns that these are unlikely to be effective unless accompanied by measures to provide soil coverage beneath crops grown on the terraces. Their construction is also labour-intensive, and they require constant maintenance if they are to be effective. Because much gully erosion is initiated on access trails and roads, special care is needed in the location of these and in maintaining soil cover on them (Lal 1990).

Use of trees in land rehabilitation

Trees are potentially important components of rehabilitation efforts in steeplands. In most areas, they formed the original vegetation cover, and when grown in forest communities can be effective in suppressing soil erosion. Deeply rooted trees can also help to stabilize soils against mass movements. Trees are often the sole source of fuel for cooking in rural areas, and Evans (1982) estimates that the average family consumes 2–5 m^3 of stacked firewood per year. Trees can also provide fruit for human consumption and forage for livestock. Unless completely deforested, most tropical steeplands contain some remnant woodland (Figure 5.9). This is often confined to the steepest slopes in headwater basins and on ridge tops where, fortuitously, it can have its largest hydrological and soil conservation benefits. Protection of these forest remnants from grazing and over-harvesting of firewood is often an important first step in watershed rehabilitation, although most can continue to be used at low intensity as a source of wood for fuel, construction and poles.

Re-establishment of a forest cover on other very steeply sloping terrain is also often a priority activity, as many of these areas have severely degraded soils. Normally some artificial tree planting is needed in these areas, with fast-growing exotics such as species of pine and *Eucalyptus* being used where indigenous forest has been totally eliminated. In semi-arid environments, species of both *Acacia* and *Eucalyptus* have been widely used (Evans 1982). These tree plantations frequently act primarily as a fuelwood source, but ideally should also act as a nurse-tree stage for the establishment of volunteer seedlings of any native trees left in the region. Whether these volunteers will establish will depend on the intensity of the fuelwood harvesting and the characteristics of the exotics; some of these, such as *Eucalyptus*, may suppress the regeneration of native species (Behmel and Neumann 1982). In contrast, Parrotta (1993) has recorded successful recruitment of native tree and shrub seedlings beneath plantations of the legume *Albizia lebbek* in Puerto Rico. After 6.7 years, seedlings of 22 species were recorded in plantations that had been established in degraded pasture, most of which were bird- or bat-dispersed. Patterns of tree plantation establishment that provide for continuity between forest remnants will maximize the benefits of tree plantations to local wildlife, and may facilitate seed dispersal among forest remnants. A dispersed but interconnected system of stream-side gallery forests and ridge-crest forests is likely to best achieve this.

Trees can be usefully incorporated into agricultural landscapes as barriers to stabilize erosion (Augustin and Nortcliff 1994), as living fences (Figure 11.6), along hedgerows and roadsides, and in home-site gardens, where they can also provide a source of firewood, fruit and livestock forage. Despite their benefits, promotion of trees in tropical land rehabilitation faces several obstacles. Tree growth is slow relative to that of annual crops, and requires a long-term commitment if it is to be successful. The slowness with which benefits accrue from tree planting may make it difficult to enlist the co-operation of local land users in tree establishment efforts. Tree seedling establishment is a labour-intensive activity, and considerable subsequent effort is needed to protect seedlings during their maturation. Production of tree seedlings is a relatively costly process that normally requires the establishment of nurseries by governments or other agencies with access to the necessary capital. For these reasons, trees should be seen as one important element of land rehabilitation, but not necessarily a panacea. For example, where livestock raising

Figure 11.6 A living fence of *Glircidia sepium*, Belize. Branches of this leguminous tree are frequently cut for firewood, and the tree is also used to provide shade in cacao plantations

is an important agricultural activity, concentration of efforts on the rehabilitation of over-grazed pastures may provide greater overall benefits than a programme aimed only at tree establishment.

Permanent land rehabilitation

The severity of the problem of land degradation in tropical steeplands is such that the best that can be hoped for in the immediate future is to slow the rates of degradation. In many degraded areas, land rehabilitation to the point where it can be used sustainably will require significant depopulation. In the interim, a realistic objective is to stabilize conditions sufficiently that some soil and biotic ingredients of the original ecosystem are preserved, with which to fashion a more permanent solution a century or two centuries later. In the mountainous areas of Ethiopia, where land degradation is severe and the evidence of an exceeded carrying capacity was manifested in the 1984–5 famine, Hurni (1993) estimates average soil erosion rates of 4 mm.yr^{-1}, and an annual decline in productivity of 2 per cent. This author estimates that only investment at two to four times the present level will result in effective rehabilitation within the next generation. When longer-term scenarios were considered, only that involving achievement of zero population growth within fifty years provided a stable solution (Hurni 1993).

Reforestation of derived grasslands

Derived tropical grasslands are of two basic types:

1 Derived savannas created by forest destruction and subsequent establishment of a frequent-fire regime. These are especially common in the Asian tropics, and are often dominated by the aggressive weedy grass *Imperata cylindrica* (Chapter 7).

2 Pastures created after forest clearance, which are subsequently abandoned owing to declining productivity. Abandoned pastureland of this sort is a relatively recent phenomenon that is frequent in the Neotropics and especially common in Brazilian Amazonia, where its creation has been encouraged by inappropriate state subsidies and land speculation (Hecht 1985).

Both types of secondary grassland tend to be maintained by frequent fires, and protection from fire for an extended time period is a prerequisite for successful forest re-establishment. This is likely to be a major task where human populations are present, as fire is widely used in the tropics to clear land, in hunting, and as a means of eliminating noxious pests such as venomous snakes.

Most natural savannas contain a fire-tolerant woody flora of trees and shrubs that expands after fire suppression and can act as a facilitator of the establishment of forest tree seedlings (Kellman 1989; Kellman and Miyanishi 1982; Figure 7.16), but relatively recently derived savannas and abandoned pastures may lack this component. Forest seedlings face a number of barriers to establishment in abandoned pastures (Nepstad *et al.* 1991; Sun and Dickinson 1996), and when transplanted without special treatment may suffer high mortality (Gerhardt 1993). However, work by Uhl *et al.* (1988a) on recently abandoned pastures in Amazonia indicates a relatively rapid re-establishment of woody vegetation on these sites, and Viera *et al.* (1994) have found that *Cordia multispicata*, a local shrub, is acting as a facilitator of forest tree seedling establishment. It seems probable that other tropical forest floras will contain at least some species with comparable properties that can act as catalysts for forest re-establishment.

Forest re-establishment is likely to be much less easily achieved by spontaneous processes in the extensive derived savannas that are common in the Asian tropics. Here, savannas are often continuous over large areas and subject to frequent fires. Prolonged fire exposure has often eliminated most woody taxa that could act as nucleii for forest seedling establishment, and extinguished most forest remnants that could act as seed sources. Successful reforestation under these conditions is likely to require the establishment of fire protection zones within which tree plantations can be initiated. If these plantations are protected from fires, and located so as to facilitate seed dispersal from any extant forest remnants, they may eventually accumulate a volunteer flora of indigenous forest plants (cf. Parrotta 1993). However, where a seed source is lacking, artificial introduction of seeds or seedlings may be necessary.

RESEARCH ON TROPICAL LAND USE

We take the objective of research on tropical land use to be the development of types of land use that are both productive and sustainable in particular regions of the tropics. Research with this objective is inherently scientific, as it seeks to understand the functioning of the real world, and how this may be best modified to benefit human populations. Consequently, we begin by considering the scientific process.

The scientific enterprise

Science is a methodology, which has evolved over several centuries, that enables persons to decipher the workings of the natural world. A necessary precursor to the practise of science is the belief by the practitioner in his or her ability to achieve this without the assistance of some 'intervenor'. In pre-scientific societies, intervenors are often members of priestly castes, who interpret the natural world by reading natural phenomena as 'signs' that correspond to supernatural forces and that require priestly or ritual intervention. Science should therefore be a profoundly liberating endeavour, as it frees the individual from the need to rely upon such persons when interpreting natural phenomena. Unfortunately, scientists have themselves often fallen into the role of intervenors when dealing with traditional societies, and in so doing have served to perpetuate a tradition of passive dependence upon externally derived solutions. We will argue below that there is an urgent need to involve local people in scientific endeavours that seek to identify better forms of land use for the areas in which they dwell. We believe that the single greatest obstacle to this involvement is a continuing perception of the need for intervenors by indigenous people, and that overcoming this perception will be an important task for those promoting improvements in land use.

The methodology of science is essentially formalized 'trial and error' or 'conjecture and refutation'. Conjectures about the properties or causes of natural phenomena are proposed ('hypotheses') and their truthfulness is evaluated by reference to real-world situations or experimentally manipulated systems. Popper (1965), who formulated much of the modern philosophy of science, has argued that, logically, hypotheses cannot be confirmed, only refuted. This requires that hypotheses be subject to repeated attempts at refutation, and that conclusions be accepted as tentative representations of reality only after attempts at refutation have failed repeatedly. Good science, which leads rapidly to clearer understanding of how the natural world works, requires both innovative hypotheses and their rigorous testing. Novel conjectures are required if one is to avoid limiting oneself to infinite reconfirmation of the same hypotheses (the logical trap of the Popperian argument), but novel ideas must also be treated sceptically and tested rigorously before being accepted as being a reasonable approximation of reality.

Hypotheses may describe the properties of certain phenomena (e.g. 'little erosion takes place beneath tree crops'), or speculate on the underlying causative processes that are responsible for the phenomena (e.g. 'erosion is less beneath tree crops because soil cover is more complete'). Property and process are clearly interconnected in that properties are the physical manifestation of one or more (usually many) underlying processes. Organized science has traditionally focused its efforts on the understanding of underlying processes, on the reasonable assumption that these are more likely to allow generalizations that cut across local idiosyncrasies. Indeed, such generalization is often seen as being a central objective of science. However, in a world such as the tropics, where generalizations are strongly modified by site-specific idiosyncrasies, it can be argued that hypotheses about properties are also important or even essential components of the scientific enterprise. Moreover, a preoccupation with process can lead to infinite reductionism as one searches for ultimate causes through increasingly detailed layers of process. In such an endeavour, an understanding of the original phenomena which initiated the search becomes increasingly unlikely. In contrast, a research focus upon system properties involves the trade-off of generality: a particular land use can be proven to be effective in one area, but an absence of knowledge about the reasons why this should be so limits one's ability to predict whether it will be equally effective elsewhere. We believe that research on both the properties and the processes of land use systems is needed in the tropics. The latter endeavour is likely to be confined to persons with formal scientific training, while the former is accessible also to land use practitioners.

There is, in theory, no reason why scientific endeavours should be the exclusive domain of the scientist. The major limitations on the practice of science by non-scientists are their more limited knowledge of the phenomena and processes being examined, and their limited training in the formal procedures of hypothesis-testing. While both may limit the sophistication of the hypotheses proposed and the tests performed, lesser hypotheses that use simple tests are within the capability of non-scientists who have accepted the proposition that the workings of the natural world are decipherable. Below we will argue that there is much room for localized testing of rudimentary hypotheses about the properties of tropical land use systems, as rural populations attempt to improve their quality of life.

Pure and applied science

Basic or pure research is curiosity-driven research which is undertaken to advance scientific knowledge, but conventionally understood to have no practical application. Applied research is undertaken to advance scientific or technological knowledge with a practical application in view. In tropical environmental science these two traditions have often involved, respectively, examinations of the functioning of natural ecosystems, and the design or modification of artificial ecosystems to yield products useful to people. Although the immediate objectives of the two endeavours differ, both invoke (or should invoke) the same principles of scientific investigation: the formulation and testing of refutable hypotheses. Moreover, in tropical environmental science, there is an increasing need for the two endeavours to be seen as complementary. The inability of many artificial ecosystems to be sustained has placed an increasing emphasis on the need to seek analogues in the more stable natural systems that these have replaced (Ewel 1986). In contrast, the pervasiveness of human activities in tropical landscapes makes an exclusive preoccupation with pristine ecosystems increasingly anachronistic.

Regrettably, in tropical environmental science the two traditions have often coexisted in relative isolation, mirroring the dichotomy between 'preservationist' and 'utilitarian' views of nature (Chapter 1). Although there has been an increasing willingness among basic researchers to recognize the pervasiveness of human influences, these are often seen as external threats to pristine ecosystems, and much of the basic research continues to be concentrated at a few research stations located in isolated natural pockets. In contrast, applied environmental scientists often see natural ecosystems as irrelevancies to be removed and replaced by 'productive' managed systems. The lessons to be learned from these natural systems have usually been ignored, sometimes with disastrous consequences.

Unfortunately, applied tropical research is often directed by policy decisions that fail to recognize that 'strategies' or 'solutions' for tropical land use problems are no more than hypotheses in need of critical evaluation. In the absence of such hypothesis-testing, policy directives become 'visions' which scientists are required to implement rather than to evaluate rigorously and possibly reject. One of the most spectacular examples of 'vision-driven' tropical applied research was the Jari forestry project in Amazonia, whose origins lay in the developer's decision that the planting of *Gmelina arborea* pulpwood plantations was the solution to land use in that region. Not only has this project proven to be without net financial benefit (Fearnside 1988), but most wood now grown by the project is of other tree species. In effect, Jari represented a gigantic experiment whose results rejected the original hypothesis. Unfortunately, the developer had the financial resources to destroy large areas of tropical forest in an experiment that could have been conducted with equal success at a far smaller scale.

A more pervasive example of 'vision-driven' research has been the enthusiastic promotion of 'agroforestry' as a panacea for the problems of small-scale farming in the tropics. The term has been given a plethora of definitions over the past two decades, but that by Nair (1984) is now widely used:

'Agroforestry is a land use that involves deliberate retention, introduction, or mixture of trees and other woody perennials in crop/animal production fields to benefit from the resultant ecological and economic interactions.' The potential benefits usually ascribed to agroforestry (e.g. MacDicken and Vergara 1990) are often simply those of diverse agricultural polycultures and do not necessarily require the involvement of woody perennials: ecological stability, nutritional variety, nitrogen fixation, and more effective use of resources for growth. Other alleged benefits remain unconfirmed, such as 'nutrient-pumping' from subsoils (see Chapter 4). There are also several theoretical reasons to be cautious about attempts to integrate trees and annual crops in the same field. Most obviously, these two life forms represent unequal competitors that are unlikely to be able to coexist stably (see Chapter 7), or necessarily result in higher total productivity of usable materials. Even when the tree component is reduced to widely spaced rows ('alley cropping'), there can be significant depression of crop yield (Lal 1991; Sanchez 1995). Finally, a tree cover often increases the potential erosivity of rainfall unless combined with a complete ground cover (Chapter 9), and the slow growth of trees makes them a less flexible crop than annuals.

These considerations tend to suggest that trees may often be more usefully and sustainably grown in tropical agricultural systems when segregated from, rather than incorporated with, annual crops and pastures, i.e. as windbreaks, living fences, woodlots and erosion barriers. We do not question that trees may be usefully incorporated into tropical agricultural systems in certain ways and under certain circumstances, but we question the wisdom of transforming this very general hypothesis into a crusade (e.g. Garrity n.d.). In crusades, there is little room for the scepticism and rigorous evaluation of assumptions and alternative hypotheses that are the essence of good science. As long ago as 1982 agroforestry was described as 'suffering from oversale and wishful thinking rather than objective analysis based on what it can and cannot offer' (Budowski 1982). Unfortunately, since then there has been a paucity of the careful comparisons between agroforestry and alternative systems that Budowski considered essential to making rational decisions about its value as a land use technique.

Applied research efforts that are driven by *ad hoc* policy decisions provide few opportunities for scientists wishing to practise their craft rigorously. When one adds to this the pressures for quick 'solutions' and the possibility of changing 'visions', it is not surprising that many tropical environmental scientists choose to pursue pure, rather than applied, research. This is unfortunate, because some of the most fascinating problems in tropical environmental science lie in the area of land use management, and solutions to them are urgently required if we are to forestall environmental collapse under the population pressures of the next century.

Towards site-specific environmental management

Throughout this book we have stressed the intrinsic heterogeneity of tropical environments, and earlier in this chapter we have emphasized the need for land uses to be sensitive to this variability. Here we address the question of how best to achieve land uses that are productive and sustainable on an environmental substrate that is very heterogeneous. Recent approaches to improving tropical land use have often involved providing national governments with improved land use 'packages' that have been developed at centralized international research stations. Implementation of these packages is left to governmental agencies in individual countries. Not only are these unitary packages unlikely to be applicable to a large subset of the habitats or social customs in each country, but these countries rarely have the resources needed to engage in the research necessary to make the systems more locally applicable. The result is usually a concentration of land use change in the more productive habitats which most closely correspond to those of the originating research stations.

This 'top-down' approach to land use change has other drawbacks that go beyond its site insensitivity. It is highly paternalistic, and perpetuates a psychology of dependency. It often proposes solutions that are dependent on large external inputs which further lock the land user into external dependency. Implementers of change (scientists, extension workers) often lack a rural background and avoid living in the areas where change is being implemented, and so are insensitive to the subtleties of the local environment, both natural and societal (Bentley and Andrews 1991). While this approach to land use can be justified in the case of land colonization by migrants with little experience in local habitats, it is unlikely to be effective in areas which already possess a resident rural population. Below we discuss some alternatives.

A logical starting point to improving land use in long-occupied areas is to accept, as a basic strategy, the need for improvement upon what already exists, rather than its replacement with entirely new land uses. Only where indigenous use has been so inappropriate as to result in effective ecosystem collapse is there justification for seeking to replace it. Local land uses represent the products of prolonged periods of trial and error by local populations, who observe and experiment continually (Howes 1980). The land uses that they engage in are usually functional, although they may possess many suboptimal solutions, and may be vulnerable to deterioration under the pressure of an expanding human population. They are also conservative in normally possessing many risk avoidance techniques. Moreover, local land users are likely to be much more receptive to the modification of known systems that they use than to the replacement of these by entirely new entities.

Plans for improvement need to address the perceived needs of the existing land users, although they cannot be confined to these. Local land users have the greatest stake in achieving better forms of land use and so may be expected to be receptive to suggestions for obtaining this. When aware of a problem such as accelerated rates of erosion, they can become active participants in its solution (e.g. Fujisaka 1989). However, their perceptions as to what is most needed may be constrained by limited experience. For example, farmers may ask for better pesticides to cope with pest outbreaks rather than a long-term plan for pest avoidance (Bentley and Andrews 1991). Also, when environmental deterioration is gradual, it may be imperceptible to an individual during his or her lifetime.

Existing land uses in tropical landscapes are likely to comprise mixtures of systems of varying degrees of sustainability and productivity. Some may represent relatively benign forms of land use but be of low productivity, while others may be temporarily productive, but be unsustainable. While some especially damaging forms of land use may need to be gradually eliminated (e.g. annual crops in steep terrain), most elements of the original are probably best preserved, even if in a peripheral form, as a means of risk-spreading. This also ensures *in situ* preservation of genetic resources (Altieri and Merrick 1987). Elements of existing land use systems that are unique or particularly promising need to be identified and explored, if necessary in an *ex situ* experimental setting, with a view to reintroduction in an elaborated form. The introduction of new land use elements, such as new crops or cropping techniques, needs to be a gradual process in which local land users are involved as 'farmer-scientists', evaluating the feasibility of these and their modification to meet local needs (Marten 1986). This is a very different research role from that traditionally engaged in by the applied environmental scientist. In this role, he or she acts as a 'window on a larger world' for the local land user, rather than as simply a prescriber of solutions. Chambers *et al.* (1989) provide a useful review of the potential for research of this sort.

While the indigenous knowledge possessed by local land users may be considerable, it is important to recognize its limitations and not to romanticize it. Because of a lack of microscopy, traditional land users are limited in what they can observe. They are often accomplished taxonomists of larger organisms, but may have no knowledge of microscopic organisms, and may have a limited understanding of process and a limited ability to predict (Howes and Chambers 1980). They may also be burdened by misconceptions about some processes (e.g. spontaneous generation; Bentley and Andrews 1991). Consequently, an important task for scientists engaged in promoting improved forms of land use is to introduce some selective understanding of the workings of local land use systems which will enable land users to test hypotheses of greater sophistication. For example, introduction of farmers to the basic elements of nitrogen fixation, including the demonstration of root nodulation and its benefits, would allow farmers to identify potential N-fixers in the local flora and to experiment with these.

Improvements in land use techniques within a region are often stimulated by the success of especially innovative local individuals who act as catalysts of change. Similarly, cross-regional observation can often lead to spontaneous adoption of new techniques, such as *Mucuna* cover-crop use in Central America (Buckles 1995), and can be used in planned improvements. For example, the use of terracing to reduce erosion in steep terrain in Mindanao was achieved by sending groups of farmers on visits to areas elsewhere in the Philippines where the technique was already in use (Fujisaka 1989). On their return to their home regions these farmers did not simply replicate the terracing techniques, but modified these to better fit the local environment. Different regions vary enormously in both natural and social conditions, so it is not possible to offer well-defined prescriptions about how to achieve better forms of land use. Rather, those seeking to implement change need to approach the endeavour with a sensitivity to local needs, an acceptance of local accomplishments, and a willingness to promote among local land users the confidence that they can participate in the solving of local problems. Stimulation of the imaginations of local land users promotes not only innovation but a commitment to seeing that their own ideas are carried to completion in their own environments.

GLOSSARY

'A' horizon The uppermost mineral soil horizon (cf. *parent material*) in a soil profile, usually characterized by some accumulation of organic material, and loss of other materials due to *leaching*.

Abiotic Unrelated to biological phenomena or processes.

Adaptation A morphological or functional property of an organism that is assumed to have been selected during the course of organic evolution. In practice, such selection cannot normally be observed, so most properties of organisms can only be inferred to be adaptations.

Adsorbed ion In soils, an *ion* held to the surface of a soil particle by electrostatic forces, but able to be removed in an exchange reaction.

Advance regeneration The population of juvenile plants already established in some plant community.

Advection The transfer of energy in the atmosphere by means of lateral air movement.

Aerenchyma tissue Spongy tissue of some aquatic plants, containing intercellular air spaces that enable gaseous exchange.

Alkaloid Member of a group of nitrogenous plant compounds, many of which are toxic to animals and are assumed to serve a defensive function.

Allelopathy The inhibition by one plant of others due to the release of toxic compounds into the surrounding environment.

Allopatric speciation Appearance of new species during the course of organic evolution following isolation of subpopulations. Also called 'geographic speciation'.

Alluvium Water-lain deposits, often stratified by particle size.

Amino acids Acids containing the amino group NH_2, which are essential constituents of *proteins* in living organisms.

Amphibian Animal of the class Amphibia, which includes frogs, toads, newts and salamanders.

Anaerobic Functioning in the absence of free oxygen.

Angiosperm A member of the Angiospermae, or true flowering plants, which, together with the *gymnosperms*, comprise the Spermatophyta, or seed plants.

Anion An *ion* bearing one or more negative charges.

Anion exchange capacity (AEC) The capacity of a soil to retain *anions* in an adsorbed state on the surfaces of mineral and organic particles.

Anoxic A deficiency of oxygen.

Anticyclone A cell of surface high pressure and its associated diverging surface airflows. As a result of the Coriolis effect, these flows are clockwise in the northern hemisphere and anticlockwise in the southern hemisphere.

Arbovirus Arthropod-borne virus.

Arthropod A member of the phylum Arthropoda; animals with articulated bodies and limbs. In terrestrial habitats, insects and spiders are the most common components of the phylum.

Asexual reproduction Reproduction without the fusion of gametes from different parent organisms, e.g. the production of vegetative suckers by a plant that later becomes independent of it.

Autecology The study of the environmental relations of individual plants (as opposed to plant communities, which is termed 'synecology').

'B' horizon The horizon of a soil profile in which various materials have accumulated as a result of movement from above. These materials commonly include clay, organic matter and mineral material such as iron oxides.

Bacteria A diverse group of unicellular micro-organisms.

Benthic Referring to the benthos, or assemblage of organisms which live on the bottom sediments of aquatic systems.

Bioassay Indirect measurement of some factor by means of the response of an organism to it.

Biomass The total mass of live organic material occupying some area, normally measured as oven-dry weight per unit area. Most biomass is composed of plant material ('*phytomass*').

Biota The total assemblage of types of organisms (normally species) occupying some region. Normally subdivided into plants (the 'flora') and animals (the 'fauna').

Blackwater river A river darkened by colloidal organic matter. In the tropics, such rivers usually drain watersheds dominated by coarse-textured quartzitic sands.

Borehole A narrow-diameter well, drilled to obtain ground water.

Bund A low embankment used in irrigated rice farming to retain standing water in fields.

C_4 plants Plants possessing a specialized photosynthetic pathway that differs from the normal 'C_3' patttern and provides superior carbon fixation efficiency at high light intensities. The C_4 pathway is especially common in tropical grasses.

Caecum A blind sac which forms part of the large intestine in the horse and some other vertebrate herbivores.

Cambium Meristematic tissue in the stems of plants. In most trees, the cambium forms a layer beneath the bark that subdivides to form woody water-conducting 'xylem' tissue on its inner surface and photosynthate-conducting 'phloem' tissue on its outer surface.

Carbohydrate Compounds produced by the process of photosynthesis, containing carbon, hydrogen and oxygen, and including sugars and starches.

Cation An *ion* bearing one or more positive charges.

Cation exchange capacity (CEC) The capacity of a soil to retain *cations* in an adsorbed state on the surfaces of mineral and organic particles.

Cellulose A relatively inert organic compound making up the cell walls of organisms.

Clay Soil mineral particles smaller than 0.002 mm in diameter.

Clear-cutting In forestry, the felling of all trees on a particular block of land (cf. *selective cutting*).

Clone A group of individuals produced vegetatively from a single parent organism and normally possessing an identical genetic composition.

Coefficient of variation A measure of the degree of variability in a set of numbers. Formally defined as the standard deviation of the set expressed as a percentage of the set's mean.

Colluvium Unsorted slope-wash or mass-movement deposits, usually found at the bases of slopes.

Conifer Member of the order Coniferales, relatively primitive seed-bearing plants that have been displaced by the more advanced *angiosperms* over much of the earth's surface since the Cretaceous, but have successfully persisted in some habitats that are more marginal for plant growth.

Convection The movement of fluids that results in heat transfer. In the tropical atmosphere the most common form of convection involves vertical movement of air parcels that have been warmed by contact with the earth's surface or by latent heat released when water vapour condenses as clouds.

Coppice Regrowth from the cut stems of woody plants.

Coriolis effect The apparent distortion of airflow patterns caused by the earth's rotation. Winds in the northern hemisphere are deflected to the right, and those in the southern hemisphere are deflected to the left.

Crustacean An animal of the class Crustacea of the *arthropod* phylum. Mostly aquatic (e.g. shrimp, crabs).

Cuirasse A hardened layer that can act as a protective shield.

Cultivar A domesticated plant variety that has been produced by horticultural techniques and is not usually found in wild populations.

Cycad A member of the Cycadaceae, a family of primitive *gymnosperms* whose gross morphologies are similar to that of palms, with which they are sometimes confused.

Density dependence A condition in which the rate of some ecological process (e.g. birth rate) is influenced by the density at which the organisms involved occur.

329

Density independence A condition in which the rate of some ecological process is unrelated to the density at which the organisms involved occur.

Deterministic Referring to any process or prediction in which a certain outcome is necessary, and chance variability is considered to play no role (cf. *stochastic*).

Detritivory The consumption of dead organic tissues.

Diaspore Any biological organ capable of producing a new organism after dispersal (e.g. a seed).

Diffusion Spontaneous movement of ions in solution from areas of high concentration to areas of low concentration.

Diversity (organic) The variety of life in some area. Most commonly used in reference to the variety of species, but can also be applied to higher taxonomic groupings (e.g. genera, families; cf. *taxonomy*) or to genetic diversity within species. A distinction is often made between local, within-community, diversity ('alpha' diversity), between-community ('beta') diversity and regional ('gamma') diversity.

El Niño Southern Oscillation (ENSO) A multi-year quasi-periodic oscillation in the patterns of atmospheric pressure in the Pacific Ocean, which is accompanied by a change in ocean currents and rainfall patterns in the Pacific and beyond.

Eluviation Loss of materials during soil formation, usually from an upper, 'A' horizon.

Embolism The appearance of a foreign body, such as an air bubble, in the vascular system of an organism.

Endemic disease A disease which prevails continuously in a region.

Epidemic A disease outbreak in which the numbers of affected animals or plants are much higher than would normally be expected.

Epiphyll An organism growing on the surface of a leaf.

Erosion The removal of particulate materials from a part of the earth's surface by wind or running water.

Eutrophy A sufficiency (or over-abundance) of nutrients (cf. *oligotrophy*).

Evapotranspiration The loss of water to the atmosphere from a plant-covered surface, comprising both direct surface evaporation and transpiration from the stomata of plant leaves.

Exchangeable ion An *ion* held to the surface of soil particles in an adsorbed state, but capable of being exchanged for another ion (or group of ions) of similar charge.

Extinction The disappearance of a type of organism (normally a species). A distinction is commonly made between the total disappearance of the species from earth ('global extinction'), and the disappearance of some subpopulation of the species ('local extinction').

Extractive reserves Forest reserves established with the intention of preserving the forest while extracting minor forest products from it, usually in the form of non-timber products.

Facilitation In vegetation, the process in which the presence of one plant encourages the establishment and growth of another plant, usually of some other species.

Fallow The period when a field is not used for crops. In the literature on shifting cultivation, the term is frequently used adjectively (e.g. 'fallow vegetation').

Fermentation The process of gradual decomposition of organic substances by the action of various agents such as micro-organisms or enzymes of plant or animal origin.

Field capacity The maximum quantity of water that can be held in soil in the face of gravitational force promoting drainage. Soil moisture at field capacity has been conventionally assumed to be held with a suction approximating –0.3 bar, but in many soils water can be retained by lower suctions.

Fingerling A young, finger-sized fish.

Floodplain The lower-lying portions of a river valley that is subject to periodic flooding during high river discharge.

Folivory The consumption of leaf tissues.

Forage Plant material that is suitable for animal consumption by either direct grazing or browsing, or for harvesting by humans for animal use.

Forb A non-*graminoid* herb.

Fragmentation The breaking up of a previously continuous entity, such as the forest cover of a region, into a series of spatially isolated fragments.

Frugivore Fruit-eating animal.

Fry Very young fish.

Full glacial The times of maximum cooling of the earth's climate, and maximum extent of glaciation, during the Pleistocene epoch.

Fungi Non-photosynthetic organisms which obtain nutrients by the absorption of organic compounds from their surroundings. Some fungi infect living organisms or engage in the decomposition of dead organic materials.

Genotype The genetic constitution of an organism, as distinct from its physical appearance, or *phenotype*.

Germ plasm The genetic constituents of an organism that are capable of transmitting inherited characteristics to its offspring.

Graminoid A grass-like plant. Normally includes grasses and sedges.

Growth ring (tree ring) A morphologically distinguishable layer of woody xylem tissue laid down during one annual cycle of tree growth. In detail, tree rings comprise low-density tissue laid down early in the growth period, and higher-density tissue laid down later. Annual growth rings are characteristic of most temperate trees but are extremely rare in tropical trees.

Gymnosperm A member of the Gymnospermae, a plant group comprising the *conifers* and their allies. Seed plants that have unenclosed seeds and more rudimentary xylem tissue than the *angiosperms*.

Gyre A large-scale oceanic circulation pattern, the best-developed of which involves basin-wide circulation around the subtropical cells of high atmospheric pressure.

Hadley circulation cells Large-scale convective circulation patterns of near-equatorial zones, comprising uplift in the *Inter-Tropical Convergence Zone*, poleward movement at high altitude, subsidence in subtropical zones, and a reverse equatorward airflow as trade winds.

Helminth A parasitic intestinal worm.

Herbivory The consumption of plant tissues by animals.

Hyphae Thread-like structures forming the vegetative mass of *fungi*.

Illuviation Accumulation of materials in a soil profile, usually as a result of translocation from higher in the profile.

Inoculation Infection of some body (usually an organism) with the *diaspores* of another organism, such as a parasite or *symbiont*, that will allow the establishment of a population of the infecting organism.

Interglacial Relatively warm period during the Pleistocene epoch, the most recent of which began about 10,000 years ago.

Inter-Tropical Convergence Zone (ITCZ) The relatively continuous near-equatorial zone where trade winds from northern and southern hemispheres converge, producing uplift and relatively heavy rainfall. Position of the zone oscillates seasonally.

Ion An electrically charged atom or group of atoms.

Isotope Any of two or more alternative forms of the same element that have the same atomic number but differ in the number of neutrons in their nuclei.

Lapse rate The rate of air temperature change with increasing altitude. Normally the change involves a decrease, but occasionally a reversal takes place to produce an 'inversion'.

Laterite *See* **Plinthite**.

Leaching The downward movement of materials held in percolating soil water.

Legume A member of the plant family Leguminosae (Fabaceae), which is sometimes treated as three separate families: Caesalpinioidae; Mimosoidae; Papilionoidae. A large, cosmopolitan family with many tropical members, many of which support root nodules containing nitrogen-fixing bacteria.

Limiting nutrient The nutrient whose availability ultimately limits growth in a given environment. For plants, nitrogen and phosphorus are the most frequent limiting nutrients.

Macro-fauna Animals of sufficient size to be visible to the naked eye.

Macro-pore A soil void too large to retain soil water by capillarity at field capacity.

Mangrove Woody vegetation of tropical coastal areas, comprising a limited number of tree and shrub species from several unrelated families. The term is also used to refer to the group of species that make up this vegetation type.

Mass flow Movement of materials in solution by the flow *en masse* of the solution (cf. *diffusion*).

Meristem Plant tissue capable of cell division to form more specialized permanent tissues.

Micro-fauna Animals visible only with the aid of microscopy.

Mineralization The freeing of nutrients from organic compounds during the decomposition process, and their release into the soil, normally as *ions* in solution.

Mollisol A soil order of the USDA System, characterized by a deep surface horizon high in organic matter, and generally of high fertility.

Monoculture The cultivation of single species (often single varieties) of crops or aquatic organisms in simplified and homogeneous artificial systems.

Monsoons Wind systems that include seasonal reversals in direction in response to major changes in the pattern of atmospheric pressure. Best-developed in Asia.

Mulch Material, normally dead plant material, spread on the soil surface of agricultural fields to reduce evaporation, inhibit *erosion*, exclude *weeds*, etc.

Mutation A spontaneous and heritable change in the genetic structure of an organism.

Mycorrhizas Complex symbiotic structures comprising plant roots and *fungi*. The host plants provide organic materials to sustain the fungi, which in turn facilitate nutrient absorption by the host plant.

Myrmecophyte A plant used by ants as a place in which to live. Usually a symbiotic relationship between ant and plant is involved.

Natural enemy Any predator or pathogen that occurs natually in an ecosystem and by its action tends to control the populations of some other organism (or group of organisms).

Nematode Roundworms of the phylum Aschelminthes. The phylum contains both free-living and parasitic species that affect both plants and animals.

Niche The functional, spatial or temporal position of an organism within an interacting biotic assemblage.

Nitrogen fixation The fixation of atmospheric nitrogen (N_2) by micro-organisms that are either free-living or exist symbiotically with higher plants.

Occluded phosphorus Phosphorus that is retained in soil in a form that is not readily available to plants (e.g. complexed with iron compounds).

Oligotrophy A deficiency of nutrients (cf. *eutrophy*).

Orographic rainfall Rainfall resulting from the vertical movement of moist air that is enforced by a mountain barrier in the air stream.

Padi (also **Paddy**) A term referring either to the cultivated rice of Asia (*Oryza sativa*) generally, or to this rice being grown under flooded-field conditions. In this book it is used in the latter context.

Palaeolatitude The latitude of some position on the earth's surface at some time in the past (usually the distant past). While the dominant pattern of past continental plate movements has been longitudinal, some continental fragments, such as Australia, have moved latitudinally, and so existed at palaeolatitudes in the past that differ from those of the present (cf. Figure 2.2).

Palaeosol An ancient soil that has been buried by overlying material but retains some of its original characteristics.

Palm Member of the family Palmae, a tropical and subtropical family of trees, miniature trees and woody vines (rattans) whose morphology is characterized by a rosette of large leaves borne terminally on the stem.

Palynology The analysis of the pollen content of sedimentary deposits, usually conducted with a view to inferring earlier vegetation of the site or its surrounding area.

Pandemic An *epidemic* which occurs over an extensive geographic area.

Parasitism The existence of one organism (often a micro-organism) at the expense of another (its host), from which it obtains sustenance.

Parent material Unconsolidated material that is transformed during soil formation into a soil profile exhibiting characteristic layers or 'horizons'.

Pathogen A *virus*, micro-organism or other agent which causes disease.

Ped A natural soil structural aggregate composed of cohering mineral or organic particles.

Percolation The downward movement of water contained in soil pores under the force of gravity.

Phenolic compound Member of a large group of plant secondary compounds including lignin and tannins.

Photosynthesis Synthesis of organic compounds from water and carbon dioxide using solar energy absorbed by chlorophyll in plant tissues.

Phytomass The total mass of live plant material occupying some piece of ground, normally measured as oven-dry weight per unit area.

Phytoplankton Small, drifting aquatic plants that form the basis of aquatic food chains.

Plant formation A large-scale category of the earth's vegetation, distinguished on the basis of gross morphology rather than on taxonomic composition (e.g. broad-leaf evergreen forest, savanna, broad-leaf deciduous forest).

Plantation Any stand of intentionally planted perennial plants (normally trees). The term is also used to refer to large-scale commercial agricultural enterprises in the tropics, irrespective of the type of crop grown.

Plate tectonics The geophysical theory proposing that the earth's surface is made up of rigid plates (both continental and oceanic) that undergo constant slow movement resulting in both coalescence and fragmentation of plates.

Plinthite Hard concretionary material containing abundant iron oxide that exists in some tropical soils. Previously, the term 'laterite' was loosely applied to this and a variety of other related phenomena in tropical soils.

Pneumatophore A specialized vertical root branch, produced by some plants grown in swampy conditions, and containing a system of inter-cellular air spaces allowing gaseous exchange with the atmosphere through pores on the aerial portion.

Podzol A distinctive soil profile, found primarily in temperate areas, comprising an intensively leached 'A' horizon of quartzitic sand overlying a 'B' horizon with an abundant accumulation of iron oxides, often complexed with colloidal organic matter. Giant podzols are occasionally found in the tropics on quartzitic sandy parent materials. In the USDA System, podzols have been renamed 'spodosols'.

Point of Zero Charge (PZC) The pH level at which a soil's *cation* and *anion exchange capacities* are equal. At pH values above this point the soil takes on net cation exchange capacity, and at values below this point the soil takes on net anion exchange capacity.

Pollen Microspore of seed plants containing a greatly reduced male gamete, whose movement between plants enables cross-fertilization.

Polyculture The cultivation of a variety of species together in the same place.

Pore space The total volume of a soil body not composed of particulate material. Normally filled by soil moisture and gases.

Predation The destruction of one organism by another, usually by consumption.

Predator satiation Production of materials consumed by a predator (e.g. seeds) in quantities sufficient fully to satiate the predator, leaving some unpredated.

Primary minerals Crystalline rock-forming structures, usually high in silica, that have formed during the cooling of molten volcanic magma or the metamorphosis of other rocks as a result of extreme pressures or temperatures (cf. *secondary minerals*).

Primary plant succession The development of vegetation on new terrain.

Primate A member of the mammalian order Primates which includes humans, monkeys, apes, lemurs, tarsiers and marmosets.

Proteins Complex organic molecules that are essential constituents of all living cells. Made up of combinations of *amino acids* and containing carbon, hydrogen, oxygen, nitrogen and sulphur.

Protozoa A subkingdom of the animal kingdom which is composed of uni-cellular and acellular animals.

Puddling The process of working soil by treading, or otherwise compacting it, to make it relatively impervious to water movement.

Radiometric dating Dating of ancient material by measuring its relative content of an unstable *isotope*, which decays at a known rate.

Recessional agriculture Agriculture practised seasonally on *floodplains* after floods have receded.

Refugia Places where organisms temporarily survive during periods of regionally inhospitable environmental conditions.

Regolith Weathered material overlying unweathered bedrock.

Reptile A member of the class Reptilia which includes lizards, snakes, turtles and crocodilians.

Root mat A thick layer of dead organic material, thoroughly impregnated with roots, that overlies the mineral soil in some tropical forests.

Salinization Progressive accumulation of salts near the soil surface due to their precipitation during the evaporation of soil solution. The process is especially common at irrigated sites in arid areas, where evaporation rates often exceed the rates of water percolation through the soil.

Sand Soil mineral particles 0.05–2.0 mm in diameter.

Saprolite Partially weathered bedrock.

Saprophyte Any organism living on the dead or dying tissues of another organism.

Secondary compounds Compounds produced by organisms that do not serve a direct biochemical function in metabolism, but may serve other functions, such as protecting the organism from predators.

Secondary minerals Crystalline or quasi-crystalline structures, usually in the form of finely divided clay-sized particles, that are synthesized during the course of chemical weathering of *primary minerals* and are usually much more stable than these. Much of the bulk of older tropical soils is made up of these materials.

Secondary plant succession The redevelopment of vegetation at a previously vegetated site.

Seed bank The reservoir of viable seeds that exists in, or on, the soil.

Selective cutting In forestry, the felling of selected individual trees in a forest stand, as opposed to all trees ('*clear-cutting*').

Sexual reproduction Reproduction involving the fusion of gametes from different parent organisms.

Shade tolerance The ability of a plant to maintain a positive net photosynthetic balance (gross *photosynthesis* minus respiration) even at low ambient light levels.

Silt Soil mineral particles 0.002–0.05 mm in diameter.

Silviculture The management of forests, usually for wood production.

Soil moisture suction The force needing to be applied to the water held in the pores of soil in order to remove this. Usually measured as the converse force (i.e. pressure) needing to be applied to the soil in order to expel moisture from it.

Soil structure The tendency for soil particles to coalesce into aggregates (*peds*), often of characteristic shapes and sizes.

Soil texture The mineral particle-size distribution of a soil, normally expressed as the proportional representation of clay-, silt- and sand-sized particles.

Solum The entire soil profile, consisting of one or more identifiable soil horizons.

Speciation The production of new species during the course of organic evolution.

Species richness Species diversity measured as the total number of species occupying some unit area (also called 'species density').

Spontaneous generation The now discredited belief that living organisms can arise spontaneously from non-living material.

Spore Microscopic detached reproductive body of lower plants and some protozoa.

Stochastic Probabilistic. Referring to any process or prediction in which an element of chance prevails (cf. *deterministic*).

Stratigraphy The study of strata in geological materials, which are usually composed of sediments that were deposited sequentially in water-filled depressions.

Symbiont An organism living symbiotically with another organism, in which both receive mutual benefit from the association.

Sympatric speciation Appearance of new species during the course of organic evolution, without geographic isolation of subpopulations.

Taxonomy The study of the natural groupings of objects. For organisms, hierarchical taxonomic classifications have usually been developed, comprising species, genera, families, orders, classes, and phyla or divisions, representing a decreasing sequence of similarity and assumed ancestral relatedness.

Termite Member of the insect order Isoptera (also called white ants) of colonial insects. Important in the decomposition of woody tissue.

Trade Wind Inversion A widespread reversal of the normal decrease in air temperatures with altitude, that occurs in the trade wind zone of the tropics. The inversion tends to be lowest on the eastern margins of tropical oceans, and to rise westward. When low, it serves to suppress convection and rainfall.

Trophic level Literally, the 'feeding level' (e.g. autotrophic plants, which synthesize organic compounds from inorganic constituents; herbivores; carnivores).

Ungulates The hoofed mammals, almost all of which are herbivorous.

Upwelling The tendency for deep (and often nutrient-enriched) ocean water to be brought to the ocean surface at some continental margins, because of ocean currents that diverge from the coastline.

Vector An agent which is capable of transmitting disease-causing organisms among animals or plants.

Vertebrate Animal containing a segmented spinal column (backbone).

Virulence The capacity of a micro-organism to cause disease in a given host.

Virus A non-cellular particle which consists of genetic material surrounded by a *protein* coat. A virus does not have a metabolism of its own and must infect a cell of a suitable host organism in order to duplicate.

Walker circulation cells Large-scale meridional convective circulation cells of equatorial latitudes that result in alternating zones of uplift and heavy rainfall, and subsidence accompanied by drought. Best-developed in the equatorial Pacific ocean, where periodic oscillations in cell positions accompany the *El Niño Southern Oscillation.*

Weathering The physical and chemical decomposition of rocks at or near the earth's surface.

Weed Any plant regarded as undesirable by people, usually because it competes with economically valuable plants.

Wilting point The level of soil moisture suction beyond which plants are assumed to be unable to extract further water and wilt irreversibly. Wilting point has conventionally been assumed to approximate −15 bar of suction, but there can be much inter-species variation in this value.

Wind shear Air movements that involve the sliding of different contiguous air parcels relative to each other along their plane of contact.

Zooplankton Small, drifting aquatic animals that form an important food source for other animals.

BIBLIOGRAPHY

Abbadie, L., Mariotti, A. and Menault, J.-C. 1992. Independence of savanna grasses from soil organic matter for their nitrogen supply. *Ecology* **73**: 608–613.

Abrams, E. and Rue, D. 1989. The causes and consequences of deforestation among the prehistoric Maya. *Human Ecology* **16**: 377–395.

Adam, P. 1992. *Australian rain forests*. Oxford: Clarendon Press.

Ahmad, N. and Jones, R.L. 1969. A plinthaquult of the Aripo savannas, north Trinidad: I. Properties of the soil and chemical composition of the natural vegetation. *Soil Science Society of America, Proceedings* **33**: 762–765.

Aide, T.M. 1992. Dry season leaf production: an escape from herbivory. *Biotropica* **24**: 532–537.

Aide, T.M. 1993. Patterns of leaf development and herbivory in a tropical understorey community. *Ecology* **74**: 455–466.

Alexander, I. 1989. Mycorrhizas in tropical forests. In *Mineral nutrients in tropical forest and savanna ecosystems* (ed. J. Proctor). Oxford: Blackwell, 169–188.

Allen, B.A. 1985. Dynamics of fallow successions and introduction of robusta coffee in shifting cultivation areas in the lowlands of Papua New Guinea. *Agroforestry Systems* **3**: 227–238.

Altieri, M.A. and Merrick, L.C. 1987. In situ conservation of crop genetic resources through maintenance of traditional farming systems. *Economic Botany* **41**: 86–96.

Amador, M.F. and Gliessman, S.R. 1990. An ecological approach to reducing external inputs through the use of intercropping. In *Agroecology* (ed. S.R. Gliessman). New York: Springer-Verlag, 146–159 .

Anderson, A.N. and Lonsdale, W.M. 1990. Herbivory by insects in Australian tropical savannas: a review. *Journal of Biogeography* **17**: 433–444.

Anderson, C. and Lee, S.Y. 1995. Defoliation of the mangrove *Avicennia marina* in Hong Kong: cause and consequences. *Biotropica* **27**: 218–226.

Anderson, J.M. and Swift, M.J. 1983. Decomposition in tropical forests. In *Tropical rain forest ecology and management* (ed. S.L. Sutton, T.C. Whitmore and A.C. Chadwick). Oxford: Blackwell, 287–309.

Anderson, J.M., Proctor, J. and Vallack, H.W. 1983. Ecological studies in four contrasting lowland rain forests in Gunung Mulu National Park, Sarawak III. Decomposition processes and nutrient losses from leaf litter. *Journal of Ecology* **71**: 503–527.

Aranguren, J., Escalante, G. and Herrera, R. 1982. Nitrogen cycle of tropical perennial crops under shade trees II. Cacao. *Plant and Soil* **67**: 259–269.

Asibey, E.O.A. 1974. Wildlife as a source of protein in Africa south of the Sahara. *Biological Conservation* **6**: 32–39.

Asibey, E.O.A. and Child, G.S. 1990. Wildlife management for rural development in sub-Saharan Africa. *Unasylva* **161** vol. 41: 3–10.

BIBLIOGRAPHY

Augspurger, C.K. 1984. Light requirements of Neotropical tree seedlings: a comparative study of growth and survival. *Journal of Ecology* 72: 777–795.

Augustin, E.O. and Nortcliff, S. 1994. Agroforestry practices to control runoff and erosion on steeplands in Ilocos Norte, Philippines. In *Soil science and sustainable land management in the tropics* (ed. J.K. Syers and D.L. Rimmer). Wallingford: CAB International, 59–72.

Ault, S.K. 1994. Environmental management: a re-emerging vector control strategy. *American Journal of Tropical Medicine and Hygiene* 50 (Suppl.): 35–49.

Bahrin, T.S., Thong, L.B. and Dorall, R.F. 1988. The Jenka Triangle: a report on research in progress. In *Agricultural expansion and pioneer settlement in the humid tropics* (ed. W. Manshard and W.B. Morgan). Tokyo: United Nations University Press, 106–116.

Bailey, N.T.J. 1975. *The mathematical theory of infectious diseases and its applications*, 2nd edition. London: Charles Griffin.

Balick, M.J. 1990. Ethnobotany and the identification of therapeutic agents from the rainforest. In *Bioactive compounds from plants* (ed. D.J. Chadwick and J. Marsh). Chichester: John Wiley, 22–31.

Banerjee, B. 1981. An analysis of the effect of latitude, age and area on the number of arthropod pest species on tea. *Journal of Applied Ecology* 18: 339–342.

Bantilan, R.T. and Harwood, R.R. 1977. Weed management in intensive cropping systems. In *Multiple cropping* (ed. R.I Papendick, P.A. Sanchez and G.B. Triplett). Special Publication No. 27. Madison, WI: American Society of Agronomy, 85–96.

Barber, R.T. and Chavez, F.P. 1983. Biological consequences of El Niño. *Science* 222: 1203–1210.

Barducci, T.B. 1972. Ecological consequences of pesticides used for the control of cotton insects in Caneta Valley, Peru. In *The careless technology: ecology and international development* (ed. M.T. Farver and J.P. Milton). Garden City, NY: Natural History Press, 423–438.

Barnes, J., Burgess, J. and Pearce, D. 1992. Wildlife tourism. In *Economics for the wilds* (ed. T.M. Swanson and E.B. Barbier). Washington, DC: Island Press, 136–151.

Barron, E.J. 1983. A warm, equitable Cretaceous: the nature of the problem. *Earth Science Reviews* 19: 305–338.

Barry, R.G. 1992. *Mountain weather and climate*, 2nd edition. London: Routledge.

Bartholomew, P.W., Khibe, T. and Ly, R. 1995. In-village studies of the use of work oxen in central Mali. *Tropical Animal Health and Production* 27: 241–248.

Baskin, Y. 1994. There's a new wildlife policy in Kenya: use it or lose it. *Science* 265: 733–734.

Batra, S.W.T. 1982. Biological control in agroecosystems. *Science* 215: 134–139.

Baur, G.N. 1968. *Ecological basis of rainforest management*. Forestry Commission of New South Wales, Sydney, Australia.

Bayliss-Smith, T. 1985. Pre-Ipomoean agriculture in the New Guinea highlands above 2000 metres: some experimental data on taro cultivation. In *Prehistoric intensive agriculture in the tropics* (ed. I.S. Farrington). BAR International Series 232. Oxford: Osney Mead, 285–320.

Beadle, L.C. 1981. *The inland waters of tropical Africa: an introduction to tropical limnology*, 2nd edition. New York: Longman.

Beard, J.S. 1946. The Mora forests of Trinidad, British West Indies. *Journal of Ecology* 33: 173–192.

Behmel, F. and Neumann, I. 1982. An example of agro-forestry for tropical mountain areas. In *Agroforestry in the African humid tropics* (ed. L.H. MacDonald). Tokyo: United Nations University Press, 92–98.

Bell, G., Lechowicz, M.J., Appenzeller, A., Chandler, M., Deblois, E., Jackson, L.,

Mackenzie, B., Preliosi, R., Schallenberg, M. and Tinker, N. 1993. The spatial structure of the physical environment. *Oecologia* **96**: 114–121.

Bell, R.H. 1982. The effect of soil nutrient availability on community structure in African ecosystems. In *Ecology of tropical savannas* (ed. B.J. Huntley and B.H. Walker). New York: Springer-Verlag, 193–216.

Bellotti, A. and Kawano, K. 1980. Breeding approaches in cassava. In *Breeding plant resistance to insects* (ed. F.G. Maxwell and P.R. Jennings). New York: John Wiley, 313–335.

Belsky, A.J., Mwonga, S.M., Amundson, R.G., Duxbury, J.M. and Ali, A.R. 1993. Comparative effects of isolated trees on their undercanopy environments in high- and low-rainfall savannas. *Journal of Applied Ecology* **30**: 143–155.

Bentley, J.W. and Andrews, K.L. 1991. Peasants, pests and publications: anthropological and entomological views of an integrated pest management program for small-scale Honduran farmers. *Human Organization* **50**: 113–122.

Beven, K. and Germann, P. 1982. Macropores and water flux in soils. *Water Resources Research* **18**: 1311–1325.

Biddulph, J. 1997. Boundary characteristics affecting savanna fire entry into forest fragments in the Gran Sabana, Venezuela. M.Sc. thesis, York University, Toronto.

Bierregaard, R.O., Lovejoy, T.O., Kapos, V., Dos Santos, A.A. and Hutchings, R.W. 1992. The biological dynamics of tropical rainforest fragments. *BioScience* **42**: 859–866.

Bigarella, J.J. and Mousinho, R.M. 1966. Slope development in southern and southeastern Brazil. *Zeitschrift für Geomorphologie* (NS) **10**: 150–160.

Bjerknes, J. 1969. Atmospheric teleconnections from the equatorial Pacific. *Monthly Weather Review* **97**: 163–172.

Black, F.L. 1992. Why did they die? *Science* **258**: 1739–1740.

Black, F.L. 1994. An explanation of high death rates among New World peoples when in contact with Old World diseases. *Perspectives in Biology and Medicine* **37**: 292–307.

Blanchard, J. and Prado, G. 1995. Natural regeneration of *Rhizophora mangle* in strip clearcuts in northwest Ecuador. *Biotropica* **27**: 160–167.

Blanton, C.M. and Ewel, J.J. 1985. Leaf-cutting ant herbivory in successional and agricultural tropical ecosystems. *Ecology* **66**: 861–869.

Blasco, F. 1983. The transition from open forest to savanna in continental southeast Asia. In *Tropical savannas* (ed. F. Bourlière). New York: Elsevier, 167–181.

Bodmer, R.E., Fang, T.G., Moya I.L. and Gill, R. 1994. Managing wildlife to conserve Amazonian forests: population biology and economic considerations of game hunting. *Biological Conservation* **67**: 29–35.

Boid, R., Hunter, A.G., Jones, T.W., Ross, C.A., Sutherland, D. and Luckins, A.G. 1996. Trypanosomosis research at the Centre for Tropical Veterinary Medicine (CTVM) 1970 to 1995. *Tropical Animal Health and Production* **28**: 5–22.

Bongaarts, J. 1994. Population policy options in the developing world. *Science* **263**: 771–776.

Bonnefille, R., Roeland, J.C. and Guiot, J. 1990. Temperature and rainfall estimates of the past 40,000 years in equatorial Africa. *Nature* **346**: 347–349.

Boo, E. 1990. *Ecotourism: the potentials and pitfalls.* vol. 2: *Country case studies.* Washington, DC: World Wildlife Fund.

Boom, B.M. 1989. Use of plant resources by the Chacabo. *Advances in Economic Botany* **7**: 78–96.

Bornemisza, E. 1982. Nitrogen cycling in coffee plantations. *Plant and Soil* **67**: 241–246.

Bos, E., Massiah, E. and Bulatao, R.A. 1994. *World population projections 1994–95.* Baltimore: The Johns Hopkins University Press.

Boserup, E. 1965. *The conditions of agricultural growth: the economics of agrarian change under population pressure.* Chicago: Aldine.

Botkin, D.B. 1990. *Discordant harmonies: a new ecology for the twenty-first century.* Oxford: Oxford University Press.

Boucher, D.H., Vandermeer, J.H., Yih, K. and Zamora, N. 1990. Contrasting hurricane damage in tropical rain forest and pine forest. *Ecology* 71: 2022–2024.

Boutton, T., Tieszen, L.L. and Imbamba, S.K. 1988. Seasonal changes in the nutrient content of East African grassland vegetation. *African Journal of Ecology* 26: 103–115.

Bowman, D.M.J.S. and Woinarski, J.C.Z. 1994. Biogeography of Australian monsoon rainforest mammals: implications for the conservation of rainforest mammals. *Pacific Conservation Biology* 1: 98–106.

Bradbury, J.P., Leyden, B., Salgada-Labouriau, M., Lewis, W.M., Schubert, C., Benford, M.W., Frey, D.G., Whitehead, D.R. and Weibezahn, F.H. 1981. Late Quaternary environmental history of Lake Valencia, Venezuela. *Science* 214: 1299–1305.

Brader, L. 1979. Integrated pest control in the developing world. *Annual Review of Entomology* 24: 225–254.

Brandani, A., Hartshorn, G.S. and Orians, G.H. 1988. Internal heterogeneity of gaps and species richness in Costa Rican tropical wet forests. *Journal of Tropical Ecology* 4: 99–119.

Brandt, J. 1988. The transformation of rainfall energy by a tropical forest canopy in relation to soil erosion. *Journal of Biogeography* 15: 41–48.

Breman, H. and de Wit, C.T. 1983. Rangeland productivity and exploitation in the Sahel. *Science* 221: 1341–1347.

Brener, G.F. and Silva, J.F. 1995. Leaf-cutting ant nests and soil fertility in a well-drained savanna in western Venezuela. *Biotropica* 27: 250–254.

Brokaw, N.V.L. 1985. Gap-phase regeneration in a tropical forest. *Ecology* 66: 682–687.

Brookfield, H. and Padoch, C. 1994. Appreciating agrodiversity. *Environment* 36: 6–11, 37–45.

Browder, J.O. 1992. Social and economic constraints on the development of market-oriented extractive reserves in Amazon rain forests. *Advances in Economic Botany* 9: 33–42.

Brown, B.J. and Marten, G.G. 1986. The ecology of traditional pest management in Southeast Asia. In *Traditional agriculture in Southeast Asia* (ed. G.G. Marten). Boulder, CO: Westview Press, 241–272.

Brown, S. and Lugo, A.E. 1990. Tropical secondary forests. *Journal of Tropical Ecology* 6: 1–32.

Bruijnzeel, L.A. 1989. Nutrient cycling in moist tropical forests: the hydrological framework. In *Mineral nutrients in tropical forest and savanna ecosystems* (ed. J. Proctor). Oxford: Blackwell, 383–415.

Brush, S.B. 1989. Rethinking crop genetic resource conservation. *Conservation Biology* 3: 19–29.

Buchanan, F. 1807. *A journey from Madras through the countries of Mysore, Canara, and Malabar,* vol. 2. St James: W. Bulmer and Co.; London: East India Co.

Buckles, D. 1995. Velvetbean: a 'new' plant with a history. *Economic Botany* 49: 13–25.

Budowski, G. 1982. Applicability of agro-forestry systems. In *Agro-forestry in the African humid tropics* (ed. L.H. MacDonald). Tokyo: United Nations University Press, 13–16.

Bukenya, G.B., Nsungwa, J.L., Makanga, B. and Salvator, A. 1994. *Schistosomiasis mansoni* and paddy-rice growing in Uganda: an emerging new problem. *Annals of Tropical Medicine and Parasitology* 88: 379–384.

Burger, J. and Gochfeld, M. 1993. Tourism and short-term behavioural responses of nesting Masked, Red-footed, and Blue-footed Bobbies in the Galápagos. *Environmental Conservation* **20**: 255–259.

Burkey, T.V. 1994. Tropical tree species diversity: a test of the Janzen–Connell model. *Oecologia* **97**: 533–540.

Buschbacher, R.J. 1990. Natural forest management in the humid tropics: ecological, social and economic considerations. *Ambio* **19**: 253–258.

Buschbacher, R.J., Uhl, C. and Serrão, E.A.S. 1987. Large-scale development in eastern Amazonia. Case study no. 10: pasture management and environmental effects near Paragominas, Pará. In *Amazonian rain forests, ecosystem disturbance and recovery* (ed. C.F. Jordan). New York: Springer-Verlag, 90–99.

Bush, M.B. 1994. Amazonian speciation: a necessarily complex model. *Journal of Biogeography* **21**: 5–17.

Bush, M.B. and Colinvaux, P.A. 1990. A pollen record of a complete glacial cycle from lowland Panama. *Journal of Vegetation Science* **1**: 105–118.

Bush, M.B., Colinvaux, P.A., Wiemann, M.C., Piperno, D.R. and Liu, K.B. 1990. Late Pleistocene temperature depression and vegetation change in Ecuadorian Amazonia. *Quaternary Research* **34**: 330–345.

Cahn, M.D., Bouldin, D.R. and Cravo, M.S. 1992. Nitrate sorption in the profile of an acid soil. *Plant and Soil* **143**: 179–183.

Cahoon, D.R., Stocks, B.R., Levine, J.S., Cofer, W.R. and O'Neil, K.P. 1992. Seasonal distribution of African savanna fires. *Nature* **359**: 812–815.

Cane, M.A. 1983. Oceanographic events during El Niño. *Science* **222**: 1189–1195.

Carroll, C.R. and Risch, S. 1990. An evaluation of ants as possible candidates for biological control in tropical annual agroecosystems. In *Agroecology* (ed. S.R. Gliessman). New York: Springer-Verlag, 30–46.

Cavelier, J. and Goldstein, G. 1989. Mist and fog interception in elfin cloud forests in Colombia and Venezuela. *Journal of Tropical Ecology* **5**: 309–322.

Chambers, R., Pacey, A. and Thrupp, L.I. (eds) 1989. *Farmer first: farmer innovation and agricultural research*. London: Intermediate Technology Publications.

Chandler, J.A., Highton, R.B. and Hill, M.N. 1976. Mosquitoes of the Kano Plain, Kenya II. Results of outdoor collections in irrigated and nonirrigated areas using human and animal bait and light traps. *Journal of Medical Entomology* **13**: 202–207.

Chang, T.T. 1989. Domestication and spread of the cultivated rices. In *Foraging and farming: the evolution of plant exploitation* (ed. D.R. Harris and G.C. Hillman). London: Unwin Hyman, 408–417.

Chantalakhana, C. 1981. A scope on buffalo breeding for draft. In *Recent advances in buffalo research and development* (ed. M.H. Tetangco). Taipei: Food and Fertilizer Technology Center, 131–140.

Chao, T.T., Harward, M.E. and Fang, S.C. 1964. Iron and aluminum coatings in relation to sulfate adsorption characteristics of soils. *Soil Science Society of America, Proceedings* **28**: 632–635.

Chapin, F.S. 1980. The mineral nutrition of wild plants. *Annual Review of Ecology and Systematics* **11**: 233–260.

Chapin, G. and Wasserstrom, R. 1981. Agricultural production and malaria resurgence in Central America and India. *Nature* **293**: 181–185.

Chapman, V.J. 1977. Introduction. In *Wet coastal ecosystems* (ed. V.J. Chapman). New York: Elsevier, 1–29.

Chenoweth, P.J. 1994. Aspects of reproduction in female *Bos indicus* cattle: a review. *Australian Veterinary Journal* **71**: 422–426.

Chrispeels, M.J and Sadava, D. 1977. *Plants, food and people*. San Francisco: W.H. Freeman.

Clapperton, C.M. 1993. Nature of environmental changes in South America at the last glacial maximum. *Palaeogeography, Palaeoclimatology and Palaeoecology* 101: 189–208.

Clawson, D.L. 1984. Harvest security and intraspecific diversity in traditional tropical agriculture. *Economic Botany* 39: 56–67.

Coale, A.J. 1983. Recent trends in fertility in less developed countries. *Science* 221: 828–832.

Coe, M.J., Cumming, D.H. and Phillipson, J. 1976. Biomass and production of large African herbivores in relation to rainfall and primary production. *Oecologia* 22: 341–354.

Cohen, M.N. 1977. *The food crisis in prehistory: overpopulation and the origins of agriculture*. New Haven, CT: Yale University Press.

Cohen, M.N. and Armelagos, G.J. (eds) 1984. *Paleopathology at the origins of agriculture*. Orlando, FL: Academic Press.

Coley, P.D. 1983. Herbivory and defensive characteristics of tree species in a lowland tropical forest. *Ecological Monographs* 53: 209–233.

Coley, P.D. 1988. Effects of plant growth rate and leaf lifetime on the amount and type of ant-herbivore defense. *Oecologia* 74: 531–536.

Coley, P.D. and Aide, T.M. 1991. Comparison of herbivory and plant defences in temperate and tropical broad-leaved forests. In *Plant–animal interactions: evolutionary ecology in tropical and temperate regions* (ed. P.W. Price, G.W. Fernandes, T.M. Lewinsohn and W.W. Benson). New York: John Wiley, 25–49.

Coley, P.D., Bryant, J.P. and Chapin, F.S. 1985. Resource availability and plant anti-herbivore defence. *Science* 230: 895–899.

Colinvaux, P.A. 1993. Pleistocene biogeography and diversity in tropical forests of South America. In *Biological relationships between Africa and South America* (ed. P. Goldblatt). New Haven, CT: Yale University Press, 473–499.

Collier, R.J. and Beede, D.K. 1985. Thermal stress as a factor associated with nutrient requirements and interrelationships. In *Nutrition of grazing ruminants in warm climates* (ed. L.R. McDowell). Orlando, FL: Academic Press, 59–71.

Colt, J. 1991. Aquacultural production systems. *Journal of Animal Science* 69: 4183–4192.

Coluzzi, M. 1994. Malaria and the Afrotropical ecosystems: impact of man-made environmental changes. *Parassitologia* 36: 223–227.

Condit, R., Hubbell, S.P. and Foster, R.B. 1992a. Recruitment near conspecific adults and the maintenance of tree and shrub diversity in a Neotropical forest. *American Naturalist* 140: 261–286.

Condit, R., Hubbell, S.P. and Foster, R.B. 1992b. Short-term dynamics of a Neotropical forest: change within limits. *BioScience* 42: 822–828.

Conklin, H.C. 1957. *Hanunoo agriculture in the Philippines*. Rome: FAO.

Connah, G. 1985. Agricultural intensification and sedentism in the Firki of NE Nigeria. In *Prehistoric intensive agriculture in the tropics* (ed. I.S. Farrington). BAR International Series 232. Oxford: Osney Mead, 765–785.

Connell, J.H. 1971. On the role of natural enemies in preventing competitive exclusion in some marine animals and rain forest trees. In *Dynamics of populations* (ed. P.J. den Boer and G.R. Gradwell). Wagenigen, the Netherlands: Center for Agricultural Publishing and Documentation, 298–310.

Connell, J.H. 1978. Diversity in tropical rain forests and coral reefs. *Science* 199: 1302–1310.

Connell, J.H. and Lowman, M.D. 1989. Low-diversity tropical rain forests: some possible mechanisms for their existence. *American Naturalist* 134: 88–119.

Conway, G.R. and Pretty, J.N. 1988. Fertilizer risks in the developing countries. *Nature* 334: 207–208.

Cook, A.G. and Perfect, T.J. 1989. The population characteristics of the brown planthopper, *Nilaparvata lugens*, in the Philippines. *Ecological Entomology* 14: 1–9.

Cornell, H.V. and Lawton, J.H. 1992. Species interaction, local and regional processes, and the limits to the richness of ecological communities: a theoretical perspective. *Journal of Animal Ecology* 61: 1–12.

Coughenour, M.B., Ellis, J.E., Swift, D.M., Coppock, D.L., Galvin, K., McCabe, J.T. and Hart, T.C. 1985. Energy extraction and use in a nomadic pastoral system. *Science* 230: 619–625.

Coutinho, L.M. 1990. Fire in the ecology of the Brazilian cerrado. In *Fire in the tropical biota* (ed. J.G. Goldammer). New York: Springer-Verlag, 82–105.

Crane, P.R. and Lidgard, S. 1989. Angiosperm diversification and paleolatitudinal gradients in floristic diversity. *Science* 246: 675–678.

Crane, P.R., Friis, E.M. and Pedersen, K.R. 1995. The origin and early diversification of angiosperms. *Nature* 374: 27–33.

Crews, T.E., Kitayama, K., Fownes, J.H., Riley, R., Herbert, D.A., Mueller-Dombois, D. and Vitousek, P.M. 1995. Changes in soil phosphorus fractions and ecosystem dynamics across a long chronosequence in Hawaii. *Ecology* 76: 1407–1424.

Cry, G.W. 1965. *Tropical cyclones of the North Atlantic ocean*. US Dept. of Commerce Weather Bureau, Technical Paper No. 55. Washington, DC: US Government Printing Office.

Cuevas, E. and Medina, E. 1986. Nutrient dynamics within Amazonian forest ecosystems. I. Nutrient flux in fine litter fall and efficiency of nutrient utilization. *Oecologia* 68: 466–472.

Cuevas, E. and Medina, E. 1988. Nutrient dynamics within Amazonian forest ecosystems. II. Fine root growth, nutrient availability and leaf litter decomposition. *Oecologia* 76: 222–235.

Culliton, B.J. 1990. Emerging viruses, emerging threat. *Science* 247: 279–280.

Damuth, J.E. and Fairbridge, R.W. 1970. Equatorial Atlantic deep-sea arkosic sands and ice-age aridity in tropical South America. *Bulletin of the Geological Society of America* 81: 189–206.

Darch, J.P. (ed.) 1983. *Drained field agriculture in Central and South America*. BAR International Series 189. Oxford: Osney Mead.

Das, P.K. 1987. Short- and long-range monsoon prediction in India. In *Monsoons* (ed. J.S. Fein and P.L. Stephens). New York: John Wiley, 549–578.

Dasmann, R.F. 1964. *African game ranching*. Oxford: Pergamon Press.

Davis, T.A.W. and Richards, P.W. 1933–4. The vegetation of Moraballi Creek, British Guiana; an ecological study of a limited area of tropical rain forest. Parts I and II. *Journal of Ecology* 21: 350–384; 22: 106–155.

de Graff, N.R. 1986. *A silvicultural system for natural regeneration of tropical rain forest in Surinam*. Wagenigen, the Netherlands: Agricultural University.

de Graff, N.R. 1991. Managing natural regeneration for sustained timber production in Surinam: the CELOS silvicultural and harvesting system. In *Rain forest regeneration and management* (ed. A. Gomez-Pompa, T.C. Whitmore and M. Hadley). Man and the Biosphere Series, vol. 6. Paris: UNESCO; Park Ridge, NJ: Parthenon Publishing Group, 393–405.

De Vos, A. 1991. Wildlife utilization on marginal lands in developing countries. In *Wildlife production: conservation and sustainable development* (ed. L.A. Renecker and R.J. Hudson). Agricultural and Forestry Experiment Station Miscellaneous Publication 91–6. Fairbanks, AL: University of Alaska Fairbanks, 8–10.

Deevey, E.S., Rice, D.S., Rice, P.M., Vaughn, H.H., Brenner, M. and Flannery, M.S. 1979. Mayan urbanism: impact on a tropical karst environment. *Science* 206: 298–306.

Demenocal, P.B. 1995. Plio-Pleistocene African climate. *Science* 270: 53–58.

Denslow, J. 1987. Tropical rain forest gaps and tree species diversity. *Annual Review of Ecology and Systematics* **18**: 431–451.

Denslow, J.S., Schultz, J.C., Vitousek, P.M. and Strain, B.R. 1990. Growth responses of tropical shrubs to treefall gap environments. *Ecology* **71**: 165–179.

Desowitz, R.S. 1978. Health and epidemiology. In *Tropical forest ecosystems: a state-of-knowledge report* (prepared by UNESCO/UNEP/FAO). Paris: UNESCO, 372–403.

Desowitz, R.S. 1980. Epidemiological–ecological interactions in savanna environments. In *Human ecology in savanna environments* (ed. D.R. Harris). London: Academic Press, 457–477.

Dezzeo, N. (ed.) 1994. *Ecología de la altiplanicie de la Gran Sabana (Guyana Venezolana) I.* Scientia Guianae vol. 4, Caracas, Venezuela.

Dhar, S.K. 1994. Rehabilitation of degraded tropical forest watersheds with people's participation. *Ambio* **23**: 216–221.

Dickson, D. and Jayaraman, K.S. 1995. Aid groups back challenge to neem patents. *Nature* **377**: 95.

Dietricht, W.E., Windsor, D.M. and Dunne, T. 1982. Geology, climate, and hydrology of Barro Colorado Island. In *The ecology of a tropical forest: seasonal rhythms and long-term changes* (ed. E.G. Leight, A.S. Rand and D.M. Windsor). Washington, DC: Smithsonian Institution Press, 21–46.

Dobzhansky, T. 1950. Evolution in the tropics. *American Scientist* **38**: 209–221.

Dodd, J.L. 1994. Desertification and degradation in sub-Saharan Africa: the role of livestock. *BioScience* **44**: 28–34.

Douglas, I., Spencer, T., Greer, T., Bidin, K., Sinun, W. and Meng, N.W. 1992. The impact of selective commercial logging on stream hydrology, chemistry and sediment loads in the Ulu Segama rain forest, Sabah, Malaysia. *Philosophical Transactions of the Royal Society of London, Series B* **335**: 397–406.

Dransfield, R.D., Williams, B.G. and Brightwell, R. 1991. Control of tsetse flies and trypanosomiasis: myth or reality? *Parasitology Today* **7**: 287–291.

Drew, M.C. 1975. Comparison of the effects of localized supply of phosphate, nitrate and potassium on the growth of the seminal root system, and the shoot, in barley. *New Phytologist* **75**: 479–490.

Durand, J.D. 1977. Historical estimates of world population: an evaluation. *Population and Development Review* **3**: 253–296.

Eiten, G. 1982. Brazilian 'savannas'. In *Ecology of tropical savannas* (ed. B.J. Huntley and B.H. Walker). New York: Springer-Verlag, 25–47.

El-Badry, M.A. 1991. The growth of world population: past, present and future. In *Consequences of rapid population growth in developing countries.* Proceedings of the United Nations/Institut national d'études demographiques Expert Group Meeting, New York, 23–26 August 1988. New York: Taylor and Francis, 15–40.

Ellis, J.E. and Swift, D.M. 1988. Stability of African pastoral systems: alternate paradigms and implications for development. *Journal of Range Management* **41**: 450–459.

Ellis, P. 1984. Health control in relation to livestock production. In *Development of animal production systems* (ed. B. Nestel). Amsterdam: Elsevier, 63–77.

El-Swaify, E.A. 1993. Soil erosion and conservation in the humid tropics. In *World soil erosion and conservation* (ed. D. Pimentel), Cambridge: Cambridge University Press, 233–255.

Eltringham, S.K. 1994. Can wildlife pay its way? *Oryx* **28**: 163–168.

Emmons, L.H. 1983. Geographic variation in densities and diversities of non-flying mammals in Amazonia. *Biotropica* **16**: 210–222.

Endler, J.A. 1982. Pleistocene forest refuges: fact or fancy? In *Biological diversification in the tropics* (ed. G.T. Prance). New York: Columbia University Press, 641–657.

BIBLIOGRAPHY

Eriksson, O. 1993. The species-pool hypothesis and plant community diversity. *Oikos* 68: 371–374.

Eussen, J.H.H. 1979. Some competition experiments with Alang-Alang (*Imperata cylindrica* (L.) Beauv.) in replacement series. *Oecologia* 40: 351–356.

Evans, J. 1982. *Plantation forestry in the tropics.* Oxford: Clarendon Press.

Evans, L.T. 1993. *Crop evolution, adaptation and yield.* Cambridge: Cambridge University Press.

Evans, T.R. 1982. Overcoming nutritional limitations through pasture management. In *Nutritional limits to animal production from pastures* (ed. J.B. Hacker). Proceedings of an International Symposium held at St Lucia, Queensland, Australia, 24–28 August, 1981. Farnham Royal: Commonwealth Agricultural Bureaux, 343–361.

Ewel, J.J. 1980. Tropical succession: manifold routes to maturity. *Biotropica* 12 (Suppl.): 2–7.

Ewel, J.J. 1986. Designing agricultural systems for the humid tropics. *Annual Review of Ecology and Systematics* 17: 245–271.

Fa, J.E., Juste, J., Del Val, J.P. and Castroviejo, J. 1995. Impact of market hunting on mammal species in Equatorial Guinea. *Conservation Biology* 9: 1107–1115.

Fairall, N. 1984. The use of non-domesticated African mammals for game farming. *Acta Zoologica Fennica* 172: 215–218.

Fairall, N. 1989. Extensive containment systems: game ranching. In *Wildlife production systems: economic utilisation of wild ungulates* (ed. R.J. Hudson, K.R. Drew, and L.M. Baskin). Cambridge: Cambridge University Press, 243–247.

FAO 1970. *Amino-acid content of foods and biological data on proteins.* FAO Nutritional Study 24. Rome: FAO.

FAO 1995. *FAO Production Yearbook 1994*, vol. 48. FAO Statistic Series No. 125. Rome: FAO.

FAO–UNESCO 1988. *Soil map of the world, revised legend.* FAO World Soil Resources Report 60. Rome: FAO.

Fearnside, P.M. 1986. *Human carrying capacity of the Brazilian rainforest.* New York: Columbia University Press.

Fearnside, P.M. 1987. Rethinking continuous cultivation in Amazonia. *BioScience* 37: 209–214.

Fearnside, P.M. 1988. Jari at age 19: lessons for Brazil's silvicultural plans at Carajas. *Interciencia* 13: 12–24.

Fein, J.S. and Stephens, P.L. (eds) 1987. *Monsoons.* New York: John Wiley.

Fink, H. and Baptist, R. 1992. Herd dynamics modelling applied to a venison ranch in Kenya. *World Animal Review* 73: 2–8.

Fischer, A.G. 1960. Latitudinal gradients in organic diversity. *Evolution* 14: 64–81.

FitzGibbon, C.D., Mogaka, H. and Fanshawe, J.H. 1995. Subsistence hunting in Arabuko-Sokoke Forest, Kenya, and its effects on mammal populations. *Conservation Biology* 9: 1116–1126.

Flaherty, M. and Karnjanakesorn, C. 1995. Marine shrimp aquaculture and natural resource degradation in Thailand. *Environmental Management* 19: 27–37.

Flemming, T.H. 1973. Numbers of mammal species in North and Central American forest communities. *Ecology* 54: 555–563.

Flenley, J.R. 1979. The Late Quaternary vegetational history of the equatorial mountains. *Progress in Physical Geography* 4: 488–509.

Forsberg, B.R., Araujo-Lima, C.A.R.M., Martinelli, L.A., Victoria, R.L. and Bonassi, J.A. 1993. Autotrophic carbon sources for fish of the central Amazon. *Ecology* 74: 643–652.

Franco, W. and Dezzeo, N. 1994. Soils and soil water regime in the terra firme–caatinga forest complex near San Carlos de Rio Negro, State of Amazonas, Venezuela. *Interciencia* 19: 305–316.

Freedman, S.M. 1980. Modifications of traditional rice production practices in the developing world: an energy efficiency analysis. *Agro-Ecosystems* **6**: 129–146.

Frost, P.G.H. and Robertson, F. 1987. The ecological effects of fires in savannas. In *Determinants of tropical savannas* (ed. B.H. Walker). Oxford: IRL Press, 93–140.

Fujisaka, S. 1989. A method for farmer-participatory research and technology transfer: upland soil conservation in the Philippines. *Experimental Agriculture* **25**: 423–433.

Furch, K. 1984. Water chemistry of the Amazon basin: the distribution of chemical elements among fresh waters. In *The Amazon, limnology and landscape ecology of a mighty tropical river and its basin* (ed. H. Sioli). Dordrecht: Dr W. Junk, 167–200.

Gajaseni, J. and Jordan, C.F. 1990. Decline of teak yield in northern Thailand: effects of selective logging on forest structure. *Biotropica* **22**: 114–118.

Ganzhorn, J.U. 1995. Low-level disturbance effects on primary production, leaf chemistry and lemur populations. *Ecology* **76**: 2084–2096.

Garlick, P.J. and Reeds, P.J. 1993. Proteins. In *Human nutrition and dietetics* (ed. J.S. Garrow and W.P.T. James). Edinburgh: Churchill Livingstone, 56–76.

Garrett, L. 1994. *The coming plague.* New York: Farrar, Straus and Giroux.

Garrity, D.P. (n.d.). *ICRAF Southeast Asia: implementing the vision.* Bogor, Indonesia: International Centre for Research in Agroforestry, Southeast Asian Regional Research Programme.

Gasser, C.S. and Fraley, R.T. 1989. Genetically engineering plants for crop improvement. *Science* **244**: 1293–1299.

Gause, G.F. 1934. *The struggle for existence.* Baltimore: Waverly Press.

Geist, V. 1988. How markets in wildlife meat and parts, and the sale of hunting privileges, jeopardize wildlife conservation. *Conservation Biology* **2**: 15–26.

Gentry, A.H. 1986. Endemism in tropical versus temperate plant communities. In *Conservation biology: the science of scarcity and diversity* (ed. M.E. Soulé). Sunderland, MA: Sinauer, 153–181.

Gentry, A.H. 1988. Changes in plant community diversity and floristic composition on environmental and geographic gradients. *Annals of the Missouri Botanical Garden* **75**: 1–34.

Gentry, A.H. 1990. Floristic similarities and differences between southern Central America and upper central Amazon. In *Four Neotropical rainforests* (ed. A.H. Gentry). New Haven, CT: Yale University Press, 141–157.

Gerhardt, K. 1993. Tree seedling development in tropical dry abandoned pasture and secondary forest in Costa Rica. *Journal of Vegetation Science* **4**: 95–102.

Giannecchini, J. 1993. Ecotourism: new partners, new relationships. *Conservation Biology* **7**: 429–432.

Gibbons, A. 1993. Where are 'new' diseases born? *Science* **261**: 680–681.

Gilbert, F.S. 1980. The equilibrium theory of island biogeography: fact or fiction? *Journal of Biogeography* **7**: 209–235.

Giller, K.E., McDonagh, J.F. and Cadisch, G. 1994. Can biological nitrogen fixation sustain agriculture in the tropics? In *Soil science and sustainable land management in the tropics* (ed. J.K. Syers and D.L. Rimmer). Wallingford: CAB International, 173–191.

Gillis, M. 1992. Economic policies and tropical deforestation. *Advances in Economic Botany* **9**: 129–141.

Gillman, G.P. 1985. Influence of organic matter and phosphate content on the point of zero charge of variable charge components in oxidic soils. *Australian Journal of Soil Research* **23**: 643–646.

Ginsberg, J.R. and Milner-Gulland, E.J. 1994. Sex-biased harvesting and population

BIBLIOGRAPHY

dynamics in ungulates: implications for conservation and sustainable use. *Conservation Biology* **8**: 157–166.

Giresse, P., Maley, J. and Brenac, P. 1994. Late Quaternary palaeoenvironments in the Lake Barombi Mbo (West Cameroon) deduced from pollen and carbon isotopes of organic matter. *Palaeogeography, Palaeoclimatology and Palaeoecology* **107**: 65–78.

Gliessman, S.R., Garcia, R. and Amador, M. 1981. The ecological basis for the application of traditional agricultural technology in the management of tropical agro-ecosystems. *Agro-Ecosystems* **7**: 173–185.

Goedert, W.J. 1983. Management of the cerrado soils of Brazil: a review. *Journal of Soil Science* **34**: 405–428.

Goldammer, J.G. 1993. *Feuer in Waldökosystemen der Tropen und Subtropen*. Basel and Boston: Birkenhauser-Verlag.

Goldammer, J.G. and Seibert, B. 1990. The impacts of droughts and forest fires on tropical lowland rain forest of east Kalimantan. In *Fire in the tropical biota* (ed. J.G. Goldammer). New York: Springer-Verlag, 11–31.

Golden, B.E. 1993. Primary protein–energy malnutrition. In *Human nutrition and dietetics* (ed. J.S. Garrow and W.P.T. James). Edinburgh: Churchill Livingstone, 440–455.

Golding, E.J. 1985. Providing energy–protein supplementation during the dry season. In *Nutrition of grazing ruminants in warm climates* (ed. L.R. McDowell). Orlando, FL: Academic Press, 129–163.

Golson, J. 1989. The origins and development of New Guinea agriculture. In *Foraging and farming: the evolution of plant exploitation* (ed. D.R. Harris and G.C. Hillman). London: Unwin Hyman, 678–687.

Goodman, D. 1987. Consideration of stochastic demography in the design and management of biological reserves. *Natural Resource Modeling* **1**: 205–234.

Gourou, P. 1958. *The tropical world*, 2nd edition (trans. E.D. Laborde). London: Longmans.

Graham, A. and Dilcher, D. 1995. The Cenozoic record of tropical dry forest in northern Latin America and the southern United States. In *Seasonally dry tropical forests* (ed. S.H. Bullock, H.A. Mooney and E. Medina). Cambridge: Cambridge University Press, 124–145.

Graham, N.E., Barnett, T.P., Wilde, R., Ponater, M. and Schubert, S. 1994. On the roles of tropical and midlatitude SSTs in forcing interannual to interdecadal variability in the winter northern hemisphere circulation. *Journal of Climate* **7**: 1416–1441.

Greenwood, D.J. 1981. Fertilizer use and food production: world scene. *Fertilizer Research* **2**: 33–51.

Greenwood, D.J. 1982. Nitrogen supply and crop yield: the global scene. *Plant and Soil* **67**: 45–59.

Griffin, P.B. 1989. Hunting, farming and sedentism in a rain forest foraging society. In *Farmers as hunters: the implications of sedentism* (ed. S. Kent). Cambridge: Cambridge University Press, 60–79.

Grigg, D. 1979. Ester Boserup's theory of agrarian change: a critical review. *Progress in Human Geography* **3**: 64–84.

Grigg, D. 1995. The pattern of world protein consumption. *Geoforum* **26**: 1–17.

Groube, L. 1989. The taming of the rain forests: a model for Late Pleistocene forest exploitation in New Guinea. In *Foraging and farming: the evolution of plant exploitation* (ed. D.R. Harris and G.C. Hillman). London: Unwin Hyman, 292–304.

Guerrero, R.D. 1990. Aquaculture in tropical Asia. *Resource Management and Optimization* **7**: 269–281.

Guilderson, T.P., Fairbanks, R.G. and Rubenstone, J.L. 1994. Tropical temperature variations since 20,000 years ago: modulating interhemispheric climate change. *Science* **263**: 663–665.

Guo, J.Y. and Bradshaw, A.D. 1993. The flow of nutrients and energy through a Chinese farming system. *Journal of Applied Ecology* **30**: 86–94.

Gwynne, M.D. and Bell, R.H.V. 1968. Selection of vegetation components by grazing ungulates in the Serengeti National Park. *Nature* **220**: 390–393.

Hackett, C. 1972. A method of applying nutrients locally to roots under controlled conditions, and some morphological effects of locally applied nitrate on the branching of wheat roots. *Australian Journal of Biological Science* **25**: 1169–1180.

Haffer, J. 1969. Speciation in Amazonian forest birds. *Science* **165**: 131–137.

Hagmann, J. and Prasad, V.L. 1995. Use of donkeys and their draught performance in smallholder farming in Zimbabwe. *Tropical Animal Health and Production* **27**: 231–239.

Hamilton, A. 1976. The significance of patterns of distribution shown by forest plants and animals in tropical Africa for the reconstruction of upper Pleistocene palaeoenvironments: a review. *Palaeoecology of Africa and the Surrounding Islands and Antarctica* **9**: 63–97.

Hammond, L.L., Chien, S.H. and Mokwunye, A.U. 1986. Agronomic value of unacidulated and partially acidulated phosphate rocks indigenous to the tropics. *Advances in Agronomy* **40**: 89–140.

Harper, J.L. 1977. *Population biology of plants*. London: Academic Press.

Harris, D.R. 1989. An evolutionary continuum of people–plant interaction. In *Foraging and farming: the evolution of plant exploitation* (ed. D.R. Harris and G.C. Hillman). London: Unwin Hyman, 11–26.

Harris, D.R. and Hillman, G.C. (eds) 1989. *Foraging and farming: the evolution of plant exploitation*. London: Unwin Hyman.

Hart, T.B., Hart, J.A. and Murphy, P.G. 1989. Monodominant and species-rich forests of the humid tropics: causes for their co-occurrence. *American Naturalist* **133**: 613–633.

Hartshorn, G.S. 1980. Neotropical forest dynamics. *Biotropica* **12** (Suppl.): 23–30.

Hartshorn, G.S. 1989. Application of gap theory to tropical forest management: natural regeneration on strip clear-cuts in the Peruvian Amazon. *Ecology* **70**: 567–569.

Hartshorn, G.S. 1990a. An overview of Neotropical forest dynamics. In *Four Neotropical rainforests* (ed. A.H. Gentry). New Haven, CT: Yale University Press, 585–599.

Hartshorn, G.S. 1990b. Natural forest management by the Yanesha Forestry Cooperative in Peruvian Amazon. In *Alternatives to deforestation: steps toward sustainable use of the Amazon rain forest* (ed. A.B. Anderson). New York: Columbia University Press, 128–138.

Hastenrath, S. 1991. *Climate dynamics of the tropics*. Dordrecht: Kluwer.

Haviland, W.A. 1967. Stature at Tikal, Guatemala: implications for ancient Maya demography and social organization. *American Antiquity* **32**: 316–325.

He, F., Legendre, P. and LaFrankie, J.V. 1996. Spatial pattern of diversity in a tropical rain forest in Malaysia. *Journal of Biogeography* **23**: 57–74.

Hecht, S.B. 1985. Environment, development and politics: capital accumulation and the livestock sector in eastern Amazonia. *World Development* **13**: 663–684.

Henwood, K. 1973. A structural model of forces in buttressed tropical rain forest trees. *Biotropica* **5**: 83–93.

Herrera, R., Tatiana, M. and Stark, N. 1978. Direct phosphorus transfer from leaf litter to roots. *Naturwissenschaften* **65**: 208–209.

Hildebrand, A.R., Pilkington, M., Connors, M., Oritz-Aleman, C. and Chavez, R.E. 1995. Size and structure of the Chicxulub crater revealed by horizontal gravity gradients. *Nature* **376**: 415–417.

Hodell, D.A., Curtis, J.H. and Brenner, M. 1995. Possible role of climate in the collapse of the Classic Maya civilization. *Nature* **375**: 391–394.

Hogberg, P. 1989. Root symbioses of trees in savannas. In *Mineral nutrients in tropical forest and savanna ecosystems* (ed. J. Proctor). Oxford: Blackwell, 121–136.

Holthuijzen, A.M.A. and Boerboom, J.H.A. 1982. The *Cecropia* seed bank in the Surinam lowland rain forest. *Biotropica* **14**: 62–68.

Homewood, K.M. and Rodgers, W.A. 1991. *Maasailand ecology, pastoralist development and wildlife conservation in Ngorongoro, Tanzania*. Cambridge: Cambridge University Press.

Homma, A.K.O. 1992. The dynamics of extraction in Amazonia: a historical perspective. *Advances in Economic Botany* **9**: 23–32.

Honey, M. 1994. Paying the price of ecotourism. *Américas* **46**: 40–47.

Hoogesteijn Reul, R. 1979. Productive potential of wild animals in the tropics. *World Animal Review* **32**: 18–24.

Hooghiemstra, H. and Van der Hammen, T. 1993. Late Quaternary vegetation history and paleoecology of Laguna Pedro Palo (subandean forest belt, Eastern Cordillera, Colombia). *Reviews of Paleobotany and Palynology* **77**: 235–262.

Hopkins, M.S., Ash, J., Graham, A.W., Head, J. and Hewett, R.K. 1993. Charcoal evidence of the spatial extent of the *Eucalyptus* woodland expansions and rainforest contractions in north Queensland during the late Pleistocene. *Journal of Biogeography* **20**: 357–372.

Horel, J.D. and Wallace, J.M. 1981. Planetary-atmosphere phenomena associated with the Southern Oscillation. *Monthly Weather Review* **109**: 813–829.

Howeler, R.H., Sieverding, E. and Saif, S. 1987. Practical aspects of mycorrhizal technology in some tropical crops and pastures. *Plant and Soil* **100**: 249–283.

Howes, M. 1980. The use of indigenous technical knowledge in development. In *Indigenous knowledge systems and development* (ed. D.W. Brokensha, D.M. Warren and O. Werner). Lanham, MD: University Press of America, 341–357.

Howes, M. and Chambers, R. 1980. Indigenous technical knowledge: analysis, implications and issues. In *Indigenous technical knowledge systems and development* (ed. D.W. Brokensha, D.M. Warren and O. Werner). Lanham, MD: University Press of America, 329–340.

Hubbell, S.P. 1980. Seed predation and the coexistence of tree species in tropical forests. *Oikos* **35**: 214–229.

Hubbell, S.P. and Foster, R.B. 1986a. Commonness and rarity in a Neotropical forest: implications for tropical tree conservation. In *Conservation biology: the science of scarcity and diversity* (ed. M.E. Soulé). Sunderland, MA: Sinauer, 205–236.

Hubbell, S.P. and Foster, R.B. 1986b. Biology, chance and history and the structure of tropical rain forest tree communities. In *Community ecology* (ed. J. Diamond and T.J. Chase). New York: Harper and Row, 314–329.

Hubbell, S.P. and Foster, R.B. 1992. Short-term dynamics of a Neotropical forest: why ecological research really matters. *Oikos* **63**: 48–61.

Hudson, N. 1981. *Soil conservation*, 2nd edition. Ithaca, NY: Cornell University Press.

Hudson, R.J., Drew, K.R. and Baskin, L.M. (eds) 1989. *Wildlife production systems: economic utilisation of wild ungulates*. Cambridge: Cambridge University Press.

Humphreys, L.R. 1991. *Tropical pasture utilization*. Cambridge: Cambridge University Press.

Huntley, B. 1990. European post-glacial forests: compositional changes in response to climate. *Journal of Vegetation Science* **1**: 507–518.

Huntley, B.J. 1982. Southern African savannas. In *Ecology of tropical savannas* (ed. B.J. Huntley and B.H. Walker). New York: Springer-Verlag, 101–119.

Hurni, H. 1993. Land degradation, famine, and land resource scenarios in Ethiopia. In *World soil erosion and conservation* (ed. D. Pimentel). Cambridge: Cambridge University Press, 27–61.

Huston, M.A. 1979. A general hypothesis of species diversity. *American Naturalist* **113**: 81–101.

Huston, M.A. 1980. Soil nutrients and tree species richness in Costa Rican forests. *Journal of Biogeography* **7**: 147–157.

Huston, M.A. 1993. Biological diversity, soils, and economics. *Science* **262**: 1676–1680.

Huston, M.A. 1994. *Biological diversity: the coexistence of species on a changing landscape*. Cambridge: Cambridge University Press.

Hutchinson, I. 1977. Ecological modelling and the stand dynamics of *Pinus caribaea* in Mountain Pine Ridge, Belize. Ph.D. thesis, Simon Fraser University, Vancouver.

Irion, G. 1989. Quaternary geological history of the Amazon lowlands. In *Tropical forests: botanical dynamics, speciation and diversity* (ed. L.B. Holm-Nielsen, I.C. Nielsen and H. Balslev). New York: Academic Press, 23–34.

Iwama, G.K. 1991. Interactions between aquaculture and the environment. *Critical Reviews in Environmental Control* **21**: 177–216.

Jablonski, D. 1993. The tropics as a source of evolutionary novelty through geological time. *Nature* **364**: 142–144.

Jackson, P.C., Cavelier, J., Goldstein, G., Meinzer, F.C. and Holbrook, N.M. 1995. Partitioning of water resources among plants of a lowland tropical forest. *Oecologia* **101**: 197–203.

Jackson, R.B., Manwaring, J.H. and Caldwell, M.M. 1990. Rapid physiological adjustment of roots to localized soil enrichment. *Nature* **344**: 58–60.

Jacobson, S.K. and Figueroa Lopez, A. 1994. Biological impacts of ecotourism: tourists and nesting turtles in Tortuguero National Park, Costa Rica. *Wildlife Society Bulletin* **22**: 414–419.

Janos, D.P. 1983. Tropical mycorrhizas, nutrient cycles and plant growth. In *Tropical rain forest: ecology and management* (ed. S.L. Sutton, T.C. Whitmore and A.C. Chadwick). Oxford: Blackwell, 327–345.

Jans, L., Poorter, L., Van Rompaey, R.S.A.R. and Bongers, F. 1993. Gaps and forest zones in tropical moist forest in Ivory Coast. *Biotropica* **25**: 258–269.

Janzen, D.H. 1966. Coevolution of mutualism between ants and acacias in Central America. *Evolution* **20**: 249–275.

Janzen, D.H. 1970. Herbivores and the number of tree species in tropical forests. *American Naturalist* **104**: 501–528.

Janzen, D.H. 1973. Sweep samples of tropical foliage insects: effects of seasons, vegetation types, elevation, time of day, and insularity. *Ecology* **54**: 687–701.

Jehne, W. and Thompson, C.H. 1981. Endomycorrhizae in plant colonization on coastal sand dunes at Cooloola, Queensland. *Australian Journal of Ecology* **6**: 221–230.

Jenkin, R.N., Rose-Innes, R., Dunsmore, J.R., Walker, S.H., Birchall, C.J. and Briggs, J.S. 1976. *The agricultural potential of the Belize valley*. Land Resources Study 24. Surbiton, Surrey: Land Resources Division.

Johns, A.D. 1988. Effects of 'selective' timber extraction on rain forest structure and composition and some consequences for frugivores and folivores. *Biotropica* **20**: 31–37.

Johns, A.D. 1992a. Species conservation in managed tropical forests. In *Tropical deforestation and species extinction* (ed. T.C. Whitmore and J.A. Sayer). London: Chapman and Hall, 15–53.

Johns, A.D. 1992b. Vertebrate response to selective logging: implications for the design of logging systems. *Philosophical Transactions of the Royal Society of London, Series B* **335**: 437–442.

Jones, J.R. 1990. *Colonization and environment: land settlement projects in Central America.* Tokyo: United Nations University Press.

Jones, R.J. 1988. The future for the grazing herbivore. *Tropical Grasslands* **22**: 97–115.

Jones, R.J. 1990. Phosphorus and beef production in northern Australia. 1. Phosphorus and pasture productivity – a review. *Tropical Grasslands* **24**: 131–139.

Jordan, C., Golley, F., Hall, J. and Hall, J. 1980. Nutrient scavenging of rainfall by the canopy of an Amazonian rain forest. *Biotropica* **12**: 61–66.

Jordan, C.F. and Escalante, G. 1980. Root productivity in an Amazonian rain forest. *Ecology* **61**: 14–18.

Joyce, C. 1988. Nature helps Indonesia cut its pesticide bill. *New Scientist* 16 June: 35.

Junk, W.J. 1989. Flood tolerance and tree distribution in central Amazonian flood-plains. In *Tropical forests: botanical dynamics, speciation and diversity* (ed. L.B. Holm-Nielsen, I.C. Nielsen and H. Balslev). New York: Academic Press, 47–64.

Jutsum, A.R., Cherrett, J.M. and Fisher, M. 1981. Interactions between the fauna of citrus trees in Trinidad and the ants *Atta cephalotes* and *Azteca* sp. *Journal of Applied Ecology* **18**: 187–195.

Karr, J.R. 1982. Population variability and extinction in the avifauna of a tropical land bridge island. *Ecology* **63**: 1975–1978.

Kattan, G.H., Alvarez-Lopez, H. and Giraldo, M. 1994. Forest fragmentation and bird extinctions: San Antonio eighty years later. *Conservation Biology* **8**: 138–146.

Katz, S.H., Heidiger, M.L. and Valleroy, L.A. 1974. Traditional maize processing techniques in the New World. *Science* **184**: 765–773.

Kavanagh, T. and Kellman, M. 1992. Seasonal pattern of fine root proliferation in a tropical dry forest. *Biotropica* **24**: 157–165.

Kellman, M. 1969. Some environmental components of shifting cultivation in upland Mindanao. *Journal of Tropical Geography* **28**: 40–56.

Kellman, M. 1970. *Secondary plant succession in tropical montane Mindanao.* Publication BG/2. Canberra: Australian National University Press.

Kellman, M. 1973. Dry season weed communities in the upper Belize Valley. *Journal of Applied Ecology* **10**: 683–694.

Kellman, M. 1974a. The viable weed seed content of some tropical agricultural soils. *Journal of Applied Ecology* **11**: 669–678.

Kellman, M. 1974b. Some implications of biotic interactions for sustained tropical agriculture. *Proceedings of the Association of American Geographers* **6**: 142–145.

Kellman, M. 1979. Soil enrichment by Neotropical savanna trees. *Journal of Ecology* **67**: 565–577.

Kellman, M. 1980. Geographical patterning in tropical weed communities and early secondary successions. *Biotropica* **12** (Suppl.): 34–39.

Kellman, M. 1984. Synergistic relationships between fire and low soil fertility in Neotropical savannas: a hypothesis. *Biotropica* **16**: 158–160.

Kellman, M. 1985a. Forest seedling establishment in Neotropical savannas: transplant experiments with *Xylopia frutescens* and *Calophyllum brasiliense*. *Journal of Biogeography* **12**: 373–379.

Kellman, M. 1985b. Nutrient retention by savanna ecosystems III. Response to artificial loading. *Journal of Ecology* **73**: 963–972.

Kellman, M. 1986. Long-term effects of cutting tap roots of *Pinus caribaea* growing on infertile savanna soils. *Plant and Soil* **93**: 137–140.

Kellman, M. 1989. Mineral nutrient dynamics during savanna–forest transformation in Central America. In *Mineral nutrients in tropical forest and savanna ecosystems* (ed. J. Proctor). Oxford: Blackwell, 137–151.

Kellman, M. 1990. Root proliferation in recent and weathered sandy soils from Veracruz, Mexico. *Journal of Tropical Ecology* **6**: 355–370.

Kellman, M. 1996. Redefining roles: plant community reorganization and species preservation in fragmented systems. *Global Ecology and Biogeography Letters* **5**: 111–116.

Kellman, M. and Adams, C.D. 1970. Milpa weeds of the Cayo District, Belize (British Honduras). *Canadian Geographer* **14**: 323–343.

Kellman, M. and Carty, A. 1986. Magnitude of nutrient influxes from atmospheric sources to a Central American *Pinus caribaea* woodland. *Journal of Applied Ecology* **23**: 211–226.

Kellman, M. and Delfosse, B. 1993. Effect of the red land crab (*Gecarcinus lateralis*) on leaf litter in a tropical dry forest in Veracruz, Mexico. *Journal of Tropical Ecology* **9**: 55–65.

Kellman, M. and Meave, J. 1997. Fire in the tropical gallery forests of Belize. *Journal of Biogeography* (in press).

Kellman, M. and Miyanishi, K. 1982. Forest seedling establishment in Neotropical savannas: observations and experiments in the Mountain Pine Ridge savanna, Belize. *Journal of Biogeography* **9**: 193–206.

Kellman, M. and Roulet, N. 1990. Nutrient influx and retention in a tropical sand dune succession. *Journal of Ecology* **78**: 664–676.

Kellman, M. and Sanmugadas, K. 1985. Nutrient retention by savanna ecosystems I. Retention in the absence of fire. *Journal of Ecology* **73**: 935–951.

Kellman, M. and Tackaberry, R. 1993. Disturbance and tree species coexistence in tropical riparian forest fragments. *Global Ecology and Biogeography Letters* **3**: 1–9.

Kellman, M., Hudson, J. and Sanmagudas, K. 1982. Temporal variability in atmospheric nutrient influx to a tropical ecosystem. *Biotropica* **14**: 1–9.

Kellman, M., Miyanishi, K. and Hiebert, P. 1985. Nutrient retention by savanna ecosystems II. Retention after fire. *Journal of Ecology* **73**: 953–962.

Kellman, M., Miyanishi, K. and Hiebert, P. 1987. Nutrient sequestering by the understorey strata of natural *Pinus caribaea* stands subject to prescription burning. *Forest Ecology and Management* **21**: 57–73.

Kellman, M., Tackaberry, R., Brokaw, N. and Meave, J. 1994. Tropical gallery forests. *National Geographic Research and Exploration* **10**: 92–103.

Kellman, M., Tackaberry, R. and Meave, J. 1996. The consequences of prolonged fragmentation: lessons from tropical gallery forests. In *Forest patches in tropical landscapes* (ed. J. Schelhas and R. Greenberg). Washington, DC: Island Press, 37–58.

Kemp, D.C. 1987. Draught animal power: some recent and current work. *World Animal Review* **63**: 7–14.

Kemp, E.M. 1981. Tertiary paleogeography and the evolution of Australian climate. In *Ecological biogeography of Australia* (ed. A. Keast). The Hague: Dr W. Junk, 31–49.

Kenchington, R.A. 1989. Tourism in the Galápagos Islands: the dilemma of conservation. *Environmental Conservation* **16**: 227–232, 236.

Kendall, H.W. and Pimentel, D. 1994. Constraints on the expansion of the global food supply. *Ambio* **23**: 198–205.

Kershaw, A.P. 1978. Record of last glacial–interglacial cycle from north-eastern Queensland. *Nature* **272**: 159–161.

Kikkawa, J. and Dwyer, P.D. 1992. Use of scattered resources in rain forests of humid tropical lowlands. *Biotropica* **24**: 293–308.

Kingsbury, N. and Kellman, M. 1997. Forest root mat depths and surface soil conditions in southeastern Venezuela. *Journal of Tropical Ecology* (in press).

Kinjo, T. and Pratt, P.F. 1971. Nitrate adsorption: I. In some acid soils from

Mexico and South America. *Soil Science Society of America, Proceedings* 35: 722–725.

Kio, P.R.O. 1979. Management strategies in the natural tropical high forest. *Forest Ecology and Management* 2: 207–220.

Kirch, P.V. 1985. Intensive agriculture in prehistoric Hawaii: the wet and dry. In *Prehistoric intensive agriculture in the tropics* (ed. I.S. Farrington). BAR International Series 232. Oxford: Osney Mead, 435–454.

Klink, C.A., Moreira, A.G. and Solbrig, O.T. 1993. Ecological impact of agricultural development in the Brazilian cerrados. In *The world's savannas* (ed. M.D. Young and O.T. Solbrig). Paris: UNESCO; Carnforth: Parthenon Publishing Group, 259–282.

Knoll, A.H. 1984. Patterns of extinction in the fossil record of vascular plants. In *Extinctions* (ed. M.H. Nitecki). Chicago: University of Chicago Press, 21–68.

Knoop, W.T. and Walker, B.H. 1985. Interactions of woody and herbaceous vegetation in a Southern African savanna. *Journal of Ecology* 73: 235–253.

Kock, R.A. 1995. Wildlife utilization: use it or lose it – a Kenyan perspective. *Biodiversity and Conservation* 4: 241–256.

Kogan, M. 1991. Contemporary adaptations of herbivores to introduced legume crops. In *Plant-animal interactions: evolutionary ecology in tropical and temperate regions* (ed. P.W. Price, T.M. Lewinsohn, G.W. Fernandes and W.W. Benson). New York: John Wiley, 591–605.

Krebs, C. 1994. *Ecology: the experimental analysis of distribution and abundance*, 4th edition. New York: Harper Collins.

Kreuter, U.P. and Workman, J.P. 1994. The comparative economics of cattle and wildlife production in the Zimbabwe Midlands. In *Wildlife ranching: a celebration of diversity*. Proceedings of the 3rd International Wildlife Ranching Symposium, October 1992, Pretoria, South Africa (ed. W. van Hoven, H. Ebedes and A. Conroy). Pretoria: Centre for Wildlife Management, University of Pretoria, 220–228.

Kumar, A., Leetmaa, A. and Ming, Ji 1994. Simulations of atmospheric variability induced by sea surface temperatures and implications for global warming. *Science* 266: 632–634.

Kummerow, J. Castillanos, J., Maas, M. and Larigauderie, A. 1990. Production of fine roots and the seasonality of their growth in a Mexican deciduous forest. *Vegetatio* 90: 73–80.

Lal, R. 1976a. Erosion on alfisols in western Nigeria, I. Effects of slope, crop rotation and residue management. *Geoderma* 16: 363–375.

Lal, R. 1976b. Soil erosion on alfisols in western Nigeria, IV. Nutrient element losses in runoff and eroded sediments. *Geoderma* 16: 403–417.

Lal, R. 1990. *Soil erosion in the tropics: principles and management*. New York: McGraw-Hill.

Lal, R. 1991. Myths and scientific realities of agroforestry as a strategy for sustainable management of soils in the tropics. *Advances in Soil Science* 15: 91–137.

Landauer, K. and Brazil, M. (eds) 1990. *Tropical home gardens*. Tokyo: United Nations University Press.

Lascano, C.E. 1991. Managing the grazing resource for animal production in savannas of tropical America. *Tropical Grasslands* 25: 66–72.

Latham, M.C. 1979. *Human nutrition in tropical Africa*, 2nd edition. Rome: FAO.

Latham, R.E. and Ricklefs, R.E. 1993. Global patterns of tree species richness in moist forests: energy–diversity theory does not account for variation in species richness. *Oikos* 67: 325–333.

Lathwell, D.J. 1990. *Legume green manures: principles for management based on recent research*. Topsoils Bulletin 90–01. Raleigh, NC: Soil Management Collaborative Research Support Program, North Carolina State University.

Laws, R.M. 1970. Elephants as agents of habitat and landscape change in East Africa. *Oikos* **21**: 1–15.

Lawton, J.H., Lewinsohn, T.M. and Compton, S.G. 1993. Patterns of diversity for insect herbivores on bracken. In *Species diversity in ecological communities: historical and geographical perspectives* (ed. R.E. Ricklefs and D. Schluter). Chicago: University of Chicago Press, 178–184.

Lawton, R.O. 1982. Wind stress and elfin stature in a montane rain forest tree: an adaptive explanation. *American Journal of Botany* **69**: 1224–1230.

Le Guenno, B. 1995. Emerging viruses. *Scientific American* **273**: 56–64.

Le Roux, X., Bariac, T. and Mariotti, A. 1995. Spatial partitioning of the soil water resource between grass and shrub components in a West African humid savanna. *Oecologia* **104**: 147–155.

Ledru, M.-P. 1993. Late Quaternary environmental and climate changes in central Brazil. *Quaternary Research* **39**: 90–98.

LeDuc, J.W. and Pinheiro, F.P. 1988. Oropouche fever. In *The arboviruses: epidemiology and ecology*, vol. 4 (ed. T.P. Monath). Boca Raton, FL: CRC Press, 1–14.

Leeuw, P.N. de and Rey, B. 1995. Analysis of current trends in the distribution patterns of ruminant livestock in tropical Africa. *World Animal Review* **83**: 47–59.

Leighton, M. and Leighton, D.R. 1983. Vertebrate responses to fruiting seasonality within a Bornean rain forest. In *Tropical rain forest: ecology and management* (ed. S.L. Sutton, T.C. Whitmore and A.C. Chadwick). Oxford: Blackwell, 181–209.

Leighton, M. and Wirawan, N. 1986. Catastrophic drought and fire in Borneo rain forests associated with the 1982–83 El Niño Southern Oscillation event. In *Tropical rain forests and the world atmosphere* (ed. G.T. Prance). American Association for the Advancement of Science, Symposium 101. Boulder, CO: Westview Press, 83–107.

Lesack, L.F.W. 1993. Water balance and hydrologic characteristics of a rain forest catchment in the central Amazon basin. *Water Resources Research* **29**: 759–773.

Leslie, A.J. 1987. A second look at the economics of natural management systems in tropical mixed forests. *Unasylva* **155** vol. 39: 46–58.

Leston, D. 1973. The ant-mosaic: tropical tree crops and the limiting of pests and diseases. *Pest Articles and News Summaries (PANS)* **19**: 311–341.

Levin, D.A. 1976. The chemical defences of plants to pathogens and herbivores. *Annual Review of Ecology and Systematics* **7**: 121–159.

Lewis, W.M., Hamilton, S.K., Jones, S.L. and Runnels, D.D. 1987. Major element chemistry, weathering and element yields for the Caura River drainage, Venezuela. *Biogeochemistry* **4**: 159–181.

Leyden, N.W. 1984. Guatemalan forest synthesis after Pleistocene aridity. *Proceedings of the National Academy of the USA* **81**: 4856–4859.

Lidgard, S. and Crane, P.R. 1988. Quantitative analysis of the early angiosperm radiation. *Nature* **331**: 344–346.

Lieberman, D., Lieberman, M., Hartshorn, G.S. and Peralta, R. 1985. Growth rates and age–size relationships of tropical wet forest trees in Costa Rica. *Journal of Tropical Ecology* **1**: 97–109.

Linares, O.F. 1976. 'Garden hunting' in the American tropics. *Human Ecology* **4**: 331–349.

Litsinger, J.A. 1993. A farming systems approach to insect pest management for upland and lowland rice farmers in tropical Asia. In *Crop protection strategies for subsistence farmers* (ed. M.A Altieri). Boulder, CO: Westview Press, 45–101.

Litsinger, J.A. and Moody, K. 1976. Integrated pest management in multiple cropping systems. In *Multiple cropping* (ed. R.I. Papendick, P.A. Sanchez and G.B. Triplett). Madison, WI: American Society of Agronomists, 293–316.

Liu, K. and Colinvaux, P.A. 1985. Forest changes in the Amazon Basin during the last glacial maximum. *Nature* **318**: 556–557.

Livingstone, D.A. 1993. Evolution of African climate. In *Biological relationships between Africa and South America* (ed. P. Goldblatt). New Haven, CT: Yale University Press, 455–472.

Lockeretz, W. 1980. Energy inputs for nitrogen, phosphorus and potash fertilizers. In *Handbook of energy utilization in agriculture* (ed. D. Pimentel). Boca Raton, FL: CRC Press, 23–24.

Loevinsohn, M.E. 1987. Insecticide use and increased mortality in rural central Luzon, Philippines. *Lancet* **336**: 1359–1362.

Longhurst, A.R. and Pauley, D. 1987. *Ecology of tropical oceans.* New York: Academic Press.

Lopes, A.S. and Cox, F.R. 1977a. A survey of the fertility status of surface soils under 'cerrado' vegetation of Brazil. *Soil Science Society of America, Journal* **41**: 742–747.

Lopes, A.S. and Cox, F.R. 1977b. Cerrado vegetation in Brazil: an edaphic gradient. *Agronomy Journal* **69**: 828–831.

Loughnan, F.C. 1969. *Chemical weathering of the silicate minerals.* New York: Elsevier.

Lovell, R.T. 1979. Fish culture in the United States. *Science* **206**: 1368–1372.

MacArthur, R. 1972. *Geographical ecology.* New York: Harper and Row.

MacArthur, R. and Wilson, E.O. 1967. *The theory of island biogeography.* Princeton, NJ: Princeton University Press.

MacDicken, K.G. and Vergara, N.T. (eds) 1990. *Agroforestry: classification and management.* New York: John Wiley.

MacDougall, A. and Kellman, M. 1992. The understorey light regime and patterns of tree seedlings in tropical riparian forest patches. *Journal of Biogeography* **19**: 667–675.

McDowell, L.R. 1985. Nutrient requirements of ruminants. In *Nutrition of grazing ruminants in warm climates* (ed. L.R. McDowell). Orlando, FL: Academic Press, 21–36.

Mace, R. 1991. Overgrazing overstated. *Nature* **349**: 280–281.

McKey, D., Waterman, P.G., Gartlan, J.S. and Struhsaker, T.T. 1978. Phenolic content of vegetation in two African rain forests: ecological implications. *Science* **202**: 61–64.

Macnab, J. 1991. Does game cropping serve conservation? A reexamination of the African data. *Canadian Journal of Zoology* **69**: 2283–2290.

McNaughton, S.J. 1985. Ecology of a grazing ecosystem: the Serengeti. *Ecological Monographs* **55**: 259–294.

McNaughton, S.J. 1988. Mineral nutrition and spatial concentrations of African ungulates. *Nature* **334**: 343–345.

McNaughton, S.J. 1990. Mineral nutrition and seasonal movements of African migratory ungulates. *Nature* **345**: 613–615.

McNaughton, S.J. and Georgiadis, N.J. 1986. Ecology of African grazing and browsing mammals. *Annual Review of Ecology and Systematics* **17**: 39–65.

McNaughton, S.J. and Sabuni, G.A. 1988. Large African mammals as regulators of vegetation structure. In *Plant form and vegetation structure* (ed. M.J.A. Werger, P.J.M. Van der Aart, H.J. During and J.T.A. Verhoeven). The Hague: SPB, Academic Publishing 339–354.

McNaughton, S.J., Sala, O.E. and Oesterheld, M. 1993. Comparative ecology of African and South American arid to subhumid ecosystems. In *Biological relationships between Africa and South America* (ed. P. Goldblatt). New Haven, CT: Yale University Press, 548–567.

McNeill, W.H. 1993. Patterns of disease emergence in history. In *Emerging viruses* (ed. S.S. Morse). Oxford: Oxford University Press, 29–36.

Macpherson, C.N.L. 1995. The effect of transhumance on the epidemiology of animal diseases. *Preventive Veterinary Medicine* 25: 213–224.

Madden, D. and Young, T.P. 1992. Symbiotic ants as an alternative defence against giraffe herbivory in spinescent *Acacia drepanolobium*. *Oecologia* 91: 235–238.

Malek, E.A. 1985. Blood flukes or schistosomes. In *Animal agents and vectors of human disease* (ed. P.C. Beaver and R.C. Jung). Philadelphia: Lea and Febiger, 97–109.

Malingreau, J.P., Stephens, G. and Fellows, L. 1985. Remote sensing of forest fires: Kalimantan and north Borneo in 1982–83. *Ambio* 14: 314–321.

Maloney, B.K. 1980. Pollen analytical evidence for early forest clearance in north Sumatra. *Nature* 287: 324–326.

Mannetje, L.'t. 1982. Problems of animal production from tropical pastures. In *Nutritional limits to animal production from pastures* (ed. J.B. Hacker). Proceedings of an International Symposium held at St Lucia, Queensland, Australia, 24–28 August 1981. Farnham Royal: Commonwealth Agricultural Bureaux, 67–85.

Manshard, W. and Morgan, W.B. (eds) 1988. *Agricultural expansion and pioneer settlement in the humid tropics*. Tokyo: United Nations University Press.

Mares, M.A. 1992. Neotropical mammals and the myth of Amazonian biodiversity. *Science* 255: 976–979.

Mariotti, A. and Peterschmitt, E. 1994. Forest savanna ecotone dynamics in India as revealed by carbon isotope ratios of soil organic matter. *Oecologia* 97: 475–480.

Marks, S.A. 1989. Small-scale hunting economies in the tropics. In *Wildlife production systems: economic utilisation of wild ungulates* (ed. R.J. Hudson, K.R. Drew, K.R. and L.M. Baskin). Cambridge: Cambridge University Press, 75–95.

Marschner, H. 1986. *Mineral nutrition of higher plants*. London: Academic Press.

Marten, G.G. 1986. Traditional agriculture and agricultural research in Southeast Asia. In *Traditional agriculture in Southeast Asia* (ed. G.G. Marten). Boulder, CO: Westview Press, 326–340.

Marten, G.G. 1990. A nutritional calculus for home garden design: case study from west Java. In *Tropical home gardens* (ed. K. Landauer and M. Brazil). Tokyo: United Nations University Press, 147–168.

Martin, P.S. 1973. The discovery of America. *Science* 179: 969–974.

Martinelli, L.A., Pessenda, L.C.R., Espinosa, E., Camargo, P.B., Telles, E.C., Cerri, C.C., Victoria, R.L., Aravena, R., Richey, J. and Trumbore, S. 1996. Carbon-13 variation with depth in soils of Brazil and climate change during the Quaternary. *Oecologia* 106: 376–381.

Matzke, G.E. and Nabane, N. 1996. Outcomes of a community controlled wildlife utilization program in a Zambezi Valley community. *Human Ecology* 24: 65–85.

Maxwell, F.G. and Jennings, P.R. (eds) 1980. *Breeding plants resistance to insects*. New York: John Wiley.

May, R.M. 1993. Ecology and evolution of host–virus associations. In *Emerging viruses* (ed. S.S. Morse). Oxford: Oxford University Press, 58–68.

Meade, J.W. 1989. *Aquaculture management*. New York: Van Nostrand Reinhold.

Meave, J. 1991. Maintenance of tropical rain forest plant diversity in riparian forests of tropical savannas. Ph.D. dissertation, York University, Toronto.

Meave, J. and Kellman, M. 1994. Maintenance of rain forest diversity in riparian forests of tropical savannas: implications for species diversity during Pleistocene drought. *Journal of Biogeography* 21: 121–135.

Medina, E. and Silva, J.F. 1990. Savannas of northern South America: a steady state regulated by water–fire interactions on a background of low nutrient availability. *Journal of Biogeography* 17: 403–413.

Meehl, G.A. 1994. Coupled land–ocean–atmosphere processes and South Asian monsoon variability. *Science* 266: 263–267.

Menault, J.-C. 1977. Evolution of plots protected from fire since 13 years in a

BIBLIOGRAPHY

Guinea savanna of Ivory Coast. *Actas del IV Symposio de Ecología Tropical, Panamá* 2: 541–558.

Menault, J.-C. 1983. The vegetation of African savannas. In *Tropical savannas* (ed. F. Bourlière). Amsterdam: Elsevier, 109–150.

Menault, J.-C., Barbault, R., Lavelle, P. and Lepage, M. 1985. African savannas: biological systems of humification and mineralization. In *Ecology and management of the world's savannas* (ed. J.C. Tothill and J.J. Mott). Canberra: Australian Academy of Science, 14–33.

Menault, J.-C., Lepage, M. and Abbadie, L. 1995. Savannas, woodlands and dry forests in Africa. In *Seasonally dry tropical forests* (ed. S.H. Bullock, H.A. Mooney and E. Medina). Cambridge: Cambridge University Press, 64–92.

Meyer, F.P. 1991. Aquaculture disease and health management. *Journal of Animal Science* 69: 4201–4208.

Miller, C.P. and Stockwell, T.G.H. 1991. Sustaining productive pastures in the tropics. 4. Augmenting native pasture with legumes. *Tropical Grasslands* 25: 98–103.

Miller, J.A. 1989. Diseases for our future: global ecology and emerging viruses. *BioScience* 39: 509–517.

Milton, K. 1984. Protein and carbohydrate resources of the Maku Indians of north-western Amazonia. *American Anthropologist* 86: 7–27.

Minson, D.J. 1980. Nutritional differences between tropical and temperate pastures. In *Grazing animals* (ed. F.H.W. Morley). Amsterdam: Elsevier, 143–157.

Miyanishi, K. and Kellman, M. 1986a. The role of fire in recruitment of two Neotropical savanna shrubs, *Miconia albicans* and *Clidemia sericea*. *Biotropica* 18: 224–230.

Miyanishi, K. and Kellman, M. 1986b. The role of root nutrient reserves in regrowth of two savanna shrubs. *Canadian Journal of Botany* 64: 1244–1248.

Miyanishi, K. and Kellman, M. 1988. Ecological and simulation studies of the responses of *Miconia albicans* and *Clidemia sericea* populations to prescribed burning. *Forest Ecology and Management* 23: 121–137.

Mohr, E.C.J. and Van Baren, F.A. 1954. *Tropical soils.* The Hague: NV Uitgeveriz.

Monath, T.P. 1993. Arthropod-borne viruses. In *Emerging viruses* (ed. S.S. Morse). Oxford: Oxford University Press, 138–148.

Monath, T.P. 1994. Vector-borne emergent disease. Disease in evolution: global changes and emergence of infectious diseases. *Annals of the New York Academy of Sciences* 740: 129–137.

Montagnini, F. and Jordan, C. 1983. The role of insects in the productivity decline of cassava (*Manihot esculenta* Crantz) on a slash and burn site in the Amazon Territory of Venezuela. *Agriculture, Ecosystems and Environment* 9: 293–301.

Moore, J.E. 1972. Palms in the tropical forest ecosystems of Africa and South America. In *Tropical forest ecosystems in Africa and South America: a comparative review* (ed. B.J. Meggers, E.S. Ayensu and W.D. Duckworth). Washington, DC: Smithsonian Institution Press, 63–88.

Moran, E.F. 1985. An assessment of a decade of colonisation in the Amazon basin. In *Change in the Amazon basin* (ed. J. Hemming). Manchester: Manchester University Press, 91–102.

Morley, R.J. 1982. A paleoecological interpretation of a 10,000 year pollen record from Danau Padang, central Sumatra, Indonesia. *Journal of Biogeography* 9: 151–190.

Morse, S.S. 1993. Examining the origins of emerging viruses. In *Emerging viruses* (ed. S.S. Morse). Oxford: Oxford University Press, 10–28.

Morse, S.S. 1995. Controlling infectious diseases. *Technology Review* 98: 54–61.

Morse, S.S. and Schluederberg, A. 1990. Emerging viruses: the evolution of viruses and viral diseases. *Journal of Infectious Diseases* 162: 1–7.

Mt Pleasant, J., McCollum, R.E. and Coble, H.D. 1990. Weed population dynamics and weed control in the Peruvian Amazon. *Agronomy Journal* **82**: 102–112.

Mueller-Dombois, D. 1986. Perspectives for an etiology of stand-level dieback. *Annual Review of Ecology and Systematics* **17**: 221–243.

Mulongoy, K., Gueye, M. and Spencer, D.S. (eds) 1990. *Biological nitrogen fixation and sustainability of tropical agriculture*. New York: John Wiley.

Murphy, P.G. and Lugo, A.E. 1986. Ecology of tropical dry forest. *Annual Review of Ecology and Systematics* **17**: 67–88.

Nair, P.K.R. 1984. *Soil productivity aspects of agroforestry*. Nairobi: International Council for Research in Agroforestry.

Nasciamento, M.T. and Proctor, J. 1994. Insect defoliation of a monodominant Amazonian rainforest. *Journal of Tropical Ecology* **10**: 633–636.

National Research Council 1993. *Sustainable agriculture and environment in the humid tropics*. Washington, DC: National Academy Press.

Naylor, E. and Scholtissek, C. 1988. Reply to 'Fish farming and aquaculture'. *Nature* **333**: 506.

Nelson, B.W., Kapos, V., Adams, J.B., Olivera, W.J., Braun, O.P.G. and Do Amaral, I.L. 1994. Forest disturbance by large blowdowns in the Brazilian Amazon. *Ecology* **75**: 853–858.

Nelson, J.G. 1994. The spread of ecotourism: some planning implications. *Environmental Conservation* **21**: 248–255.

Nepstad, D.C. and Schwartzman, S. (eds) 1992. *Non-timber products from tropical forests: evaluation of a conservation and development strategy. Advances in Economic Botany*, vol. 9. New York: New York Botanical Garden.

Nepstad, D.C., Uhl, C. and Serrao, E.A. 1991. Recuperation of a degraded Amazonian landscape: forest recovery and agricultural restoration. *Ambio* **20**: 248–255.

Nepstad, D.C., Brown, F., Luz, L., Alechandra, R. and Viana, V. 1992. Biotic impoverishment of Amazonian forests by rubber tappers, loggers, and cattle ranchers. *Advances in Economic Botany* **9**: 1–14.

Nepstad, D.C., De Carvalho, C.R., Davidson, E.A., Jipp, P.H., Lefebvre, P.A., Negrerios, G.H., Da Silva, E.D., Stone, T.A., Tumbore, S.E. and Viera, S. 1994. The role of deep roots in the hydrological and carbon cycles of Amazonian forests and pastures. *Nature* **372**: 666–669.

New, T.R. 1994. Butterfly ranching: sustainable use of insects and sustainable benefit to habitats. *Oryx* **28**: 169–172.

Newsome, J. and Flenley, J.R. 1988. Late Quaternary vegetational history of the central highlands of Sumatra. II. Palaeopalynology and vegetational history. *Journal of Biogeography* **15**: 555–578.

Ng, F.S.P. 1978. Strategies of establishment in Malayan forest trees. In *Tropical trees as living systems* (ed. P.B. Tomlinson and M.H. Zimmerman). New York: Cambridge University Press, 129–162.

Ngatunga, E.L.N., Lal, R. and Uriyo, A.P. 1984. Effects of surface management on runoff and soil erosion from some plots at Mlingano, Tanzania. *Geoderma* **33**: 1–12.

Nicholls, N. 1993. El Niño–Southern Oscillation and vector-borne disease. *Lancet* **342**: 1284–1285.

Nichols, D.J., Jarzen, D.M., Orth, C.J. and Oliver, P.Q. 1986. Palynological and iridium anomalies at the Cretaceous–Tertiary boundary, south-central Saskatchewan. *Science* **231**: 714–717.

Niederberger, C. 1979. Early sedentary economy in the Basin of Mexico. *Science* **203**: 131–142.

Nieuwolt, S. 1977. *Tropical climatology: an introduction to the climates of low latitudes.* New York: John Wiley.

Norman, M.J.T., Pearson, C.J. and Searle, P.G.E. 1995. *The ecology of tropical food crops,* 2nd edition. Cambridge: Cambridge University Press.

Nye, P.H. and Greenland, D.J. 1960. *The soil under shifting agriculture.* Technical Communication 51, Commonwealth Bureau of Soils. Farnham Royal: Commonwealth Agricultural Bureaux.

O'Callaghan, J.R. and Wyseure, G.C.L. 1994. Proposals for quantitative criteria in the management of sustainable agricultural systems. In *Soil science and sustainable land management in the tropics* (ed. J.K. Syers and D.L. Rimmer). Wallingford: CAB International, 27–39.

Oesterheld, M., Sala, O.E. and McNaughton, S.J. 1992. Effect of animal husbandry on herbivore-carrying capacity at a regional scale. *Nature* **356**: 234–236.

Omar, M.A. 1992. Health care for nomads too, please. *World Health Forum* **13**: 307–310.

Owen-Smith, N. 1982. Factors influencing the consumption of plant products by large herbivores. In *Ecology of tropical savannas* (ed. B.J. Huntley and B.H. Walker). New York: Springer-Verlag, 359–404.

Palmen, E. 1948. On the formation and structure of tropical hurricanes. *Geophysika* (Helsinki) **3**: 26–38.

Panicker, K.N. and Dhanda, V. 1992. Community participation in the control of filariasis. *World Health Forum* **13**: 177–181.

Parrotta, J.A. 1993. Secondary forest regeneration on degraded tropical lands: the role of plantations as 'foster ecosystems'. In *Restoration of tropical forest ecosystems* (ed. H. Leith and M. Lohmann). Dordrecht: Kluwer, 63–73.

Parsons, J.J. 1955. The Miskito pine savannas of Nicaragua and Honduras. *Annals of the Association of American Geographers* **45**: 36–63.

Parsons, J.J. 1972. Spread of African pasture grasses to the American tropics. *Journal of Range Management* **25**: 12–17.

Pathak, M.D. and Saxena, R.C. 1980. Breeding approaches in rice. In *Breeding plants resistance to insects* (ed. F.G. Maxwell and P.R. Jennings). New York: John Wiley, 421–455.

Pemadasa, M.A. 1990. Tropical grasslands of Sri Lanka and India. *Journal of Biogeography* **17**: 395–400.

Peoples, M.B. and Craswell, E.T. 1992. Biological nitrogen fixation: investments, expectations and actual contributions to agriculture. *Plant and Soil* **141**: 13–39.

Peres, C.A. 1990. Effects of hunting on western Amazonian primate communities. *Biological Conservation* **54**: 47–59.

Perfecto, I. 1990. Indirect and direct effects in a tropical agroecosystem: the maize–pest–ant system in Nicaragua. *Ecology* **71**: 2125–2134.

Peters, C.M. 1992. The ecology and economics of oligarchic forests. *Advances in Economic Botany* **9**: 15–22.

Phillips, O.L. and Gentry, A.H. 1994. Increasing turnover through time in tropical forests. *Science* **263**: 954–958.

Philpot, C.S. 1970. Influence of mineral content on the pyrolysis of plant materials. *Forest Science* **16**: 461–471.

Pianka, E.R. 1966. Latitudinal gradients in species diversity: a review of concepts. *American Naturalist* **100**: 33–46.

Pimentel, D. (ed.) 1980. *Handbook of energy utilization in agriculture.* Boca Raton, FL: CRC Press.

Pimentel, D. 1993. *World soil erosion and conservation,* Cambridge: Cambridge University Press.

Pimentel, D., Da Zhong, W. and Giampietro, M. 1990. Technological changes in energy use in US agricultural production. In *Agroecology* (ed. S.R. Gliessman). New York: Springer-Verlag, 305–321.

Pimentel, D., Harvey, C., Resosudarmo, P., Sinclair, K., Kurz, D., McNair, M., Crist, S., Shapritz, L., Fitton, L., Saffouri, R. and Blair, R. 1995. Environmental and economic costs of soil erosion and conservation benefits. *Science* 267: 1117–1123.

Pimm, S.C. 1984. The complexity and stability of ecosystems. *Nature* 307: 321–326.

Piperno, D.R. 1989. Non-affluent foragers: resource availability, seasonal shortages, and the emergence of agriculture in Panamian tropical forests. In *Foraging and farming: the evolution of plant exploitation* (ed. D.R. Harris and G.C. Hillman). London: Unwin Hyman, 538–554.

Plumptre, A.J. and Reynolds, V. 1994. The effect of selective logging on the primate populations in the Budongo Forest Reserve, Uganda. *Journal of Applied Ecology* 31: 631–641.

Ponce, V.M. and Da Cunha, C.N. 1993. Vegetated earthmounds in tropical savannas of central Brazil: a synthesis. *Journal of Biogeography* 20: 219–225.

Popper, K.R. 1965. *Conjectures and refutations: the growth of scientific knowledge.* New York: Harper and Row.

Power, A.G. 1990. Cropping systems, insect movement, and the spread of insect-transmitted diseases in crops. In *Agroecology* (ed. S.R. Gliessman). New York: Springer-Verlag, 47–69.

Prance, G.T. (ed.) 1982. *Biological diversification in the tropics.* New York: Columbia University Press.

Price, P.W. 1991. Patterns in communities along latitudinal gradients. In *Plant–animal interactions: evolutionary ecology in tropical and temperate regions* (ed. P.W. Price, G.W. Fernandes, T.M. Lewinsohn and W.W. Benson). New York: John Wiley, 51–69.

Putz, F.E. 1983. Treefall pits and mounds, buried seeds, and the importance of soil disturbance to pioneer trees on Barro Colorado Island, Panama. *Ecology* 64: 1069–1074.

Putz, F.E. 1991. Silvicultural effects of lianas. In *The biology of vines* (ed. F.E. Putz and H.A. Mooney). Cambridge: Cambridge University Press, 493–501.

Qasim, S.Z. and Wafar, M.V.M. 1990. Marine resources in the tropics. *Resource Management and Optimization* 7: 141–169.

Radulovich, R., Sollins, P., Baveye, P. and Solorzano, E. 1992. Bypass water flow through unsaturated microaggregated tropical soils. *Soil Science Society of America, Journal.* 56: 721–726.

Rai, H. and Hill, G. 1984. Primary production in the Amazon aquatic system. In *The Amazon: limnology and landscape ecology of a mighty tropical river and its basin* (ed. H. Sioli). Dordrecht: Dr W. Junk, 311–336.

Rankin-de-Merona, J.M., Hutchings, R.W. and Lovejoy, T.E. 1990. Tree mortality and recruitment over a five-year period in undisturbed upland rainforest of the central Amazon. In *Four Neotropical rainforests* (ed. A.H. Gentry). New Haven, CT: Yale University Press, 573–584.

Rasmussen, E.M. and Wallace, J.M. 1983. Meteorological aspects of the El Niño/Southern Oscillation. *Science* 222: 1195–1202.

Redford, K.H. and da Fonseca, G.A.B. 1986. The role of gallery forests in the zoogeography of the cerrado non-volant mammal fauna. *Biotropica* 18: 126–135.

Redford, K.H. and Robinson, J.G. 1991. Subsistence and commercial uses of wildlife in Latin America. In *Neotropical wildlife use and conservation* (ed. J.G. Robinson and K.H. Redford). Chicago: University of Chicago Press, 6–23.

Renecker, L.A. and Hudson, R.J. (eds) 1991. *Wildlife production: conservation and sustainable development.* Agricultural and Forestry Experiment Station

BIBLIOGRAPHY

Miscellaneous Publication 91–6. Fairbanks, AL: University of Alaska, Fairbanks.

Richards, P.W. 1952. *The tropical rain forest: an ecological study.* Cambridge: Cambridge University Press.

Richter, D.D. and Babbar, L.T. 1991. Soil diversity in the tropics. *Advances in Ecological Research* 21: 315–389.

Ricklefs, R.E. 1987. Community diversity: relative roles of local and regional processes. *Science* 235: 161–171.

Ricklefs, R.E. and Schluter, D. (eds) 1993. *Species diversity in ecological communities: historical and geographical perspectives.* Chicago: University of Chicago Press.

Riehl, H. 1979. *Climate and weather in the tropics.* London: Academic Press.

Risch, S.J., Andow, D. and Altieri, M.A. 1983. Agroecosystem diversity and pest control: data, tentative conclusions and new research directions. *Environmental Entomology* 12: 625–629.

Roa Morales, P. 1979. Estudio de los medanos de los Llanos Centrales de Venezuela: evidencias de un clima desertico. *Acta Biología Venezolano* 10: 19–49.

Roberts, H.A. 1970. Viable weed seeds in cultivated soils. *Report of the National Vegetable Research Station* 9: 81–109.

Robinson, J.G. 1996. Hunting wildlife in forest patches: an ephemeral resource. In *Forest patches in tropical landscapes* (ed. J. Schelhas and R. Greenburg). Washington, DC: Island Press, 111–130.

Robinson, J.G. and Redford, K.H. (eds) 1991a. *Neotropical wildlife use and conservation.* Chicago: University of Chicago Press.

Robinson, J.G. and Redford, K.H. 1991b. Sustainable harvest of Neotropical forest mammals. In *Neotropical wildlife use and conservation* (ed. J.G. Robinson and K.H. Redford). Chicago: University of Chicago Press, 415–429.

Robinson, J.G. and Redford, K.H. 1994. Measuring the sustainability of hunting in tropical forests. *Oryx* 28: 249–256.

Roger, P.A. and Ladha, J.K. 1992. Biological N_2 fixation in wetland rice fields: estimation and contribution to nitrogen balance. *Plant and Soil* 141: 41–55.

Rogers, D.J. and Randolph, S.E. 1988. Tsetse flies in Africa: bane or boon? *Conservation Biology* 2: 57–65.

Rohde, K. 1992. Latitudinal gradients in species diversity: the search for the primary cause. *Oikos* 65: 514–527.

Root, R.B. 1973. Organization of plant–arthropod association in simple and diverse habitats: the fauna of collards (*Brassica oleracea*). *Ecological Monographs* 43: 95–124.

Rosenzweig, M.L. 1995. *Species diversity in space and time.* Cambridge: Cambridge University Press.

Rull, V. 1992. Successional patterns of the Gran Sabana (southeastern Venezuela) vegetation during the last 5000 years, and its responses to climatic fluctuations and fire. *Journal of Biogeography* 19: 329–338.

Rundel, P.W., Smith, A.P. and Meinzer, F.C. (eds) 1994. *Tropical alpine environments: plant form and function.* Cambridge: Cambridge University Press.

Russell-Smith, J. 1991. Classification, species richness and environmental relations of monsoon rain forest in northern Australia. *Journal of Vegetation Science* 2: 259–278.

Sagers, C.L. and Coley, P.D. 1995. Benefits and costs of defence in Neotropical shrubs. *Ecology* 76: 1835–1843.

St John, T.V. 1983. Response of tree roots to decomposing organic matter in two lowland Amazonian rain forests. *Canadian Journal of Forest Research* 13: 346–349.

St John, T.V. and Anderson, A.B. 1982. A re-examination of plant phenolics as a source of tropical black water rivers. *Tropical Ecology* 23: 151–154.

Saito, T., Yamanoi, T. and Kaiho, K. 1986. End-Cretaceous devastation of terrestrial flora in the boreal Far East. *Nature* 323: 253–255.

Salafsky, N., Dugelby, B.L. and Terborgh, J.W. 1993. Can extractive reserves save the rain forest? An ecological and socioeconomic comparison of nontimber forest product extraction systems in Peten, Guatemala and west Kalimantan, Indonesia. *Conservation Biology* 7: 39–52.

Salati, E. and Vose, P.B. 1984. Amazon basin: a system in equilibrium. *Science* 225: 129–138.

Salick, J. 1989. Ecological basis of Amuesha agriculture, Peruvian upper Amazon. *Advances in Economic Botany* 7: 189–212.

Salo, J., Kalliola, R., Hakkinen, I., Makinen, Y., Niemela, P., Puhakka, M. and Coley, P.D. 1986. River dynamics and the diversity of Amazon lowland forest. *Nature* 322: 254–258.

San José, J.J. and Fariñas, M.R. 1983. Changes in tree density and composition in a protected *Trachypogon* savanna, Venezuela. *Ecology* 64: 447–453.

San José, J.J., Montes, R.A. and Florentino, A. 1995. Water flux through a semi-deciduous forest grove in the Orinoco savannas. *Oecologia* 101: 141–150.

Sánchez, M. 1995. Integration of livestock with perennial crops. *World Animal Review* 82: 50–57.

Sanchez, P.A. 1976. *Properties and management of soils in the tropics.* New York: John Wiley.

Sanchez, P.A. 1995. Science in agroforestry. *Agroforestry Systems* 30: 5–55.

Sanchez, P.A. and Benites, J.R. 1987. Low-input cropping of acid soils of the humid tropics. *Science* 238: 1521–1527.

Sanchez, P.A. and Buol, S.W. 1975. Soils of the tropics and the world food crisis. *Science* 188: 598–603.

Sanchez, P.A., Bandy, D.E., Villachia, J.H. and Nicholades, J.J. 1982. Amazon basin soils: management for continuous crop production. *Science* 216: 821–827.

Sanders, J.O. 1980. History and development of Zebu cattle in the United States. *Journal of Animal Science* 50: 1188–1200.

Sanders, W.T., Parsons, J.R. and Santley, R.S. 1979. *The Basin of Mexico. Ecological processes in the evolution of a civilization.* New York: Academic Press.

Sanford, R.L. 1987. Apogeotropic roots in an Amazon rain forest. *Science* 235: 1062–1064.

Santana, M.B.M. and Cabala-Rosand, P. 1982. Dynamics of nitrogen in a shaded cacao plantation. *Plant and Soil* 67: 271–281.

Sarmiento, G. 1983. The savannas of tropical America. In *Tropical savannas* (ed. F. Bourlière). Amsterdam: Elsevier, 245–288.

Sarmiento, G. 1984. *The ecology of tropical savannas* (trans. by O. Solbrig). Cambridge, MA: Harvard University Press.

Sarmiento, G. and Monasterio, M. 1975. A critical consideration of the environmental conditions associated with the occurrence of savanna ecosystems in tropical America. In *Tropical ecological systems* (ed. F.B. Golley and E. Medina). New York: Springer-Verlag, 223–250.

Saulei, S. and Lamb, D. 1991. Regeneration following pulpwood logging in lowland rain forest in Papua New Guinea. In *Rain forest regeneration and management* (ed. A. Gomez-Pompa, T.C. Whitmore, and M. Hadley). Man and the Biosphere Series vol. 6. Paris: UNESCO; Park Ridge, NJ: Parthenon Publishing Group, 313–322.

Schelhas, J. and Greenberg, R. (eds) 1996. *Forest patches in tropical landscapes.* Washington, DC: Island Press.

Scholes, R.J. 1990. The influence of soil fertility on the ecology of Southern African dry savannas. *Journal of Biogeography* 17: 415–419.

Scholtissek, C. and Naylor, E. 1988. Fish farming and influenza pandemics. *Nature* 331: 215.

Scholz, U. 1988. Types of spontaneous pioneer settlement in Thailand. In *Agricultural expansion and pioneer settlement in the humid tropics* (ed. W. Manshard and W.B. Morgan). Tokyo: United Nations University Press, 44–61.

Schupp, E.W. 1986. *Azteca* protection of *Cecropia*: ant occupation benefits juvenile trees. *Oecologia* 70: 379–385.

Schwartz, D., Mariotti, A., Lafranchi, R. and Guillet, B. 1986. $^{13}C/^{12}C$ ratios of soil organic matter as indicators of vegetation changes in the Congo. *Geoderma* 39: 97–103.

Scott, G.A.J. 1978. *Grassland development in the Gran Pajonal of eastern Peru*. Hawaii Monographs in Geography 1. Imprint Series, Monograph Publishing. Ann Arbor, MI: University Microfilms International.

Seavoy, R.E. 1973. The transition to continuous rice cultivation in Kalimantan. *Annals of the Association of American Geographers* 63: 218–225.

Seavoy, R.E. 1975. The origin of tropical grasslands in Kalimantan, Indonesia. *Journal of Tropical Geography* 40: 48–52.

Serrão, E.A. and Toledo, J.M. 1990. The search for sustainability in Amazonian pastures. In *Alternatives to deforestation: steps toward sustainable use of the Amazon rain forest* (ed. A.B. Anderson). New York: Columbia University Press, 195–214.

Sharkey, M.J. 1970. The carrying capacity of natural and improved land in different climatic zones. *Mammalia* 34: 564–572.

Shukla, J., Nobre, C. and Sellers, P. 1990. Amazon deforestation and climate change. *Science* 247: 1322–1325.

Shuttleworth, W.J. 1988. Evaporation from the Amazon rainforest. *Proceedings of the Royal Society of London, Series B* 223: 321–346.

Silberbauer-Gottsberger, I., Morawetz, W. and Gottsberger, W. 1977. Frost damage of cerrado plants in Botucatu, Brazil, as related to the geographical distribution of the species. *Biotropica* 9: 253–261.

Simberloff, D., Farr, J.A., Cox, J. and Mehlman, D.W. 1992. Management corridors: conservation bargains or poor investments? *Conservation Biology* 6: 493–504.

Simpson, B. and Haffer, J. 1978. Speciation patterns in the Amazon forest biota. *Annual Review of Ecology and Systematics* 9: 497–518.

Simpson, R.H. and Riehl, H. 1981. *The hurricane and its impact*. Baton Rouge, LA: Louisiana State University Press.

Sinclair, A.R.E. and Fryxell, J.M. 1985. The Sahel of Africa: ecology of a disaster. *Canadian Journal of Zoology* 63: 987–994.

Singh, B.R. and Kanehiro, Y. 1969. Adsorption of nitrate in amorphous and kaolinitic Hawaiian soils. *Soil Science Society of America, Proceedings* 33: 681–683.

Sioli, H. 1973. Recent human activities in the Brazilian Amazon region and their ecological effects. In *Tropical forest ecosystems in Africa and South America: a comparative review* (ed. B.J. Meggers, E.S. Ayensu and W.D. Duckworth). Washington, DC: Smithsonian Institution Press, 321–334.

Skinner, J.D. 1989. Game ranching in Southern Africa. In *Wildlife production systems: economic utilisation of wild ungulates* (ed. R.J. Hudson, K.R. Drew and L.M. Baskin). Cambridge: Cambridge University Press, 286–306.

Smiet, A.C. 1992. Forest ecology on Java: human impact and vegetation of montane forest. *Journal of Tropical Ecology* 8: 129–152.

Smith, A.G., Hurley, A.M. and Briden, J.C. 1981. *Phanerozoic paleocontinental world maps*. Cambridge: Cambridge University Press.

Smith, C.E. 1985. Agricultural intensification in the Mexican highlands. In *Prehistoric intensive agriculture in the tropics* (ed. I.S. Farrington). BAR International Series 232. Oxford: Osney Mead, 501–519.

Sobrado, M.A. and Medina, E. 1980. General morphology, anatomical structure, and nutrient content of sclerophyllous leaves of the 'bana' vegetation of Amazonas. *Oecologia* 45: 341–345.

Soil Survey Staff 1960. *Soil classification, a comprehensive system: 7th approximation.* Washington, DC: Soil Conservation Service, United States Department of Agriculture.

Soil Survey Staff 1975. *Soil taxonomy.* Washington, DC: Soil Conservation Service, United States Department of Agriculture.

Sollins, P. and Radulovich, R. 1988. Effects of soil physical structure in solute transport in a weathered tropical soil. *Soil Science Society of America, Journal* 52: 1168–1173.

Sollins, P., Robertson, G.P. and Uehara, G. 1988. Nutrient mobility in variable- and permanent-charge soils. *Biogeochemistry* 6: 181–199.

Sommerlatte, M. and Hopcraft, D. 1992. The economics of game cropping on a Kenyan ranch. Paper presented to the 3rd International Wildlife Ranching Symposium, October 1992, Pretoria, South Africa.

Southgate, D.A.T. 1993. Meat, fish, eggs and novel proteins. In *Human nutrition and dietetics* (ed. J.S. Garrow and W.P.T. James). Edinburgh: Churchill Livingstone, 305–316.

Spain, A.V. and McIvor, J.G. 1988. The nature of herbaceous vegetation associated with termitaria in north-eastern Australia. *Journal of Ecology* 76: 181–191.

Stahl, A.B. 1989. Plant-food processing: implications for dietary quality. In *Foraging and farming: the evolution of plant exploitation* (ed. D.R. Harris and G.C. Hillman). London: Unwin Hyman, 171–194.

Stark, N.M. and Jordan, C.F. 1978. Nutrient retention by the root mat of an Amazonian rain forest. *Ecology* 59: 434–437.

Stark, N.M. and Spratt, M. 1977. Root biomass and nutrient storage in rain forest oxisols near San Carlos de Rio Negro. *Tropical Ecology* 18: 1–9.

Stehli, F.G. 1968. Taxonomic diversity gradients in pole location: the recent model. In *Evolution and environment* (ed. E.T. Drake). New Haven, CT: Yale University Press, 163–227.

Stehli, F.G. and Wells, J.W. 1971. Diversity and age patterns in hermatypic corals. *Systematic Zoology* 20: 115–126.

Stehli, F.G., Douglas, R.G. and Newell, N.D. 1969. Generation and maintenance of gradients of taxonomic diversity. *Science* 164: 947–949.

Stelfox, J.B. 1986. Effects of livestock enclosures (bomas) on the vegetation of the Athi Plains, Kenya. *African Journal of Ecology* 24: 41–45.

Stelfox, J.B. and Hudson, R.J. 1986. Body condition of male Thomson's and Grant's gazelles in relation to season and resource use. *African Journal of Ecology* 24: 111–120.

Stewart, H. and Kellman, M. 1982. Nutrient accumulation by *Pinus caribaea* in its native savanna habitat. *Plant and Soil* 69: 105–118.

Stickney, R.R. 1994. *Principles of aquaculture.* New York: John Wiley.

Stott, P. 1990. Stability and stress in the savanna forests of mainland South-East Asia. *Journal of Biogeography* 17: 373–383.

Street, F.A. 1981. Tropical paleoenvironments. *Progress in Physical Geography* 5: 157–185.

Strong, D.R. 1974. Rapid asymptotic species accumulation in phytophagous insect communities: the pests of cacao. *Science* 185: 1064–1066.

Strong, D.R. 1992. Are trophic cascades all wet? Differentiation and donor-control in speciose ecosystems. *Ecology* 73: 747–754.

Strong, D.R., McCoy, E.D. and Rey, J.R. 1977. Time and the number of herbivore species: the pests of sugarcane. *Ecology* 58: 167–175.

Struhsaker, T.T., Lwanga, J.S. and Kasenene, J.M. 1996. Elephants, selective logging and forest regeneration in the Kibale Forest, Uganda. *Journal of Tropical Ecology* 12: 45–64.

Stute, M., Forster, M., Frischkorn, H., Serejo, A., Clark, J.F., Schlosser, P., Broecker, W.S. and Bonnani, G. 1995. Cooling of tropical Brazil (5°C) during the last Glacial Maximum. *Science* 269: 379–383.

Summerfelt, R.C. 1982. Practice and prospects of fish farming for food production. In *Animal products in human nutrition* (ed. D.C. Beitz and R.G. Hansen). New York: Academic Press, 81–120.

Sun, D. and Dickinson, G.R. 1996. The competition effect of *Brachiaria decumbens* on early growth of direct-seeded trees of *Alphitonia petriei* in tropical north Australia. *Biotropica* 28: 272–276.

Swaine, M.D. and Whitmore, T.C. 1988. On the definition of ecological species groups in tropical rain forests. *Vegetatio* 75: 81–86.

Swaine, M.D., Lieberman, D. and Putz, F.E. 1987. The dynamics of tree populations in tropical forests: a review. *Journal of Tropical Ecology* 3: 359–366.

Swaine, M.D., Hawthorne, W.D. and Orgle, T.K. 1992. The effect of fire exclusion on savanna vegetation in Kpong, Ghana. *Biotropica* 24: 166–172.

Swarbrick, J.T. 1987. *Weed science and weed control in Southeast Asia.* Rome: FAO.

Swift, M.J., Heal, O.W. and Anderson, J.M. 1979. *Decomposition in terrestrial ecosystems.* Berkeley, CA: University of California Press.

Tackaberry, R. and Kellman, M. 1996. Patterns of tree species richness along peninsular extensions of tropical forests. *Global Ecology and Biogeography Letters* 5: 85–90.

Tallis, J.H. 1990. *Plant community history: long-term changes in plant distribution and diversity.* London: Chapman and Hall.

Tanner, E.V.J., Kapos, V. and Healey, J.R. 1991. Hurricane effects on forest ecosystems. *Biotropica* 23: 513–521.

Taylor, R.D. and Walker, B.H. 1978. Comparisons of vegetation use and herbivore biomass on a Rhodesian game and cattle ranch. *Journal of Applied Ecology* 15: 565–581.

Teitzel, J.K. 1992. Sustainable pasture systems in the humid tropics of Queensland. *Tropical Grasslands* 26: 196–205.

Teramura, A.H., Gold, W.G. and Forseth, I.N. 1991. Physiological ecology of mesic, temperate woody vines. In *The biology of vines* (ed. F.E. Putz and H.A. Mooney). Cambridge: Cambridge University Press, 245–285.

Terborgh, J. 1973. On the notion of favourableness in plant ecology. *American Naturalist* 107: 481–501.

Terborgh, J. 1992. Maintenance of diversity in tropical forests. *Biotropica* 24: 283–292.

Tesh, R.B. 1994. The emerging epidemiology of Venezuelan hemorrhagic fever and Oropouche fever in tropical South America. Disease in evolution: global changes and emergence of infectious diseases. *Annals of the New York Academy of Sciences* 740: 129–137.

Thomas, D.H.L. 1994. Socio-economic and cultural factors in aquaculture development: a case study from Nigeria. *Aquaculture* 119: 329–343.

Thomas, R.J. 1995. Role of legumes in providing N for sustainable tropical pasture systems. *Plant and Soil* 174: 103–118.

Thompson, C.H. 1981. Podzol chronosequences on coastal dunes of eastern Australia. *Nature* 291: 59–61.

Tiessen, H., Mermut, A.R. and Nyameke, A.L. 1991. Phosphorus sorption and properties of ferruginous nodules from semiarid soils from Ghana and Brazil. *Geoderma* 48: 373–390.

Tilman, D., Wedin, D. and Knops, J. 1996. Productivity and sustainability influenced by biodiversity in grassland ecosystems. *Nature* 379: 718–720.

Toledo, V.M. 1982. Pleistocene changes of vegetation in tropical Mexico. In *Biological diversification in the tropics* (ed. G.T. Prance). New York: Columbia University Press, 93–111.

Tolsma, D.J., Ernst, W.H.O. and Verwey, R.A. 1987. Nutrients in soil and vegetation around two artificial waterpoints in eastern Botswana. *Journal of Applied Ecology* 24: 991–1000.

Trenbath, B.R. 1989. The use of mathematical models in the development of shifting cultivation systems. In *Mineral nutrients in tropical forest and savanna ecosystems* (ed. J. Proctor). Oxford: Blackwell, 353–369.

Trewartha, G.T. 1981. *The earth's problem climates*, 2nd edition. Madison: University of Wisconsin Press.

Triomphe, B.L. 1996. Seasonal nitrogen dynamics and long-term changes in soil properties under the *Mucuna*/maize cropping system on the hillsides of northern Honduras. Ph.D. thesis, Ithaca, NY: Cornell University Press.

Tschudy, R.H., Pillmore, C.L., Orth, C.J., Gilmore, J.S. and Knight, J.D. 1984. Disruption of the terrestrial plant ecosystem at the Cretaceous–Tertiary boundary, western interior. *Science* 225: 1030–1032.

Turner, B.L. and Denevan, W.M. 1985. Prehistoric manipulation of wetlands in the Americas: a raised field perspective. In *Prehistoric intensive agriculture in the tropics* (ed. I.S. Farrington). BAR International Series 232. Oxford: Osney Mead, 11–30.

Uehara, G. and Gillman, G. 1981. *The mineralogy, chemistry and physics of tropical soils with variable charge clays.* Boulder, CO: Westview Press.

Uhl, C. 1987. Factors controlling succession following slash-and-burn agriculture in Amazonia. *Journal of Ecology* 75: 377–407.

Uhl, C. and Buschbacher, R. 1985. A disturbing synergism between cattle ranch burning practices and selective tree harvesting in the eastern Amazon. *Biotropica* 17: 265–268.

Uhl, C. and Jordan, C.F. 1984. Succession and nutrient dynamics following forest cutting and burning in Amazonia. *Ecology* 65: 1476–1490.

Uhl, C. and Kauffman, J.B. 1990. Deforestation, fire susceptibility, and potential tree responses to fire in the eastern Amazon. *Ecology* 71: 437–449.

Uhl, C. and Murphy, P. 1981. A comparison of productivities and energy values between slash and burn agriculture and secondary succession in the upper Rio Negro region of the Amazon basin. *Agro-Ecosystems* 5: 63–83.

Uhl, C. and Viera, I.C.G. 1989. Ecological impacts of selective logging in the Brazilian Amazon: a case study from the Paragominas region of the State of Pará. *Biotropica* 21: 98–106.

Uhl, C., Clark, K., Clark, H. and Murphy, P. 1981. Early plant succession after cutting and burning in the upper Rio Negro region of the Amazon basin. *Journal of Ecology* 69: 631–649.

Uhl, C., Clark, H. and Clark, K. 1982. Successional patterns associated with slash-and-burn agriculture in the upper Rio Negro region of the Amazon basin. *Biotropica* 14: 249–254.

Uhl, C., Buschbacher, R. and Serrão, E.A.S. 1988a. Abandoned pastures in eastern Amazonia I. Patterns of plant succession. *Journal of Ecology* 76: 663–681.

Uhl, C., Clark, K., Dezzeo, N. and Maquirino, P. 1988b. Vegetation dynamics in Amazonian treefall gaps. *Ecology* 69: 751–763.

Uhl, C., Kauffman, J.B. and Cummings, D.L. 1988c. Fire in the Venezuelan Amazon 2: Environmental conditions necessary for forest fires in the evergreen rainforest of Venezuela. *Oikos* 53: 176–184.

Uhlig, H. 1988. Spontaneous and planned settlement in South-East Asia. In *Agricultural expansion and pioneer settlement in the humid tropics* (ed. W. Manshard and W.B. Morgan). Tokyo: United Nations University Press, 7–43.

UNEP 1991. *United Nations Environment Programme Environmental Data Report 1991–92*. Oxford: Blackwell.

UNEP 1993. *United Nations Environment Programme Environmental Data Report 1993–94*. Oxford: Blackwell.

UNESCO 1978. *Tropical forest ecosystems: a state-of-knowledge report*. Natural Resources Research XIV, Paris: UNESCO.

Upchurch, G.R. and Wolfe, J.A. 1987. Mid-Cretaceous to Early Tertiary vegetation and climate: evidence from fossil leaves and woods. In *The origins of angiosperms and their biological consequences* (ed. M.E. Friis, W.G. Chalnor and P.R. Crane). Cambridge: Cambridge University Press, 75–105.

Van Wagner, C.E. 1988. Effect of slope on fires spreading downhill. *Canadian Journal of Forest Research* 18: 818–820.

Van Wilgen, B.W., Everson, C.S. and Trollope, W.S.W. 1990. Fire management in Southern Africa: some examples of current objectives, practices and problems. In *Fire in the tropical biota* (ed. J.G. Goldhammer). New York: Springer-Verlag, 179–215.

Vandermeer, J., Mallona, M.A., Boucher, D., Yih, K. and Perfectos, I. 1995. Three years of ingrowth following catastrophic hurricane damage on the Caribbean coast of Nicaragua: evidence in support of the direct regeneration hypothesis. *Journal of Tropical Ecology* 11: 465–471.

Vasey, D.E. 1985. Nitrogen fixation and flow in experimental bed gardens: implications for archaeology. In *Prehistoric intensive agriculture in the tropics* (ed. I.S. Farrington). BAR International Series 232. Oxford: Osney Mead, 233–246.

Vaz, J.E. and Miragaya, J.C. 1989. Thermoluminescence dating of fossil sand dunes in Apure, Venezuela. *Acta Científica Venezolana* 40: 80–81.

Viera, I.C.G., Uhl, C. and Nepstad, D. 1994. The role of the shrub *Cordia multispicata* as a 'succession facilitator' in an abandoned pasture, Paragominas, Amazonia. *Vegetatio* 115: 91–99.

Villachia, H., Silva, J.E., Peres, J.R. and Da Rocha, C.M.C. 1990. Sustainable agricultural systems of the humid tropics of South America. In *Sustainable agricultural systems* (ed. C.A. Edwards, R. Lal, P. Madden, R.H. Miller and G. House). Ankeny, IA: Soil and Water Conservation Society, 391–437.

Vitousek, P.M. 1984. Litterfall, nutrient cycling and nutrient limitation in tropical forests. *Ecology* 65: 285–298.

Vitousek, P.M. and Denslow, J.S. 1986. Nitrogen and phosphorus availability in treefall gaps of a lowland tropical rainforest. *Journal of Ecology* 74: 1167–1178.

Vitousek, P.M. and Sanford, R.L. 1986. Nutrient cycling in moist tropical forest. *Annual Review of Ecology and Systematics* 17: 137–167.

Vogt, K.A., Grier, C.S., Meier, C.E. and Edmonds, R.L. 1982. Mycorrhizal role in net primary production and nutrient cycling in *Abies amabalis* ecosystems in western Washington. *Ecology* 63: 370–380.

Vuilleumier, B.S. 1971. Pleistocene changes in fauna and flora of South America. *Science* 173: 771–780.

Wade, M.K. and Sanchez, P.A. 1983. Mulching and green manure applications for continuous crop production in the Amazon basin. *Agronomy Journal* 75: 39–45.

Walker, J., Thompson, C.H., Fergus, I.F. and Tunstall, B.R. 1981. Plant succession and soil development in coastal sand dunes of subtropical eastern Australia. In *Forest succession: concepts and applications* (ed. D.C. West, H.H. Shugart and D.B. Botkin). New York: Springer-Verlag, 107–131.

Walker, S.H. 1973. *Summary of climatic records for Belize*. Land Resources Division Supplementary Report 3. Surbiton, Surrey: Land Resources Division.

Walsh, J. 1988. Rift Valley fever rears its head. *Science* 240: 1397–1399.

Walsh, J.A. 1990. Estimating the burden of illness in the tropics. In *Tropical and geographical medicine* (ed. K.S. Warren and A.A.F. Mahmoud). New York: McGraw-Hill, 185–196.

Walter, H. 1973. *Vegetation of the earth in relation to climate and the eco-physiological conditions* (trans. J. Wieser). New York: Springer-Verlag.

Warren, S.D., Black, H.L., Eastmond, D.A. and Whaley, W.H. 1988. Structural function of buttresses of *Tachigalia versicolor*. *Ecology* **69**: 532–536.

Wastie, R.L. 1975. Diseases of rubber and their control. *Pest Abstracts and News Summaries (PANS)* **21**: 268–288.

Webb, L.J. 1958. Cyclones as an ecological factor in tropical lowland rainforest, north Queensland. *Australian Journal of Botany* **6**: 220–228.

Webb, S.D. 1978. A history of savanna vertebrates in the New World. Part II: South America and the great interchange. *Annual Review of Ecology and Systematics* **9**: 393–426.

Webb, T. 1988. Eastern North America. In *Vegetation history* (ed. B. Huntley and T. Webb). The Hague: Kluwer, 385–414.

Webster, P.J. 1987a. The elementary monsoon. In *Monsoons* (ed. J.S. Fein and P.L. Stephens). New York: John Wiley, 3–32.

Webster, P.J. 1987b. The variable and interactive monsoon. In *Monsoons* (ed. J.S. Fein and P.L. Stephens). New York: John Wiley, 269–330.

Wellman, F.L. 1968. More diseases on crops in the tropics than in the temperate zone. *Ceiba* **14**: 17–28.

Wendland, W.M. 1977. Tropical storm frequencies related to sea surface temperatures. *Journal of Applied Meteorology* **16**: 477–481.

Went, F.W. and Stark, N.M. 1968. Mycorrhiza. *BioScience* **18**: 1035–1039.

Western, D. and Finch, V. 1986. Cattle and pastoralism: survival and production in arid lands. *Human Ecology* **14**: 77–94.

Whelan, R.J. 1995. *The ecology of fire*. Cambridge: Cambridge University Press.

White, J. 1989. Ethnoecological observations on wild and cultivated rice and yams in northeastern Thailand. In *Foraging and farming: the evolution of plant exploitation* (ed. D.R. Harris and G.C. Hillman). London: Unwin Hyman, 152–158.

White, L.J.T. 1994. The effect of commercial mechanized selective logging on a transect in lowland rainforest in the Lope Reserve, Gabon. *Journal of Tropical Ecology* **10**: 313–322.

White, R.E. 1987. *Introduction to the principles and practice of soil science*, 2nd edition. Oxford: Blackwell.

Whitmore, T.C. 1984. *Tropical rain forests of the Far East*, 2nd edition. Oxford: Clarendon Press.

Whitmore, T.C. 1989. Canopy gaps and the two major groups of forest trees. *Ecology* **70**: 536–538.

Whitmore, T.C. 1990. *An introduction to tropical rain forests*. Oxford: Clarendon Press.

Whitmore, T.C. and Prance, G.T. 1987. *Biogeography and Quaternary history in tropical America*. Oxford: Oxford University Press.

Whittaker, R.H. and Likens, G.E. 1975. The biosphere and man. In *Primary productivity of the biosphere* (ed. H. Leith and R.H. Whittaker). Ecological Studies vol. 14. New York: Springer-Verlag, 305–328.

Whitten, A.J. 1987. Indonesia's transmigration program and its role in the loss of tropical rain forests. *Conservation Biology* **1**: 239–246.

WHO 1985a. The control of schistosomiasis. Report of a WHO expert committee. *World Health Organization Technical Report Series* **728**. Geneva: WHO.

WHO 1985b. Viral haemorrhagic fevers. Report of a WHO expert committee. *World Health Organization Technical Report Series* **721**. Geneva: WHO.

WHO 1994. Emerging infectious diseases: memorandum from a WHO meeting. *Bulletin of the World Health Organization* **72**: 845–850.

Wiersum, K.F. 1984. Surface erosion under various tropical agroforestry systems. In *Proceedings of the symposium on the effects of forest land use on erosion and slope stability* (ed. C.L. O'Loughlin and A.J. Pearce). Vienna: IUFRO; Honolulu: East-West Center, 231–239.

Wiersum, L.K. 1958. Density of root branching as affected by substrate and separate ions. *Acta Botanica Neerlandica* 7: 174–190.

Wijmstra, T.A. and Van der Hammen, T. 1966. Palynological data on the history of tropical savannas in northern South America. *Overdruk uit Leidse Geologische Mededelingen* 38: 71–90.

Wikkramatileke, R. 1972. The Jenka Triangle, west Malaysia: a regional development project. *Geographical Review* 62: 479–500.

Wiklander, L. 1980. Interactions between cations and anions influencing adsorption and leaching. In *Effects of acid precipitation on terrestrial ecosystems* (ed. T.C. Hutchinson and M. Havas). New York: Plenum Press, 239–254.

Wilkes, G. 1989. Maize: domestication, varietal evolution, and spread. In *Foraging and farming: the evolution of plant exploitation* (ed. D.R. Harris and G.C. Hillman). London: Unwin Hyman, 440–455.

Wilkie, D.S., Sidle, J.G. and Boundzanga, G.C. 1992. Mechanized logging, market hunting, and a bank loan in Congo. *Conservation Biology* 6: 570–580.

William, R.D. and Chiang, M.Y. 1980. Weed management in Asian vegetable cropping systems. *Weed Science* 28: 445–451.

Williams-Linera, G. 1990. Origin and development of forest edge vegetation in Panama. *Biotropica* 22: 235–241.

Wilson, J.B. and Agnew, A.D.Q. 1992. Positive feedback switches in plant communities. *Advances in Ecological Research* 23: 263–336.

Windsor, D.M. 1990. *Climate and moisture variability in a tropical forest: long-term records from Barro Colorado Island, Panama.* Washington, DC: Smithsonian Institution Press.

Wolfe, J.A. 1978. A paleobotanical interpretation of tertiary climates in the northern hemisphere. *American Scientist* 66: 694–703.

Wolfe, J.A. 1985. The distribution on major vegetational types during the Tertiary. In *The carbon cycle and atmospheric CO_2 natural variations Archean to present.* American Geophysical Union, Geophysical Monograph 32, 357–375.

Wolfe, J.A. and Upchurch, G.R. 1986. Vegetation, climatic and floral changes at the Cretaceous–Tertiary boundary. *Nature* 324: 148–152.

Wolffsohn, A.L.A. 1967. Post-hurricane fires in British Honduras. *Commonwealth Forestry Review* 46: 233–238.

Wong, M., Wright, S.J., Hubbell, S.P. and Foster, R. 1990. The spatial pattern and reproductive consequences of outbreak defoliation in *Quararibea asterolepis*, a tropical tree. *Journal of Ecology* 78: 579–588.

Woods, P. 1989. Effects of logging, drought, and fire on structure and composition of tropical forests in Sabah, Malaysia. *Biotropica* 21: 290–298.

Wright, S.J. 1992. Seasonal drought, soil fertility and species density of tropical forest plant communities. *Trends in Ecology and Evolution (TREE)* 7: 260–263.

Yavitt, J.B., Battles, J.J., Lang, G.E. and Knight, D.H. 1995. The canopy gap regime in a secondary Neotropical forest in Panama. *Journal of Tropical Ecology* 11: 391–402.

Young, T.P. and Hubbell, S.P. 1991. Crown asymmetry, treefalls, and repeat disturbance of broad-leaved forest gaps. *Ecology* 72: 1464–1471.

Young, T.P. and Perkocha, V. 1994. Treefalls, crown asymmetry, and buttresses. *Journal of Ecology* 82: 319–324.

Zimmerer, K.S. 1995. The origins of Andean irrigation. *Nature* 378: 481–483.

INDEX

Note: Page numbers in **bold** type refer to figures. Page numbers in *italic* type refer to tables.

373